THE ENCYCLOPEDIA OF
WEATHER
and CLIMATE CHANGE

THE ENCYCLOPEDIA OF
WEATHER
and CLIMATE CHANGE
A Complete Visual Guide

UNIVERSITY OF CALIFORNIA PRESS
BERKELEY   LOS ANGELES

University of California Press, one of the most distinguished
university presses in the United States, enriches lives around the
world by advancing scholarship in the humanities, social sciences,
and natural sciences. Its activities are supported by the UC Press
Foundation and by philanthropic contributions from individuals
and institutions. For more information, visit www.ucpress.edu.

University of California Press
Berkeley and Los Angeles, California

Published by arrangement with
Weldon Owen Pty Ltd
59–61 Victoria Street, McMahons Point,
Sydney NSW 2060, Australia
Copyright © 2010 Weldon Owen Pty Ltd

BONNIER PUBLISHING LTD
Group Publisher John Owen

WELDON OWEN PTY LTD
Chief Executive Officer Sheena Coupe
Creative Director Sue Burk
Art Manager Trucie Henderson
Senior Vice President, International Sales Stuart Laurence
Vice President, Sales: United States and Canada Amy Kaneko
Vice President Sales: Asia and Latin America Dawn Low
Administration Manager, International Sales Kristine Ravn
Production Manager Todd Rechner
Production Coordinators Lisa Conway, Mike Crowton
Publishing Coordinator Gina Belle

Managing Editor Jennifer Taylor
Senior Editor Barbara Sheppard
Project Editors Lachlan McLaine, Carol Natsis
Senior Designer Michelle Cutler
Designers Jacqueline Richards, Mark Thacker
Jacket Design Lia Tjandra
Picture Research Joanna Collard, Barbara Sheppard
Copy Editor Shan Wolody
Proofreader Amanda Burdon
Index Jo Rudd
Editorial Assistants Hunnah Jessup, Natalie Ryan

ISBN 978-0-520-26101-3 (cloth : alk. paper)

16 15 14 13 12 11 10
10 9 8 7 6 5 4 3 2 1

Color reproduction by Chroma Graphics (Overseas) Pte Ltd
Printed by Tien Wah Press
Manufactured in Singapore

A WELDON OWEN PRODUCTION

AUTHORS

Dr. Juliane L. Fry
Assistant Professor
Chemistry and Environmental Studies
Reed College
Oregon, U.S.A.

Dr. Hans-F Graf
Professor
Environmental Systems Analysis
Centre for Atmospheric Science
University of Cambridge
Cambridge, U.K.

Dr. Richard Grotjahn
Professor
Atmospheric Science Program
Department of Land, Air,
and Water Resources
University of California, Davis
California, U.S.A.

Dr. Marilyn N. Raphael
Professor
Department of Geography
University of California, Los Angeles
California, U.S.A.

Dr. Clive Saunders
Senior Lecturer
School of Earth, Atmospheric,
and Environmental Sciences
University of Manchester
Manchester, U.K.

Richard Whitaker
Senior Meteorologist
The Weather Channel
Sydney, Australia

CONTENTS

FOREWORD

Mark Twain once said, "The weather is always doing something." That is probably why it is never far from our thoughts. The moods and vagaries of weather fascinate and frighten us, thrill and threaten us. Weather is a perennial topic of conversation and conjecture. It affects our physical and psychological wellbeing and impacts constantly on the living conditions, livelihoods, and lifestyles of people everywhere.

In recent decades there has been a growing consciousness, among scientists and the public at large, of dramatic shifts in global weather patterns and climatic conditions. Predictions vary about the extent and severity of these changes, and there are numerous, often contradictory, theories to explain their causes. There is now significant scientific consensus that these developments are, to a large extent, human induced and that only urgent and timely human intervention can avert potentially disastrous consequences worldwide.

With clear text, written by a team of international experts, and vivid color photographs, detailed illustrations, maps, graphs, and diagrams, *The Encyclopedia of Weather and Climate Change* uncovers the processes and mechanics of weather and climate in a form that will engage and inform. Divided into six sections, this book contains the most recent scientific research, ranging in scale from planetary to molecular. Topics such as Earth's atmosphere and its global systems, extreme weather events, meteorology, and continental climate zones are examined in depth. The final section presents the evidence for climate change, outlines the likely outcomes, and spells out the ways in which governments, corporations, and individuals can adapt their long-established practices to cope with and mitigate the conditions that now confront us all.

HOW TO USE THIS BOOK

This book is divided into six sections: Engine, Action, Extremes, Watching, Climate, and Change. Engine gives an overview of Earth's atmosphere and its global systems. Action explains the workings of general weather phenomena, such as clouds, rain, and snow. Extremes looks at devastating weather events, including tornadoes, hurricanes, and drought. Watching covers the science of meteorology, from ancient times to today. Climate tours the climate zones of the world. The final section, Change, provides a compelling portrait of our relationship with Earth and the effects of climate change. Each section is broken down into chapters devoted to particular subjects. Each chapter begins with an introduction to the subject (right), providing a general overview, then the subject is explored in detail in the pages that follow (examples are below). Special features, called "Insights" (far right), look at the evolution of our knowledge about weather and climate through text, illustrations, charts and graphs, maps, and photographs.

TRACKING WEATHER

Section and chapter heading
This indicates the broad theme and specific area under discussion.

Global locator map
This pinpoints the location of the key regional examples discussed below it.

Timeline
This provides information about key developments through the ages.

Diagrams
Where appropriate, diagrams are included to illustrate complex concepts.

FACT FILE

Glaciers

In mountainous and polar regions, so much winter snowfall can build up that the snow remains frozen even after the following summer's sunshine. Under the weight of snow, glacial ice slides downhill over wet ground beneath. Deep water-filled inlets in coastlines (fjords) are created by the scouring action of a valley glacier on its way to the sea.

1. Lambert Glacier This is the largest glacier in the world. Located in Antarctica, it measures about 250 by 60 miles (400 x 100 km) and has a depth of around 1.5 miles (2.5 km). It flows into the Amery Ice Shelf, an extension of the Antarctic ice shelf that overlies part of Prydz Bay.
Lambert Glacier, Antarctica

2. Kangerdlugssuaq Glacier This glacier, in the southeast of Greenland, has lately accelerated its journey to the ocean. After a century of slow progress, it now travels up to 124 feet (38 m) each day. Global warming may be the cause
Kangerdlugssuaq Glacier, Greenland

3. Jostedal Glacier Located near the west coast of Norway, this is Europe's largest glacier. Measuring 40 by 5 miles (64 x 8 km), it is up to 1,800 feet (548 m) thick. The glacier feeds more than 50 glacier branches that reach down into the fertile valleys in all directions.
Jostedal Glacier, Norway

LOUIS AGASSIZ
Swiss–American scientist Louis Agassiz is known as the father of glaciology. Although others had studied glaciers before him, he was the first to see evidence of an ice age, when Earth had been largely covered by a thick ice sheet.

Antarctic dry valleys (below) The volume of the Antarctic Ice Sheet has fluctuated greatly over geological time. The ice-free valleys in this photo were scoured by glaciers millions of years ago. They are kept dry by the action of constant wind

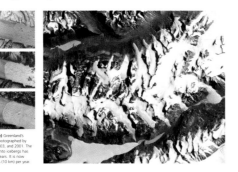

Glacial retreat (above) Greenland's Helheim Glacier was photographed by NASA in 2005 (top), 2003, and 2001. The break up of the front into icebergs has accelerated in recent years. It is now retreating about 6 miles (10 km) per year.

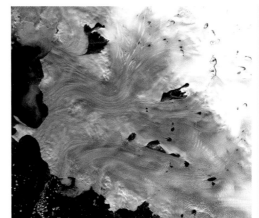

Glacier breakup (above) This satellite image of Greenland's west coast shows glaciers flowing around mountaintops and into Baffin Bay. The Greenland Ice Cap is the birthplace of countless icebergs that then drift out to sea on ocean currents.

FACT FILE

Shaping the land (below) As they flow, glaciers carve the underlying rocks into shapes an expert eye can recognize thousands of years later.

Glacial plucking As a glacier moves over uneven ground, it breaks the bedrock into pieces that are transported downhill until they are deposited as a terminal moraine.

Before glaciation Between ice ages, the climate is warm and mountains are covered by vegetation. Valleys are V-shaped.

During glaciation During an ice age, the climate is too cold for plants and they disappear. Glaciers flow through the valleys.

After glaciation When the ice retreats, the valleys that remain are U-shaped, with finger lakes in deeper places.

Rock, debris, and water | Precipitation, such as snow | Net accumulation zone | Energy loss through long-wave radiation | Solar energy | Net ablation zone | Evaporation | Meltwater and deposited debris

Geothermal heat

How a glacier flows A valley glacier is fed from above with ice, water, rock, and other debris. Geothermal heat warms the base of the glacier, causing it to flow downhill. At the upper levels, snow continues to accumulate in the net accumulation zone, but water is also lost by evaporation, sublimation, and wind (deflation). In the net ablation zone, below the equilibrium line, the glacier loses mass faster than it receives it.

Geothermal heat

Feature box
Photographs or illustrations, and text, highlight an interesting aspect of the topic being explored.

Illustration
A graphic cutaway illustration shows the inner workings of a physical phenomenon.

INSIGHTS

Fact file
This panel explains processes or profiles several examples of the subject under discussion.

Introductory text
This provides a general overview of the subject.

Charts and graphs
These group data and present statistics and forecasts in an easy-to-understand format.

Photograph
An evocative photograph shows a landform or feature that is representative of the subject matter under discussion.

Global Changes

FACT FILE

Mount Kilimanjaro Retreating glacial cover on Mount Kilimanjaro provides highly visible evidence of a warming planet. Permanent snow and ice on the summit have almost completely disappeared. The ice cap formed over 11,000 years ago.

Approximate glacier extent in 1912

Glacier extent in 2003

Rim of summit plateau

Retreat Satellite photographs of Mount Kilimanjaro's crater capture 10 years of glacial shrinkage. Over 80 percent of ice has been lost over the past century, and the summit may be ice-free by 2015.

February 17, 1993

February 21, 2000

June 2, 2003

Evidence of global warming can be seen today around the world in temperature trends, retreating ice caps, shrinking mountain glaciers, warming oceans, rising sea levels, biodiversity loss, and failing human health. Erratic weather patterns have also been observed, such as heat waves, droughts, floods, and coastal storms of escalating frequency and intensity. These events are harbingers of things to come as global temperatures rise.

Severe weather (below) A common feature in the predictions of all global climate models is an increase in severe weather events as the climate warms. There will be droughts in some regions and floods in others. Heat waves and melting glaciers also attest to the warming trend.

Heat waves Europe has experienced severe heat waves in recent years, most notably in the summer of 2003, when tens of thousands lost their lives to heat-related maladies.

Polar ice cap Average Arctic temperatures have increased at twice the rate of global temperatures over the past century, resulting in the rapid melting of the polar ice cap, one of the most visible signs of climate change.

Glacial melting Glaciers are melting at a rapid pace around the world. The Himalayan glaciers, situated in densely populated South Asia, feed the Indus, Ganges, Yangtze, and other rivers, providing fresh water for one-sixth of global population.

Rising seas Rising seas pose the greatest threat to low-lying island nations, such as the many atolls in the South Pacific. For example, Tuvalu's highest elevation is only 15 feet (4.5 m) above sea level.

FACT FILE

Warming The Intergovernmental Panel on Climate Change predicts a global average warming trend of 3.6–7.2°F (2–4°C) over the coming century. This will cause sea levels to rise, more frequent extreme weather, and the spread of disease.

Rising seas Sea levels have risen at a rate of 0.12 inches (0.3 cm) per year since 1993. Sea levels are projected to rise 7–23 inches (18–58 cm) over the next century, submerging low-lying islands.

Extreme weather Tropical cyclones are common in the Caribbean, South Pacific, and Bay of Bengal, but in June 2007, Cyclone Gonu formed in an unusual location: the Arabian Sea.

Health risks A warming climate and more frequent flooding contribute to the spread of disease, especially in areas with poor sanitation and difficult living conditions, such as Bangladesh.

Glacier

Typical cyclone tracks

Winter ice extent

Summer ice extent

Extent of iceberg drifts

Increase in drought

Increased rainfall

Increase in mean temperature

Flood hazard

Coastal areas at greatest risk

Islands and archipelagos at risk

Areas of low-lying islands

Hurricanes The intensity of Atlantic hurricanes has increased markedly over recent decades, a trend that is tied to warming sea surface temperatures in the Gulf of Mexico.

Arctic Ocean

NORTH AMERICA

EUROPE

ASIA

Atlantic Ocean

Pacific Ocean

AFRICA

Pacific Ocean

SOUTH AMERICA

Indian Ocean

AUSTRALIA

Southern Ocean

ANTARCTICA

Flooding Increasingly strong El Niño conditions have been connected to global warming. These events cause severe flooding along the western coast of the Americas, particularly in Peru.

Calving ice The ice sheets surrounding the Antarctic Continent, especially the Antarctic Peninsula, are calving icebergs into the Southern Ocean as they disintegrate. Though the Antarctic is not warming as rapidly as the Arctic, the margins of its ice sheet are thinning quickly as surrounding waters warm.

Drought The Sahel region of sub-Saharan Africa experiences frequent droughts, caused by overgrazing and poor land management. Drought frequency has increased in recent years due to climate change, human-produced atmospheric aerosols, and warming seas shifting regional precipitation patterns.

Coral bleaching Coral reefs are highly sensitive to changes in ocean temperature and acidity. Widespread bleaching death of reef ecosystems, such as the Great Barrier Reef, is an indication of destructive changes happening beneath the surface

Satellite photography
Images taken from space provide unique perspectives on Earth.

World map
This shows the global distribution of a feature being profiled, and is accompanied by text that discusses the feature in more detail.

ENGINE

ENGINE

WHAT IS WEATHER?

Earth is surrounded by a thin envelope of gases, the atmosphere. The term weather is used to describe the state of the atmosphere at any given moment. Weather conditions include the distribution and intensity of winds, air pressure, humidity, clouds, and numerous kinds of precipitation. Climate is the statistics of these conditions, including their variability, mean, and extremes, measured over an extended period of time. Earth receives a constant stream of energy from the Sun. This energy heats the planet. Earth also emits energy, a process that cools the planet. The equilibrium of the incoming and outgoing radiation determines Earth's mean temperature.

Global patterns This compilation illustration depicts various weather events around the world. Large-scale cloud clusters provide the energy to maintain weather systems. Converging cold, dry and warm, moist air masses create huge midlatitude storms, while warm tropical waters spawn hurricanes. Differential heating of air leads to pressure gradients that generate winds strong enough to transport sand and dust long distances.

Stormy skies A thunderstorm producing torrential rain and lightning redistributes massive amounts of energy. While a small thunderstorm releases 10 million kilowatts, supercell thunderstorms can generate 10–100 times as much. Most powerful thunderstorms occur in the tropics, but occasionally they are observed in the midlatitudes.

The tilt of Earth's axis and the planet's elliptical rotation around the Sun creates differences in the global distribution of incoming radiation energy, leading to seasons. The yearly growth and melting of ice sheets, long-term changes in Earth's orbital parameters around the Sun over thousands of years, and shifts over millennia in the position of the continents all affect Earth's climate.

Long ago, humans started to change the face of the planet, which has had an impact on weather and climate. Converting natural vegetation into arable land and building cities have changed surface reflectivity and water evaporation rates. Carbon formerly stored in coal and oil is emitted into the atmosphere as carbon dioxide, an effective trap for capturing outgoing long-wave radiation. Atmospheric carbon dioxide levels have reached concentrations approaching twice the natural amount in only a few decades.

ATMOSPHERIC LAYERS

Earth's atmosphere is composed of a mixture of nitrogen, oxygen, and argon, plus a number of variable trace gases and particles. Extending from Earth's surface to a height of 4.4 miles (7 km) at the poles, 9.3 miles (15 km) in the tropics, and 10.6 miles (17 km) near the Equator, the troposphere is the lowest part of the atmosphere. This layer is mixed by turbulent motion and temperature decreases with altitude, dropping to −76°F (−60°C) at the tropopause, the upper limit of the troposphere. The layer above is the stratosphere, where ozone absorbs radiation and creates a warm, stable layer that suppresses turbulence. Temperatures at the top of the stratosphere, approximately 31 miles (50 km) above sea level, reach around 32°F (0°C). Higher in the atmosphere, temperatures fall again within the mesosphere, the next layer up.

The presence of water vapor in Earth's atmosphere creates the conditions needed to sustain life and produces moisture required to make weather. Because the tropopause is so cold, nearly all water vapor is trapped within the troposphere. Therefore weather phenomena are most intense and occur almost exclusively in the troposphere. Evaporated from lakes, oceans, rivers, soil, and vegetation, water vapor is mixed with the other gases in the troposphere in varying amounts. Many processes that transform energy take place near the surface and involve phase transitions of water—changing from solid ice to liquid water to water vapor.

The maximum ratio of water vapor in air is determined by temperature: the lower the air temperature, the less moisture can be mixed with dry air. Any surplus water vapor will condense and form clouds made of liquid or solid forms of water. Condensation and freezing releases energy when the water droplets eventually fall to Earth as precipitation.

Hurricanes Intense, rotating storms develop near the Equator over tropical oceans with water temperatures above 80°F (26.5°C). Hurricanes are fueled by large amounts of condensing water, which release latent heat.

Air currents Jet streams are bands of high-speed wind that flow 5–10 miles (8–18 km) above sea level, near the top of the troposphere. Most prominent are the jet streams in the Northern Hemisphere's midlatitudes.

North Atlantic storm These storms develop at the interface between cold polar and warm subtropical air masses. They have a cold core and form cloudy bands that spiral around each other, guided by the jet stream.

Airborne particles A combination of fierce winds and prolonged dry weather can produce massive dust storms. About one-tenth the size of sand grains, dust can remain aloft longer and be carried farther than sand.

Wildfires Often caused by lightning strikes, and sometimes by human negligence or arson, wildfires are a common phenomenon in arid regions. They emit large quantities of carbon monoxide, soot, and other aerosols that affect atmospheric chemistry.

Conditions Meteorologists measure the temperature and pressure of Earth's atmosphere, the amount of moisture it holds, precipitation, wind strength, and the presence or absence of clouds. This data is compiled into maps and forecasts.

Temperature Warm surfaces radiate more energy than cold. A thermographic image shows a warm waterfall releasing more heat than its wintry surroundings.

Pressure Fluctuations in air pressure often bring a change of weather conditions. The lowest air pressure is measured within a hurricane's eye.

Humidity The amount of water vapor contained in the air is dependent upon temperature. On cold surfaces, water vapor will condense and form dew.

Earth's Systems

Earth is a system consisting of five distinct components that permanently interact through the exchange of mass and energy: atmosphere (air), lithosphere (land), hydrosphere (liquid water), cryosphere (ice), and biosphere (life). As a whole, this system is in quasi-equilibrium, fueled by incoming energy from the Sun (climate), but local disorder creates weather fluctuations.

Climate system (below) From vast oceans spanning thousands of miles and huge cities housing millions of people to clouds containing microscopic water droplets, natural and man-made features of varying size work in tandem, and in opposition, to influence Earth's climate over time.

Sun The star in the center of our solar system provides Earth with a permanent supply of energy in the form of short-wave radiation.

Snow Global snow coverage is variable from year to year. The white surface of powdery snow reflects approximately 80 percent of sunlight back to space.

Glaciers Large ice sheets in Greenland, Antarctica, and on high mountains contain most of Earth's fresh water.

Volcanoes Gases from volcanic activity and impacting comets created Earth's early atmosphere. Strong eruptions can disturb climate and weather for years.

Atmosphere This mixture of gases allowed life to develop on Earth. Most weather takes place in the troposphere, close to the surface.

Wildfires Vegetation fires, frequently caused by humans, are common around the globe. Carbon dioxide and aerosols released by burning affect atmospheric composition.

Biosphere Regions of vegetation over Europe appear as dark green in this enhanced satellite image. Brown indicates minimal vegetation. High chlorophyll concentrations off the coast are red and yellow.

Irrigation In dry areas or during drought, water is used to improve agricultural productivity. Increasing evaporation from soil and plant transpiration, and reducing reflectance, change the energy balance.

Precipitation Water droplets and ice crystals large enough to fall from the sky as rain, snow, frost, hail, mist, fog, or dew are known as precipitation.

Wind As wind moves over a body of water, surface friction causes the water to pile up, creating waves. The stronger the wind blows, the higher the waves.

Clouds Formed by condensed water vapor that evaporated from oceans and the soil, clouds reflect sunlight. Precipitation releases the heat generated by evaporation back into the atmosphere.

Sea ice Large areas of floating ice at both Poles reflect solar radiation and prohibit direct interaction between the oceans and the atmosphere.

Farms Cultivation and land-use changes impact energy exchange by modifying transpiration and surface texture and reflectivity.

Oceans The biggest reservoir of water on Earth are the oceans. Their capacity to retain heat and slow-moving currents influences weather and climate parameters.

Rivers Networks of waterways collect rain and meltwater from snow and glaciers and channel the water to the oceans.

Smog Air pollution develops from the burning of fossil fuels and industrial activities. Cities generate and retain more heat than suburban and rural areas.

Groundwater Precipitation stored in soil and underground reserves eventually makes its way back to the oceans.

EARTH'S WATER SUPPLY

Although more than two-thirds of Earth's surface is covered by water, 97.5 percent is salty from dissolved minerals. Nearly 70 percent of the fresh water is stored in frozen form in glaciers, ice caps, and permanent snow, so only a tiny and unevenly distributed fraction of fresh water is readily available.

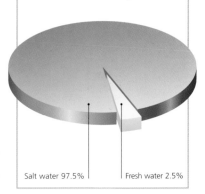

Salt water 97.5%   Fresh water 2.5%

The Atmosphere

Within the atmosphere—a mixture of gases that surrounds and protects Earth—air density and pressure decrease with height. The composition of the atmosphere is constant up to about 62 miles (100 km) from the surface, above which Earth's gravitational effects dominate, turbulent mixing decreases, and gas molecules separate based on their mass.

Moist swirls (right) Water vapor is unevenly distributed in the atmosphere. Dark colors in this infrared satellite image indicate areas of dense water vapor concentrated above cloud tops and storms in the upper troposphere along the Intertropical Convergence Zone. The brightest colors define areas of dry, descending air over the Pacific Ocean.

Noctilucent clouds (below) Although very little water exists in the upper atmosphere, high-latitude temperatures in the mesosphere may drop below −184°F (−120°C) in summer, causing water vapor microparticles to condense. Tiny ice crystals can be observed at twilight as they reflect sunlight, even when lower atmospheric layers are already dark.

FACT FILE

Planetary atmosphere All planets in our solar system have gravitational forces strong enough to maintain an atmosphere. Composition varies due to gravitational differences, planetary temperatures, and the effects of solar wind on each planet.

Great Red Spot For the last 300 years, a giant anticyclone has been observed in Jupiter's atmosphere. It is the largest known vortex in the solar system.

Venus Temperatures on Venus reach 870°F (465°C). Its dense atmosphere, composed mainly of carbon dioxide and nitrogen, traps heat like a greenhouse.

Mars Martian atmosphere is 95 percent carbon dioxide. During winter, its poles become so cold that one-fourth of the carbon dioxide condenses into dry ice.

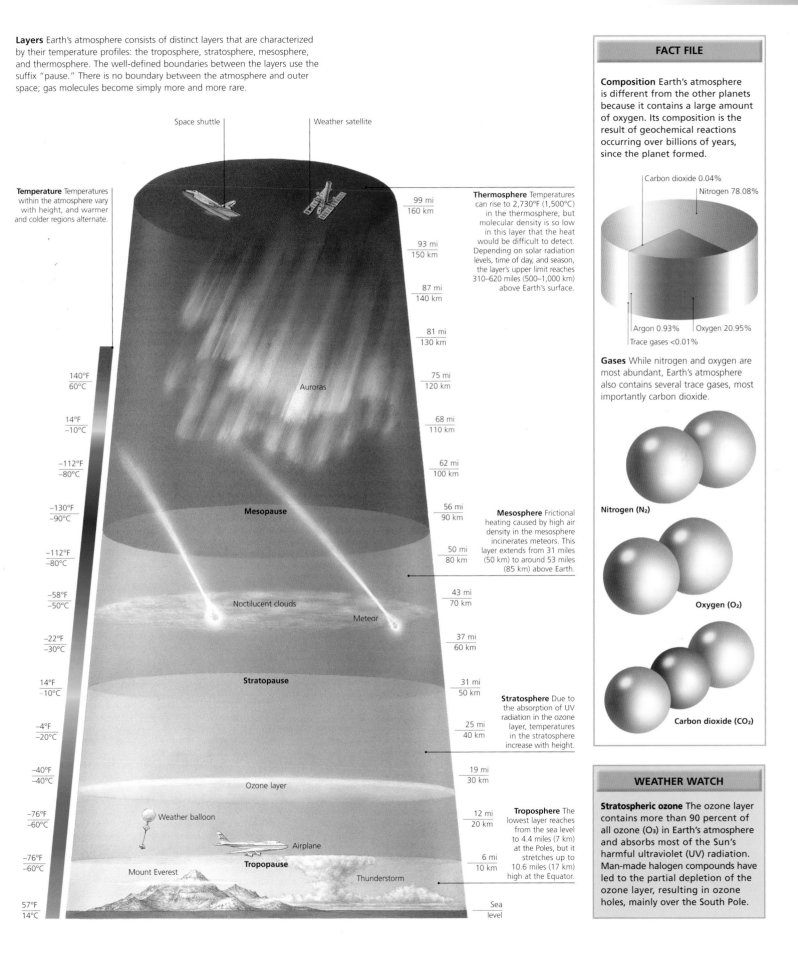

Layers Earth's atmosphere consists of distinct layers that are characterized by their temperature profiles: the troposphere, stratosphere, mesosphere, and thermosphere. The well-defined boundaries between the layers use the suffix "pause." There is no boundary between the atmosphere and outer space; gas molecules become simply more and more rare.

Space shuttle

Weather satellite

Temperature Temperatures within the atmosphere vary with height, and warmer and colder regions alternate.

99 mi / 160 km

93 mi / 150 km

87 mi / 140 km

81 mi / 130 km

75 mi / 120 km

68 mi / 110 km

62 mi / 100 km

56 mi / 90 km

50 mi / 80 km

43 mi / 70 km

37 mi / 60 km

31 mi / 50 km

25 mi / 40 km

19 mi / 30 km

12 mi / 20 km

6 mi / 10 km

Sea level

140°F / 60°C
14°F / −10°C
−112°F / −80°C
−130°F / −90°C
−112°F / −80°C
−58°F / −50°C
−22°F / −30°C
14°F / −10°C
−4°F / −20°C
−40°F / −40°C
−76°F / −60°C
−76°F / −60°C
57°F / 14°C

Auroras

Mesopause

Noctilucent clouds

Meteor

Stratopause

Ozone layer

Weather balloon

Airplane

Tropopause

Mount Everest

Thunderstorm

Thermosphere Temperatures can rise to 2,730°F (1,500°C) in the thermosphere, but molecular density is so low in this layer that the heat would be difficult to detect. Depending on solar radiation levels, time of day, and season, the layer's upper limit reaches 310–620 miles (500–1,000 km) above Earth's surface.

Mesosphere Frictional heating caused by high air density in the mesosphere incinerates meteors. This layer extends from 31 miles (50 km) to around 53 miles (85 km) above Earth.

Stratosphere Due to the absorption of UV radiation in the ozone layer, temperatures in the stratosphere increase with height.

Troposphere The lowest layer reaches from the sea level to 4.4 miles (7 km) at the Poles, but it stretches up to 10.6 miles (17 km) high at the Equator.

FACT FILE

Composition Earth's atmosphere is different from the other planets because it contains a large amount of oxygen. Its composition is the result of geochemical reactions occurring over billions of years, since the planet formed.

Carbon dioxide 0.04%
Nitrogen 78.08%
Argon 0.93%
Oxygen 20.95%
Trace gases <0.01%

Gases While nitrogen and oxygen are most abundant, Earth's atmosphere also contains several trace gases, most importantly carbon dioxide.

Nitrogen (N_2)

Oxygen (O_2)

Carbon dioxide (CO_2)

WEATHER WATCH

Stratospheric ozone The ozone layer contains more than 90 percent of all ozone (O_3) in Earth's atmosphere and absorbs most of the Sun's harmful ultraviolet (UV) radiation. Man-made halogen compounds have led to the partial depletion of the ozone layer, resulting in ozone holes, mainly over the South Pole.

Composition Atmospheric gases are consistent to 62 miles (100 km) above Earth's surface, except for the presence of tropospheric water vapor and pollutants. Water vapor is responsible for 36 percent of the natural greenhouse effect.

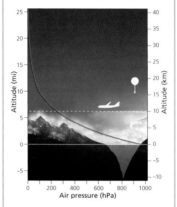

Molecule Water (H_2O) is formed from one oxygen and two hydrogen atoms. Essential to support life on Earth, water can exist as a solid, liquid, or gas.

Pressure gradient Mean air pressure is 1,013.25 hectopascals (hPa) at sea level. Air pressure (indicated by red line) decreases exponentially with height.

TURBULENCE

Commercial planes are built to fly most economically at altitudes around 33,000 feet (10 km), the level of the tropopause in the midlatitudes. Pilots avoid turbulence produced by convective clouds by flying above that altitude.

The Troposphere

The lowest level of the atmosphere, the troposphere, contains 90 percent of its mass and the vast majority of atmospheric water. Nearly all weather develops in the troposphere, which is characterized by strong convection, a heat-driven process that causes warmer air to rise and cooler air to sink, and turbulent mixing of sensible and latent heat from Earth's surface.

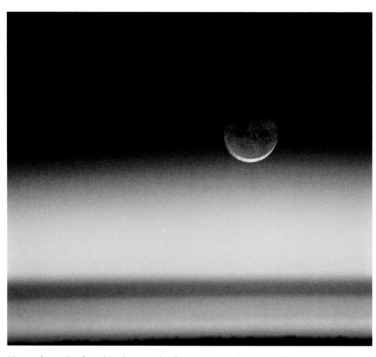

Moon above Earth In this photograph taken at sunrise, the troposphere is seen as an orange band capped with a denser, thin brown layer. The colors are due to light-scattering aerosols concentrated in the stable tropopause, the upper boundary of the troposphere.

Smog (above) Mexico City, surrounded on three sides by mountains, has one of the world's worst air pollution problems. A stable temperature inversion layer traps vehicle emissions, ground-level ozone, and soot near the surface, especially in winter.

Thunderstorms (right) Deep, convective cells, seen from a space shuttle in 1984, develop in the moist, warm air over Brazil. Most of the clouds stop at the tropopause, but some storms are so vigorous they are able to penetrate the stratosphere.

FACT FILE

Airborne Aerosols are tiny particles of various origins suspended in air. Found mostly in the troposphere, they influence cloud properties and solar energy transfer. The images below are false-colored, scanning electron microscope photographs.

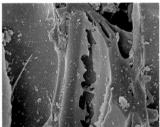

Ash Volcanic ash, released during the 1980 eruption of Mount St. Helens, Washington, U.S.A., consists of glass shards produced by exploding magma.

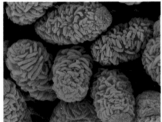

Mold Many varieties of mold fungi reproduce by toxic spores that can become airborne and cause serious health problems when inhaled.

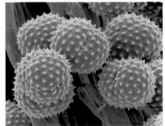

Pollen Grains of pollen, a biogenic aerosol, are responsible for allergic reactions like hay fever and are found worldwide, especially during springtime.

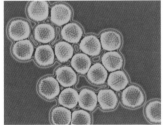

Virus More than 200 kinds of virus can cause the common cold. Although viruses become airborne easily, most infections result from direct contact.

The Sun

The Sun is the center of our solar system, around which all planets orbit. Like all stars, the Sun is a giant ball of superheated hydrogen gas. Containing 98.6 percent of the solar system's total mass, the Sun provides nearly all of Earth's energy as electromagnetic radiation. The absorption, reflection, and redistribution of this energy determines Earth's weather and climate.

Magnetic activity Most of the Sun's features are caused by vigorous magnetic field activity. Photographs taken with ultraviolet instruments show the surface and its corona, the outermost region of the star's thin atmosphere.

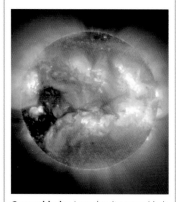

Coronal holes Low-density areas (dark regions) permit solar material to flow out to space along open magnetic lines in a constant stream called solar wind.

Loops Clustered around sunspots, arc-like phenomena develop when hot gas, channeled by magnetic field lines, joins two active places on the surface.

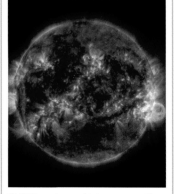

Field lines Areas of intense magnetic activity appear as bands of bright plumes, where charged particles spiral around arcing magnetic field lines.

Solar power The Sun's surface is not firm like Earth's; it is composed of churning, active gases, hot enough to vaporize any solid. The energy produced and emitted by the Sun provides warmth and light for the planets. Its core, containing only seven percent of the Sun's volume but half its mass, fuels the star with nuclear fusion.

Prominence Long-lasting arcs of gas erupt from the Sun's surface and are held in place by strong magnetic fields. One of the largest prominences ever observed extended more than 365,000 miles (587,000 km).

Photosphere The visible surface of the Sun, the photosphere, is about 9,900°F (5,500°C). Rising hot gas produces an effect called granulation. Typical granules are 600 miles (1,000 km) in diameter.

Flares Magnetic storms on the surface release bursts of high-energy particles, gas, and radiation thousands of miles into space.

Spicules Jet-like flames of hot gas and particles 310 miles (500 km) in diameter stretch up to 6,200 miles (10,000 km) above the photosphere. At any time, there are 60,000–70,000 active spicules.

Radiative zone Energy is transported from the core toward the surface by electromagnetic waves. Since the density is so high in this area, energy is absorbed and re-emitted many times, making the process very slow.

Core Energy is produced by the fusion of hydrogen into helium at 27 million °F (15 million °C) in the Sun's intensely hot core.

Auroras When charged particles from the Sun's thin atmosphere collide with Earth's atmosphere they vibrate. As the molecules return to their original state, they release different colors of light. Most common around the Poles, auroras occur 50–370 miles (80–600 km) above Earth's surface.

Convective zone In this outer portion of the Sun, energy is carried to the surface by a boiling, convective motion induced by density differences.

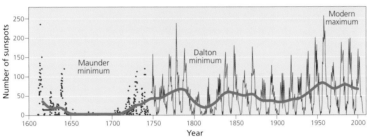

Sunspots Dark spots indicate regions of magnetic activity that inhibit convection. Gases trapped within sunspots cool by about 3,000°F (1,600°C).

Cycles Observed with telescopes since the 1600s, sunspot activity fluctuates in 11-year cycles, during which solar intensity varies by 0.1 percent. It is speculated that sunspot cycles, especially periods of prolonged minimum activity, impact climate.

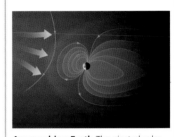

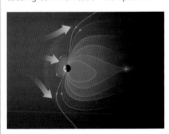

The Energy Cycle

Energy produced by fusion in the Sun's core reaches Earth as short–wave electromagnetic radiation. After encountering Earth's atmosphere, this energy is reflected, absorbed, transformed, and radiated back to space in a complex process known as the energy cycle. The balance between incoming and outgoing radiation results in Earth's stable temperature.

Solar radiation The Sun, the only notable energy source for Earth's atmosphere, emits radiation at all wavelengths, but most intensely in the wavelengths we see as visible light. Solar radiation is distributed unevenly around our planet.

Production Nuclear fusion in the Sun's core manufactures all elements heavier than hydrogen and releases energy.

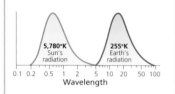

5,780°K Sun's radiation   255°K Earth's radiation

0.1 0.2   0.5 1   2   5 10 20   50 100
Wavelength

Emissions The Sun emits radiation at short wavelengths around the center of visible light. Earth emits much longer wavelengths in the infrared spectrum.

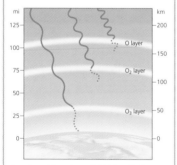

mi   km
125 —   — 200
                             O layer
100 —   — 150
75 —   O₂ layer — 100
50 —
                         O₃ layer — 50
25 —
0 —   — 0

Barriers Most harmful UV radiation (purple waves) is absorbed by oxygen atoms (O), molecular oxygen (O_2), and ozone (O_3) in the atmosphere.

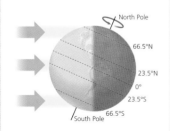

North Pole
66.5°N
23.5°N
0°
23.5°S
66.5°S
South Pole

Tilt Because Earth's axis is tilted 23.5 degrees, solar radiation hits the surface at varying angles, creating inconsistent energy densities at different latitudes.

Radiation balance This diagram illustrates solar radiation as it enters and leaves Earth's atmosphere. About 30 percent of the Sun's energy is immediately reflected back to space. Clouds and the lower atmosphere absorb another 19 percent. The remaining 51 percent is absorbed by land and oceans.

Incoming solar radiation: 342 units The Sun emits high-intensity, short-wave radiation toward Earth.

Absorbed by atmosphere: 67 units Incoming short-wave radiation is absorbed by air molecules in the atmosphere.

Reflected by atmosphere: 77 units Earth's atmosphere and clouds reflect short-wave radiation back to space.

Absorbed by surface: 168 units Albedo determines the amount of incoming solar radiation that is absorbed and available to heat the ground or evaporate water.

Reflected by surface: 30 units Depending on the reflectivity, or the albedo, of different surfaces, Earth's land and oceans reflect on average 30 units of solar radiation.

Albedo The proportion of radiation that is reflected by Earth varies depending on surface color and texture, an effect known as albedo. Incoming energy is absorbed by dark surfaces; light, smooth surfaces reflect it back to space. If the Sun's rays strike certain surfaces, such as large bodies of water, at a low angle, the amount of reflection increases. Overall, Earth's albedo averages approximately 30 percent.

Ocean Short-wave radiation can penetrate water. Oceans, therefore, have a very low albedo of 3–10 percent.

Forest Dark green leaves effectively absorb short-wave radiation, so forests have a very low albedo (5–20 percent).

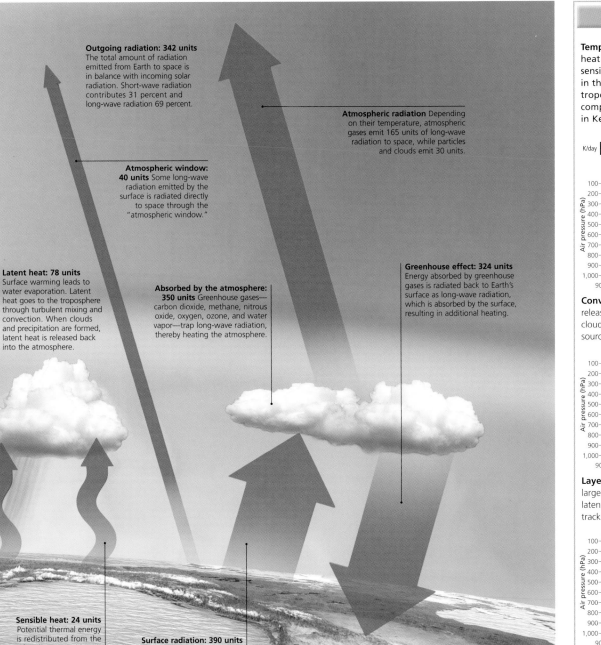

Outgoing radiation: 342 units
The total amount of radiation emitted from Earth to space is in balance with incoming solar radiation. Short-wave radiation contributes 31 percent and long-wave radiation 69 percent.

Atmospheric radiation Depending on their temperature, atmospheric gases emit 165 units of long-wave radiation to space, while particles and clouds emit 30 units.

Atmospheric window: 40 units Some long-wave radiation emitted by the surface is radiated directly to space through the "atmospheric window."

Latent heat: 78 units
Surface warming leads to water evaporation. Latent heat goes to the troposphere through turbulent mixing and convection. When clouds and precipitation are formed, latent heat is released back into the atmosphere.

Absorbed by the atmosphere: 350 units Greenhouse gases—carbon dioxide, methane, nitrous oxide, oxygen, ozone, and water vapor—trap long-wave radiation, thereby heating the atmosphere.

Greenhouse effect: 324 units
Energy absorbed by greenhouse gases is radiated back to Earth's surface as long-wave radiation, which is absorbed by the surface, resulting in additional heating.

Sensible heat: 24 units
Potential thermal energy is redistributed from the surface by warm, rising air and turbulent motion.

Surface radiation: 390 units
The global mean surface temperature is 57°F (13.9°C), and Earth's surface emits long-wave thermal radiation.

Fields Farmland reflectivity depends on crop color. Meadows and young plants have an albedo from 12–30 percent.

Desert The albedo for sand, which may consist of very different mineral types and grain sizes, varies between 15–40 percent.

Snow Ageing snow develops small bubbles, making it less reflective (40–70 percent) than freshly fallen snow (75–90 percent).

FACT FILE

Temperatures Determined by latent heat transfer, radiation levels, and sensible heat release, temperatures in the lower stratosphere and the troposphere can be theorized by computer models as heating rates in Kelvin degrees per day.

K/day
-2 -1 0 1 2

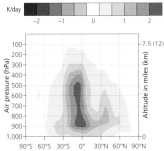

Convective clouds Latent heat released by rain-producing, tropical clouds is the most important heat source in the troposphere.

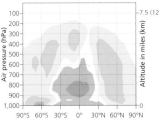

Layered clouds Rain produced by large-scale, layered clouds releases latent heat, especially in the storm tracks of middle and high latitudes.

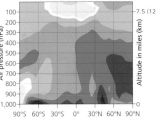

Radiation Radiation processes overall lead to cooling, except where ozone in the lower stratosphere heats the air by absorbing UV radiation.

Turbulent heat The sensible heat transferred by turbulent motion from the surface can only reach the lowest layers of Earth's atmosphere.

SEASONS

Seasons are distinct periods of a year characterized by annual recurring weather changes. The varying intensity and duration of incoming solar radiation as Earth orbits the Sun, caused by the tilt of Earth's rotational axis, creates seasons. Generally, four seasons that differ mainly by temperature—winter, spring, summer, and fall—are observed in mid and polar latitudes. In the tropics and subtropics, where solar radiation does not fluctuate much throughout the year, changing precipitation levels produce a dry and a wet season.

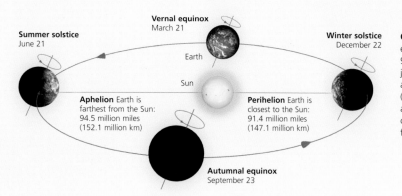

Summer solstice June 21

Vernal equinox March 21

Winter solstice December 22

Earth

Sun

Aphelion Earth is farthest from the Sun: 94.5 million miles (152.1 million km)

Perihelion Earth is closest to the Sun: 91.4 million miles (147.1 million km)

Autumnal equinox September 23

Orbital path Earth orbits the Sun in an elliptical path, at an average distance of 93 million miles (149.7 million km). The journey takes one year, or 365.25 days, at a mean speed of 18.5 miles per second (29.8 km/s). Earth also spins around its axis and completes one rotation every 24 hours, creating night and day. Dates are shown for the Northern Hemisphere.

REASON FOR SEASONS

Earth rotates in a counterclockwise direction around an imaginary line, its axis, running between the North and South Poles. Because the axis is tilted at an angle of 23.5 degrees from the vertical, different parts of the planet receive solar radiation in different amounts; this tilt is why we experience seasons.

Depending on the time of year, some latitudes are tilted toward the Sun, while other regions are tilted away. For half the year, sunlight falls most directly on the Northern Hemisphere; during the other half,

the Southern Hemisphere. In the Northern Hemisphere, the North Pole is tilted away from the Sun in December. Less light reaches the hemisphere then, resulting in short days and low temperatures—in other words, winter. If Earth's axis was not tilted at all, the Poles would be cold and dark year-round. Were the axis tilted more, the seasons would be more extreme.

The angle that the Sun's rays hit Earth affects the intensity of solar radiation that reaches the ground. When the Sun is in zenith, meaning directly overhead, solar energy is reduced only by the amount that is reflected back to space or absorbed by the atmosphere. If the Sun's rays hit at a low angle, they illuminate a broader section of Earth's surface and

distribute energy over a larger area. Low-angled rays have to penetrate the thickest part of the atmosphere and, therefore, they are weakened by the time they reach the surface.

The distribution of landmasses around the planet also has an effect on seasonal temperatures. Because water has a very high heat capacity, meaning it requires considerable energy to warm, large bodies of water have a dampening effect on temperature extremes. This causes smaller seasonal variations near oceans than in continental areas, where solar radiation quickly heats the surface. More continents are located in the Northern Hemisphere, so mean summer temperatures are higher there than those in the Southern Hemisphere.

Solar positions At noon, an observer on Earth's surface will see the Sun at different heights depending on the latitude and the time of the year. When the Sun is high in one hemisphere, it is low in the other. Navigators used this feature for centuries to determine latitudinal position.

June 21 | March 21; September 23

June 21 | March 21; September 23 | December 22

December 22 | March 21; September 23 | June 21

At the North Pole From March 21 until September 23, the Sun is visible above the horizon, reaching its highest position on the June solstice. The Sun stays below the horizon from the September equinox to the March equinox, causing "polar night."

At midlatitudes At 45°N or 45°S, the Sun reaches a noon position on the equinoxes of 45 degrees above the horizon and stays in the sky for 12 hours. On the summer solstice, the Sun is above the horizon for about 15 hours; in winter, less than nine.

At the Equator On both equinoxes, the Sun is exactly overhead, while on the June solstice it is slightly to the north and on the December solstice slightly to the south. The Sun remains essentially above the horizon for 12 hours throughout the year.

Winter Due to colder temperatures, deciduous trees lose their leaves in winter. The same woodland scene and buds for a European beech are shown in the four different seasons (clockwise, from above).

Spring Rising temperatures and increased sunlight bring the biosphere back to life in spring. Buds and new growth appear on branches and develop into leaves and blossoms.

Fall Decreased temperatures and sunlight in fall slows biological activity. Trees withdraw chlorophyll and nutrients from their leaves, causing them to change color before they drop. Pods open and release seeds.

Summer Consistently warm weather encourages growth. Photosynthesis converts atmospheric carbon dioxide into sugars using the energy from sunlight. Leaves open fully and seed pods form.

Camouflage While some species hibernate or migrate to warm climates, animals that remain active in winter adjust their coloring to blend into the changing environment. The mottled feathers of a willow ptarmigan in summer (far left) whiten in winter (left), providing protection from predators.

Almanac Seasons dictate prime agricultural cycles for sowing and harvesting crops. Before meteorological information was easily accessed, farmers often relied on almanacs, like this one from 1795, to help predict seasonal changes. Temperatures and common weather observations were noted alongside astronomical facts.

Heat Heat is the process of energy transfer from one object or area to another because of temperature differences. Molecular vibrations, or the excitation of an electron's energy level, produce internal changes that affect temperature.

Fire Burning is the result of a complex chain of chemical reactions when fuel is combined with oxygen, or oxidized, and heat and/or light is set free.

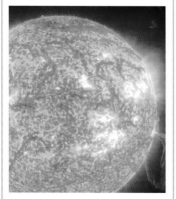

Nuclear reaction In the Sun's core, energy is created by fusion of hydrogen into helium. The star's temperature rises as its gases absorb surplus energy.

Currents Electrical energy is converted into heat energy when a current flows through a resistor. This principle is used by electric stoves and lightbulbs.

Temperature

Temperature is a measure of the heat content, or average kinetic energy, of molecules in a substance. It is one of the most important weather and climate parameters and determines the heat flow between two different areas or objects. Temperature controls biological activity, the phase state of water (ice, liquid, or water vapor), and many other physical processes.

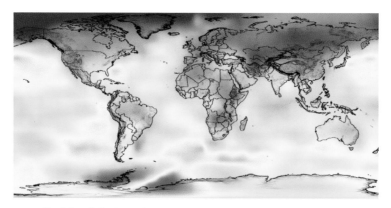

Anomalies (above) This image depicts temperature anomalies for 2002–2006 and shows that global atmospheric temperatures are inhomogeneously distributed. Strongest positive anomalies (red) are observed over the Arctic and Asia, while colder temperatures (blue) are found scattered over the oceans and parts of Antarctica.

Optical illusion This photo taken in the Namibian Desert shows a mirage, a natural phenomenon where distant light rays are refracted at the interface between hot surface air and the colder air above it.

FACT FILE

Scales Temperature is measured with thermometers that can be calibrated to a variety of scales. While most countries use the Celsius scale, a few still use Fahrenheit. Kelvin is the basic thermodynamic scale in the International System of Units (SI).

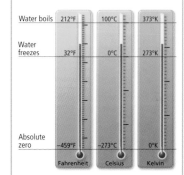

Water boils	212°F	100°C	373°K
Water freezes	32°F	0°C	273°K
Absolute zero	−459°F	−273°C	0°K
	Fahrenheit	Celsius	Kelvin

Conversion In Celsius, water freezes at 0°C (32°F) and water boils at 100°C (212°F) at sea level. Kelvin is offset from Celsius by 273.15°. To convert between the Celsius and Fahrenheit scales, use the following equations.

From Celsius to Fahrenheit:
$$°F = (1.8 \times °C) + 32$$

From Fahrenheit to Celsius:
$$°C = 0.56 \times (°F - 32)$$

COMPARISON	°F	°C
Absolute zero (0°K)	−459.67	−273.15
Lowest recorded temperature on Earth	−128.2	−89
Fahrenheit and Celsius are equal	−40	−40
Water freezes (at sea level)	32	0
Mean surface temperature on Earth	57	13.9
Average human body temperature	98.2	36.8
Highest recorded temperature on Earth	136.4	58
Water boils (at sea level)	212	100

Over the Poles (far left) Very dry air, which reduces the greenhouse effect, combined with diminished sunlight and high snow albedo, leads to cold weather. Temperatures rarely rise above 32°F (0°C) in most of the Arctic and Antarctic.

In deserts (left) Dry, virtually cloud-free, desert air allows solar radiation to heat Earth's surface at full strength, resulting in extremely high daytime temperatures. At night, the same conditions permit warmth to escape, causing quick cooling.

AIR IN MOTION

Changes in air pressure and temperature usually start and maintain winds. Generated by differential heating, pressure differences may develop on all scales, from local to global. Warm air is less dense than cold air and tends to rise, reducing the number of air molecules at ground level, leaving behind a mass deficit, or an area of low pressure. As air cools, it sinks and increases the number of surface air molecules, forming a region of high pressure. To balance the densities, air flows from high to low pressure, creating wind.

Floating (left) The skin of a hot-air balloon inhibits the mixing of the heated air inside with the cooler air outside. The balloon achieves buoyancy because the volume of air inside the balloon weights less than an equal amount of surrounding air.

Clouds (right) Surface moisture is carried aloft as warm air rises. Temperature usually decreases with altitude, so as it travels up, the air begins to cool. Clouds form when water vapor cools and condenses around tiny particles, such as dust, in the air.

Sailing (below) Invented around 5,500 years ago, sailboats use wind to propel them forward. Sails harness the energy of relative wind and water speed. Due to lessened surface friction, winds are much stronger over sea than land.

BLOWING WINDS

Wind generally refers to horizontal airflow and is measured by direction and speed. If high- and low-pressure areas are close, they trigger a strong pressure gradient, creating powerful winds. In certain circumstances, high-density air, such as frigid air collecting over a glacier, can create a gravity-driven, downslope flow, known as katabatic wind.

Depending upon the amount of incoming and outgoing solar energy, and the ability of surfaces to absorb or reflect radiation, atmospheric and surface temperature variations occur. These anomalies impact airflow; wind transports energy from warmer

to cooler regions. Midlatitude storms accomplish most of the heat transfer outside the tropics. Trade winds, monsoons, and hurricanes (tropical cyclones) are the main heat-carrying systems in the tropics.

Large-scale, global wind systems are influenced by Earth's rotation around its axis because the air is moving in relation to a surface that is also moving below it. Discovered by French engineer Gustave-Gaspard de Coriolis in 1835, the Coriolis effect, abbreviated as CorF, explains that a moving object or atmospheric or ocean current is deflected from a straight course by rotational forces. This means that air does not flow directly from high to low pressure, but is deflected to the right in the Northern Hemisphere and to the left in the Southern Hemisphere. Accordingly, in the Northern Hemisphere, wind flows clockwise around high-pressure areas and counterclockwise for low-pressure areas. In the Southern Hemisphere, the directions are reversed.

Near Earth's surface, ground friction slows winds, reducing the Coriolis effect in the lowest layers of the atmosphere. Warm air near the Equator rises and moves toward the Poles, where cold air subsides and spreads toward lower latitudes, creating a system of circulation cells that transport air around the globe.

Wind power The earliest examples of windmills were recorded in Persia between A.D. 500–900. The motion of blades turning atop a tower provided energy to pump water or grind grain. Modern wind turbines convert the kinetic energy of wind into mechanical energy, which drives electrical generators.

Atmospheric Pressure

Atmospheric pressure at any particular point is caused by the weight of air molecules above it. Differences in pressure create wind as the air moves from areas of high to low pressure. Weather experienced on the ground is greatly influenced by atmospheric pressure: high pressure usually produces fine weather, and low pressure creates unsettled, often stormy conditions.

Coastal Sea breezes develop during the day when land is heated more than water. Strong but shallow onshore winds can form high waves.

Monsoon The moist air masses of the Indian monsoon make thick clouds and produce torrential rain when they hit the Himalayas near Kathmandu, Nepal.

Trade winds Northeast trade winds from a subtropical high strike the Koolau Range on Oahu, Hawaii, forming a band of topographic clouds.

Pressure systems (below) A warm surface, such as an ocean, creates an area of low pressure. Because warm air is less dense than cold air, it starts to rise. While rising, the air cools and pressure decreases with height. Sinking air generates an area of high pressure. Winds form and attempt to balance pressure differences.

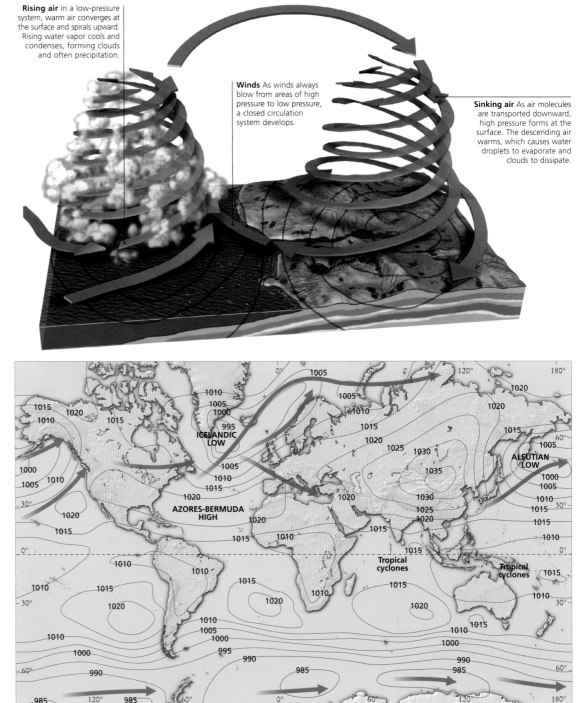

Rising air In a low-pressure system, warm air converges at the surface and spirals upward. Rising water vapor cools and condenses, forming clouds and often precipitation.

Winds As winds always blow from areas of high pressure to low pressure, a closed circulation system develops.

Sinking air As air molecules are transported downward, high pressure forms at the surface. The descending air warms, which causes water droplets to evaporate and clouds to dissipate.

Global pressure Isobars, lines on a weather map that connect points of equal sea-level pressure, identify large-scale pressure systems. Numbers are pressure readings in hectopascals (hPa). Red arrows indicate primary storm tracks.

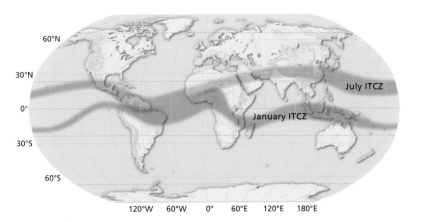

Intertropical Convergence Zone
In the tropics, where the northeast and southeast trade winds meet, a belt of low pressure circles the globe. This is a band of high precipitation known as the ITCZ. It follows the Sun's position during the seasons, moving to the north in the northern summer and south in winter.

Midlatitude storm (below) Taken in May 2003 over the Gulf of Alaska, this satellite image of a low-pressure system shows polar and subtropical air masses converging and spinning counterclockwise in a huge vortex.

FACT FILE

Comparison The relative pressure and density of air molecules are compared to the pressure and density of water (below). The white balls represent pressure and the background color of each segment is an indication of density.

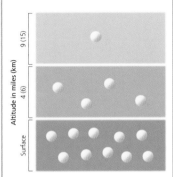

Altitude Atmospheric pressure decreases non-linearly with height. Because air is a compressible gas, lower layers in the atmosphere have a higher density.

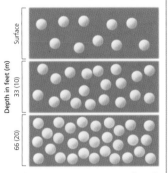

Depth Water pressure increases linearly with depth by one unit every 33 feet (10 m). Water is not compressible, so its density remains the same at all depths.

SCUBA DIVING

Divers experience higher pressure the deeper they descend, causing dissolved gases to accumulate in their blood and tissues. When surfacing, the gases form bubbles, especially in large body joints. Rapid decompression may cause severe illness, paralysis, or death.

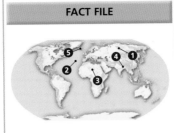
1. Siberian high A very stable, shallow high-pressure system, the Siberian high, forms each winter by extreme surface cooling over the vast, snow-covered region of northern Asia. This high is

the source of the winter monsoon that transports very cold air to eastern China, sometimes as far as Southeast Asia.

Siberia

2. Azores high This semi-permanent subtropical high is named after the archipelago of the Azores. The region of high pressure gains strength and

shifts farther north in summer, occasionally expanding to the northeast, preventing Atlantic storms from reaching Europe.

The Azores

3. Saharan high Rain-bearing storms are kept away from the Sahara by a dominant high-pressure system over North Africa. Very stable temperature

inversion over the world's largest desert stops cloud formation, and conditions range from slight frost to extremely hot.

The Sahara

4. Everest low On the summit of Mount Everest, 29,029 feet (8,848 m) high, atmospheric pressure is around 300 hPa, approximately one-third of

the pressure at sea level. Oxygen density is also reduced to one-third of normal levels; most climbers rely on canned air.

Mount Everest, Himalayas

5. Icelandic low The Icelandic low is a semi-permanent area of low pressure located between Iceland and southern Greenland. It is a center of atmospheric circulation associated with frequent

storm activity and is stronger in winter. The Icelandic low and the Azores high anchor the North Atlantic Oscillation.

Iceland

Atmospheric Pressure continued

Areas of common high and low pressure encircle Earth in well-defined bands. In equatorial regions, low pressure dominates, but in midltatitudes, large areas of high pressure exist in both hemispheres. Belts of low-pressure systems are found toward the polar regions, except that in winter localized Siberian and polar highs develop from strong surface cooling.

Bavaria High pressure usually leads to fair conditions in the Bavarian Alps. Sometimes, sinking air creates a temperature inversion, when temperature increases with height, and traps a layer of valley fog, while blue skies are found at higher altitudes.

Pattern The North Atlantic Oscillation (NAO) alters European climate. Weak differences between the Azores high and the Icelandic low indicate a negative phase with cold conditions. Strong differences in the positive phase cause mild European winters.

Negative phase Northern Europe and eastern North America experience colder than normal conditions (blue) during the NAO negative phase, while rain-bearing storm clouds enter the Mediterranean.

Death Valley (above) A lack of moisture in deserts hinders cloud formation and evaporative cooling, leading to intense surface heating. Rising hot air results in a low-pressure area called a thermal low.

Tropics (below) Over many tropical islands, a temperature inversion inhibits deep convective clouds. Evaporated ocean water concentrates at the inversion layer, where it condenses and forms wispy clouds.

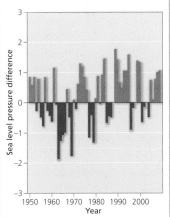

Positive phase In its positive phase, NAO causes a concentration of storm clouds in the Arctic, leading to warm, wet conditions (orange) over northern Europe and eastern North America.

NAO Index The normalized pressure differences between the Icelandic low and the Azores high create an oscillating pattern of extended phases of positive and negative values.

Global Winds

A system of global winds, also called the general circulation of the atmosphere, develops because of the interaction of large-scale pressure differences, the release of latent heat in precipitating clouds, and the effects of Earth's rotation. The Sun's rays hit the Equator directly, causing a consistent flow of warm, rising air in the tropics. Cold, dense air sinks at the Poles. These air circulation patterns transport heat around the globe.

Air cells (below) Air circulation creates three major types of cells. Those nearest the Equator are called Hadley cells, for George Hadley, the English scientist who described them in 1753. Ferrel cells, which circulate between 30 and 60 degrees, were first noted in 1856 by American scientist William Ferrel. At both Poles, dense, cold air subsides in the circulation cells.

Speed Wind speed is dependent on pressure differences over distance, known as the pressure gradient, and on local circumstances. Friction over a large body of water is low, allowing for higher wind speeds over oceans than over rough land.

Doldrums The doldrums are regions of low pressure with light winds. They form over warm tropical water in the ITCZ near the Equator.

Blowing snow Blizzards, which have sustained winds greater than 35 miles per hour (56 km/h), might reduce visibility to less than 500 feet (150 m).

Shrieking sixties No large land masses in the Southern Ocean near Antarctica hinder the path of strong winds circumnavigating the globe.

Polar cell At the Poles, circulation cells are driven by cold air sinking and spreading to lower latitudes.

Ferrel cell Ferrel cells are established by the mixing of warm and cold air along midlatitude storm tracks.

Hadley cell Energy produced by large thunderstorms along the Intertropical Convergence Zone (ITCZ) fuel Hadley cells.

Hadley cell

Ferrel cell

Polar cell

Walker circulation Three cells along the Equator are driven by convection over tropical water in the western Pacific and near Africa and the Amazon region. Transported by easterly trade winds, warm, moist air rises and forms deep convective clouds, which release rain and latent heat. After the moisture is discharged, the now-dry air warms during descent and flows back toward the east.

Polar winds Easterly winds blow from the North Pole to 60°N, and, likewise, from the South Pole to 60°S. Airflow is deflected by the Coriolis effect.

Westerlies A belt of strong, often gusty winds blowing from west to east is found in the midlatitudes of both hemispheres. Westerlies originate from air flowing from subtropical heights toward the Poles.

Subtropical high A belt of high-pressure develops at the descending branch of Hadley cells in both hemispheres. Westerly winds, as well as the trade winds that carry moisture toward the ITCZ, originate in these regions.

Trade winds Trade winds blow steadily from subtropical heights toward the Equator. They have a strong easterly component and are the most constant and predictable of all winds.

Intertropical Convergence Zone Defined by the convergence of trade winds from both hemispheres, the ITCZ is an area of deep convection that acts as the main driver of the atmosphere's general circulation.

Trade winds

Subtropical high

Westerlies

Polar winds

Deflection (above) A plane starting at the North Pole and flying on a straight course toward Miami (dashed line) would end up somewhere over the Pacific (curved line) as Earth turns beneath the plane. Rotational velocity increases farther from the Poles.

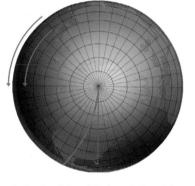

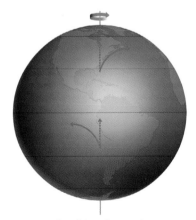

Coriolis effect (above) As Earth rotates, global winds are deflected between low- and high-pressure areas. The Coriolis effect is directed to the right in the Northern Hemisphere. The force is reversed in the Southern Hemisphere.

Visible effects Evidence of blowing winds can be seen in different ways. Steady winds along coasts bend growing trees. Strong winds can transform landscapes. Extreme winds might produce devastating conditions that destroy structures.

Sastrugi Irregular furrows and ridges made from the deposition and erosion of snow by strong surface winds are called *sastrugi,* "groove" in Russian.

Wind at sea Strong and steady winds generate ocean waves. Winds greater than 40 miles per hour (65 km/h) whip away wave tops and create whitecaps.

Dunes Winds stronger than 10 miles per hour (16 km/h) can lift fine sand particles. Dunes stretch perpendicular to the prevailing wind direction.

Planetary waves Global-scale winds meander at high altitudes, affect jet stream formation, and influence weather. These large, curving flow patterns form in the troposphere, at the border between air masses with contrasting temperatures.

Polar night jet Planetary waves barely affect the stratosphere, so the polar night jet stream remains symmetrical.

Polar jet Planetary waves are strongest around 60°N and create a meandering jet consisting of ridges and troughs.

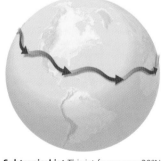

Subtropical jet This jet forms near 30°N and 30°S along Hadley cells, which determine the jet's shape and strength.

Pineapple Express The subtropical jet stream known as the Pineapple Express brings warm, moisture-laden air across the Pacific from Hawaii toward California. Strong rainfall (red) results when the jet stream interacts with coastal mountains.

Jet Streams

Jet streams are narrow bands of strong winds blowing west to east that form at the interface between air masses of different density. They travel at speeds greater than 62 miles per hour (100 km/h) and may stretch over thousands of miles in the upper troposphere, near the tropopause. Meteorologists track storms flowing along jet stream paths as a tool for weather forecasting.

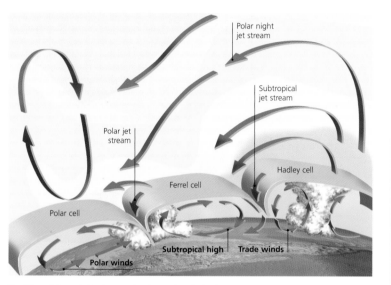

Positions (above) Jet streams develop at breaks in the tropopause between the polar and Ferrel cells and between the Ferrel and Hadley cells. Polar night jet streams form in winter in the stratosphere.

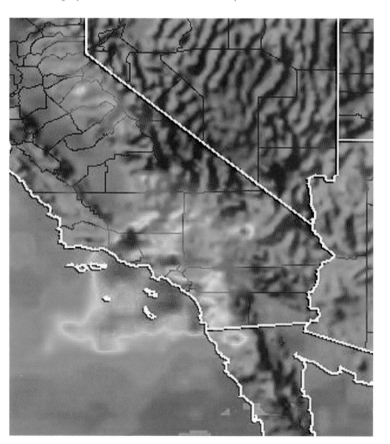

FACT FILE

Shifting flow Variable weather in the midlatitudes is generated by low-pressure storms traveling from west to east. Sometimes blocking occurs when air masses stall and weather does not change for days, or even weeks, at a time.

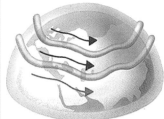

Normal Regular, slightly meandering polar jet streams bring variable weather patterns to the midlatitudes without causing extreme conditions.

Troughs Powerful planetary waves can distort jet streams and carry cold air to low latitudes and warm air to high latitudes, creating severe weather.

Blocking High- and low-pressure areas may be cut off. Stationary systems may cause flooding, extended cold spells, heat waves, and drought.

Visible jets Jet streams produce strong atmospheric turbulence, often resulting in long bands of clouds. In this photo taken by the Gemini 12 spacecraft, jet stream winds moving faster than 100 miles per hour (160 km/h) carry streaks of cirrus clouds above Egypt and the Red Sea.

Local Winds

While large-scale wind systems are generally driven by pressure differences, local circulation patterns can be caused by winds encountering physical barriers, such as mountain ranges, valleys, and fjords. Local temperature contrasts or the proximity between landmasses and large bodies of water may also produce specific conditions that lead to a huge range of local winds.

Windswept trees (right) The Canterbury northwester, a foehn-type wind, occurs when a low-pressure system drives moist air from the Tasman Sea over the Southern Alps on New Zealand's South Island. The ascending air loses moisture at the mountains, then it blows as a warm, dry northwesterly wind over the Canterbury Plains. Trees grow bent in the prevailing flow direction.

Khamsin (far right) The khamsin is an oppressively hot, dusty southerly or southeasterly desert wind that blows for some 50 days a year. The Arabic word *khamsin* means "fifty." Enforced in late winter and early summer by depressions moving eastward in the Mediterranean or across North Africa, this frontal wind causes frequent sandstorms in Egypt.

Breezes Small temperature variations between nearby places can cause local wind patterns. Day and night breezes result from local differences in daily heating and cooling. They are common in valleys and develop near coasts as sea breezes.

Daytime During the day, sunlit slopes warm up, driving a slight breeze uphill. The air eventually subsides and warms during descent.

Nighttime At night, cooled soil chills the overlaying air, which then flows down the slopes and accumulates in the valley, creating a downward breeze.

Bora A cold northerly or northeasterly wind gusts along the Adriatic Coast as cold air collects in a high-pressure system over the snow-covered plateau east of the Dinaric Mountains. The bora, most common in winter, can be felt throughout the Adriatic regions of Italy, Slovenia, and Croatia.

FOEHN WIND

A foehn wind develops when air is forced to cross a mountain. During ascent on the windward flank, air is cooled and moisture condenses, forming clouds. If precipitation is released, the now-dry air descends on the leeward side and warms.

Extreme Winds

Some winds can either be extremely hot or cold, depending on where and how they form. When the air source is very cold, such as near the poles, the winds can be freezing, even when experienced far from their origin. In contrast, strong winds blowing across sandy midlatitude deserts bring oppressively hot, dusty weather to normally moist, temperate regions.

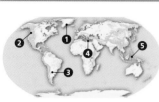

Fiery funnel In 2002, a rare, tornado-like whirlwind, or pyronado, formed during wildfires fomented by hot, dry Santa Ana winds in Southern California. Intensified by channeling through valleys, these winds blow from the high-pressure area of the Great Basin toward the Californian coast.

Harmattan Driven by high pressure, this hot, dry wind blows over West Africa, the Sahara, and the Sahel from the northeast or east toward the Gulf of Guinea. It carries large amounts of dust thousands of miles over the Atlantic, sometimes as far as North America. Tornadoes may form when the harmattan interacts with monsoon winds.

Polar wind Cold, dense air accumulates over Arctic plateaus. Downslope winds create a shallow flow of cold air, which, if channeled by topography, may reach hurricane force. As snow is relatively light, it is easily swept off the mountain peaks towering above an Inuit village in Greenland.

FACT FILE

Katabatic wind Intense winds form when cold air, which is denser than warm air, flows downhill. Driven by gravity, not pressure gradients, this downward flow is called katabatic wind. Wind strength depends on the height of the drop.

Cold air accelerates going downhill.

Air warms slightly and is lifted by turbulence.

Cold flow Air over the glaciers covering Antarctica is permanently cooled by radiation processes. The cold, dense air forms a reservoir and gravitational forces generate a strong downslope wind.

Location Because of the high elevation and distance from the warming ocean, very low temperatures over Antarctica's interior fuel a consistent katabatic flow. Winds become stronger near the coast.

WEATHER WATCH

Inversion winds With an elevation of nearly 10,000 feet (3,000 m), Antarctica's Polar Plateau offers a constant source of extremely cold air. Accumulated close to the ground due to gravitational forces, the bitterly cold air creates a strong temperature inversion that inhibits vertical mixing. Inversion winds are driven down the gentle slopes of the high interior ice plateau by gravity, much like katabatic winds, at extremely low temperatures.

Frontal Systems

Large air masses with similar temperatures and humidity form in different parts of the world. They are classified by temperature as arctic, polar, tropical, or equatorial, and by their humidity level as continental (dry) or maritime (moist). The sharp boundary between air masses is called a front. There are three main types of fronts: cold fronts, warm fronts, and occluded fronts.

Squall A sudden drop in air temperature and the appearance of smooth, rotating cloud features indicate that a swift and possibly violent wind change might be imminent. Squall lines are often associated with the passing of a cold front.

FACT FILE

Frontal cycle When two air masses of different temperatures meet, a low-pressure system can develop. A small wave disturbance along the front slowly grows into a frontal depression by the release of latent heat from precipitating clouds.

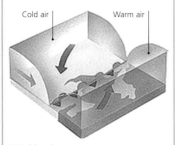

Cold air | Warm air

1 Cold and warm air masses converge and move side by side in opposing directions. Small instabilities, or waves, form at the intersection of the fronts.

2 The warm air mass begins to rise over the cold air, creating an area of low pressure. Clouds and precipitation form. The air masses begin to rotate.

3 The cold front pushes under and lifts, or occludes, the warm front. Cold air catches up to the warm air mass, which creates windy, unsettled weather.

4 Warm air is completely lifted off the surface. The cold air cuts off the supply of warm air, and precipitation and winds subside. The system dissipates.

Cold front In a cold front, dense cold air undercuts a warm air mass, causing strong lift near the surface just ahead of the cold air. Often defined by a band of heavy precipitation, the air in a cold front is frequently unstable and conducive to cumulonimbus cloud formation.

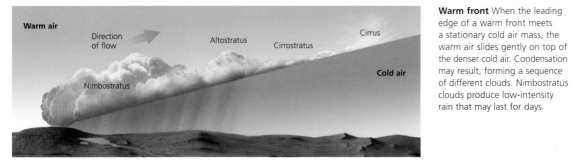

Warm front When the leading edge of a warm front meets a stationary cold air mass, the warm air slides gently on top of the denser cold air. Condensation may result, forming a sequence of different clouds. Nimbostratus clouds produce low-intensity rain that may last for days.

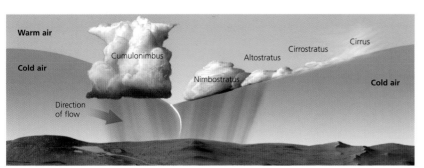

Occluded front Usually formed around a mature low-pressure area, an occluded front develops when a cold front overtakes a warm front and lifts the warm air completely off the ground. While thunderstorms sometimes appear at the frontal boundary, occluded fronts are associated with less severe weather.

FACT FILE

"The perfect storm" In October 1991, a remarkable low-pressure system formed along the North American East Coast when strongly contrasting air masses converged: cold air from the Midwest crashed into a warm, moist air mass over the Atlantic.

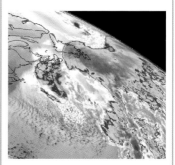

October 29 A massive low developed along a cold front east of Nova Scotia. Within two days it completely absorbed approaching Hurricane Grace.

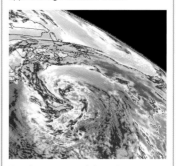

October 30 At peak intensity, pressure dropped to 972 hPa and the storm had sustained winds of 68 miles per hour (110 km/h) and 30 foot (9 m) waves.

October 31 The storm turned south and weakened to 998 hPa. Nearing warm water over the Gulf of Mexico, it re-intensified to hurricane strength.

Lightning If enough moisture is present, cold fronts often bring thunderstorms, heavy rain, and, sometimes, hail. Lightning is the result of a buildup of opposing electrical charges inside a towering thunderstorm, or cumulonimbus, cloud.

MONSOONS

Monsoons, from the Arabic word *mausim* meaning "season," are seasonal changes in wind experienced across the tropics and subtropics. Originally used by seafarers traveling from the Arabian Peninsula to India and back, the term now applies to the two main tropical seasons. Occurring over the continents of Asia, Africa, and Australia, monsoonal winds and rainfall are most dominant over southern and Southeast Asia. The word monsoon is often associated with strong rainfall, but this is not necessarily the case.

Floodwater Torrential downpours cause frequent flooding in Jakarta. Straddling the Equator, the Indonesian archipelago is alternately dominated by dry winter and rainy summer monsoon seasons.

MONSOON SEASONS

Caused by the increased amplitude of land temperature variations as compared to ocean temperatures, monsoons are seasonal prevailing winds that last several months. The southwest monsoon of the Northern Hemisphere's summer months forms when an intense heat low develops over central Asia. Rising air cools and condenses, leading to heavy clouds and precipitation. This area of low-pressure also creates a steady onshore flow that draws in moisture-laden trade winds.

In the Northern Hemisphere winter, land cools quickly and the Siberian high-pressure system intensifies. This produces strong northeasterly winds that blow toward the Equator. These winds remain dry until they build up moisture over the South China Sea.

In the case of Indian monsoons, the steep Himalayas significantly enhance precipitation levels as moist air masses are forced to rise higher and cool more drastically. The Himalayan foothills have most of the highest recorded rainfall totals. Credited as the wettest location on Earth, Cherrapunjee, in northeastern India, receives an incredible 323 inches (8,204 mm) of rain from May to August. The town's average in December is only 0.5 inch (13 mm), but the average in June, when the monsoon is at its peak, is 106 inches (2,692 mm).

Essential rain (right) In India, more than 75 percent of the annual rainfall occurs during the summer monsoon. Bhil tribesmen herd cattle during a downpour, which brings much-needed water to their land. Half the world's population relies on monsoon rains for vital water supplies.

Typhoon (below) Monsoon winds may induce tropical storms. This satellite image, taken on June 29, 2004, shows Typhoon Mindulle, a Category 4 storm that developed from a monsoon gyre over the warm Pacific Ocean waters near the Philippines.

Seasonal change Summer monsoons blow from the Indian Ocean toward Asia (front globe). Due to the Coriolis effect, the winds shift direction as they cross the Equator. In winter, wind direction reverses and monsoons originate in the north.

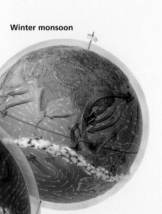

Winter monsoon

Summer monsoon

Structure (right) Data obtained from the Tropical Rainfall Measuring Mission (TRMM) satellite creates a 3-D view of Typhoon Mindulle. Rain intensity is shown from light (green) to heavy (red). The storm's cloud-free eye and spiral rain bands are visible.

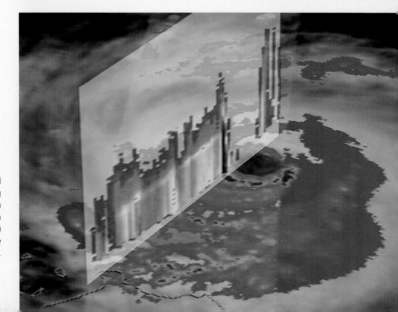

Celebration (above) The onset of the summer monsoon is heralded by festivals in Punjab, India, at the end of July. Monsoon rains are essential for farmers. If the rains arrive late or are weaker or more sporadic than normal, cattle die, crops fail, and millions of people face starvation.

Dry wind (below) In Zimbabwe, the summer monsoon season lasts from November to March, but the winds do not always bring rain. There have been consecutive years of drought in the Hwange National Park, which stresses the reserve's large elephant population.

Formation Sea breezes form when coastal land heats up more quickly than the adjacent water. Land has a lower heat capacity than water, so the same amount of solar energy causes different surface temperatures over land and sea.

Day breeze A closed circulation forms as cooled air from over the water rushes onshore to replace rising warm air, which then descends back toward the sea.

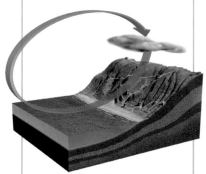

Night breeze Overnight, land cools quickly, but sea surface temperatures remain virtually unchanged. Cooler air drains off the land and over the sea.

Roll cloud So-called Morning Glory clouds may be 620 miles (1,000 km) long and 0.6–1.2 miles (1–2 km) high. These clouds develop when a frontal system interacts with a strong sea breeze, causing a strong localized upward motion of moist air.

Cloud forecast Heated surface air usually rises 1–2 miles (1.6–3.2 km) from the shoreline. Therefore, beaches often remain cloud-free during the day, but farther inland the rising air may condense and form convective clouds. At night, these clouds dissipate, but shallower clouds develop as air ascends and cools over the warmer water.

Sea Breezes

Land warms faster during the day than water and it also cools faster at night, creating a daily cycle of pressure differences that drives the development of local winds called sea breezes. They can develop anywhere there is land adjacent to a large body of water, even beside a lake. During the day, the breeze blows from the cooler water onto land. The flow reverses at night.

Mixing Funneled through mountain gaps, an offshore breeze intensifies by the time it reaches the Gulf of Papagayo, Nicaragua. The wind churns the usually warm surface water, drawing up cold deep water, creating patches of surface temperature anomalies.

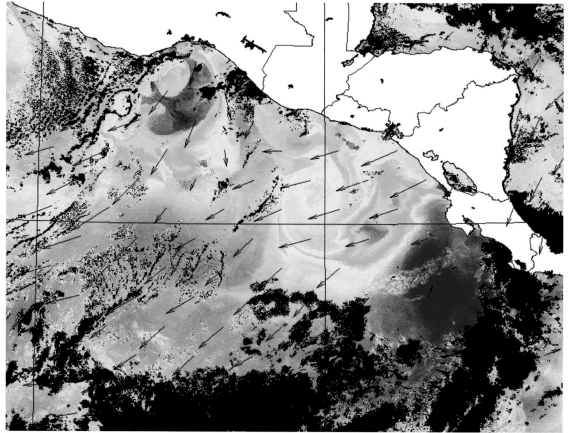

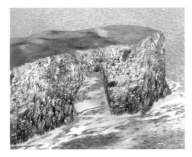

Sea cliff When steady waves, often caused by coastal winds, hit a rocky cliff they erode the lowest part, creating caves and holes. Eventually, the cliff becomes unstable and large rock masses fall into the sea.

Sea arch Over time, caves carved by wave erosion may break through the tip of the headland, resulting in an arch. Cracks and areas of softer rock in the cliff determine the position of the initial cave.

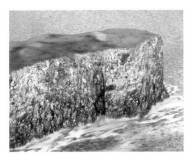

Sea stack As waves continue to erode the rock, the arch widens and pieces of rock crumble into the sea. The bridge of the arch may collapse, leaving an island-like chunk of rock, called a sea stack.

Natural sculptures The Twelve Apostles, off the shore of Port Campbell National Park, Australia, were formed by coastal erosion that began 10–20 million years ago. The erosion rate is about 0.8 inch (2 cm) per year. Only eight remain standing after the collapse of one of the sea stacks in 2005.

FACT FILE

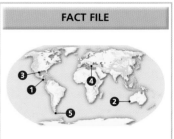

1. Papagayo The Papagayo wind is caused by a surge of cool, dry air from North America that establishes a strong pressure gradient between the cooler air over the Gulf of Mexico and the warmer, moister air in the Pacific. Such pressure systems may occur throughout the year but are more common in winter.

Papagayo, Central America

2. Fremantle Doctor A very strong, cool sea breeze moving 18–22 miles per hour (29–35 km/h) brings welcome relief to Western Australia from the high summertime temperatures. The wind is strongest in December and January, when the temperature contrast between land and sea is greatest.
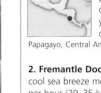
Fremantle Doctor, Western Australia

3. Norte Causing a drastic drop in air temperatures, this wind results from an outbreak of cold air from the U.S. Plains states, especially Texas. As the northeasterly wind moves toward shore, it creates large waves and good surfing conditions along the Gulf Coast.

Norte, Gulf of Mexico

4. Etesians The annually recurring Etesian winds are the prevailing, sometimes gusty summer winds that blow over large parts of Greece and the Aegean and Mediterranean seas. They bring clear skies and cool continental air between May and October.

Etesians, Aegean and Mediterranean seas

5. Williwaw squall A Williwaw squall is a sudden outburst of cold wind that blows seaward from high mountains. When funneled by fjords or inlets, they can reach hurricane force. Also termed squamish or outflow winds, they occur frequently in the Strait of Magellan.

Williwaw squall, Strait of Magellan

OCEAN WEATHER

Earth is the only planet in the solar system that has large volumes of surface water. Oceans cover 71 percent of Earth's surface and contain 97.5 percent of all water. Most water that falls as precipitation over land has evaporated from ocean surfaces, hence the oceans, although salty, are also the main source of fresh water. Oceans absorb more solar energy than land and form a huge heat reservoir. This heat is then transported around the world by ocean currents, affecting climate and local weather conditions.

WATER CYCLES

Oceans determine Earth's climate but also have a direct influence on weather. It is water's high heat capacity that is particularly important. For the same amount of heat exchange, the temperature change in water is only one-fifth of the temperature change for a similar mass of sand. This means that oceans heat and cool much slower than land, creating daily and annual temperature cycles that cause sea breezes and monsoons.

Heat is stored mainly in the turbulent, shallow upper layer that lies above the cold, deeper ocean. These warm and cold oceanic layers are separated by a zone of rapid temperature change, called the thermocline. The mean temperature of the world ocean is only 38.8°F (3.8°C).

Oceans are also important for the balance of atmospheric gases. About one-third of the man–made, or anthropogenic, carbon dioxide is absorbed and stored in oceans. Since solubility of gases depends on temperature and pressure, the gas accumulates in the deep, cold waters. If the ocean waters warm, the gas is released back into the atmosphere.

Ocean currents—along with atmospheric processes—transport heat from low to higher latitudes, keeping the global climate in balance. In the Pacific Ocean, about equal amounts of heat are transported toward the North and South Poles, but in the Indian Ocean the northward transport is blocked by land. The Atlantic Ocean is unique because here heat flows northward across all latitudes. This movement of water around the world is called thermohaline circulation, or the great ocean conveyor. Because salty and cold water is heavier than fresher and warmer water, the salt-rich subtropic surface water masses cool during their transport toward the Poles and sink to deeper layers due to gravitational forces. This formation of deep water masses is concentrated in small areas in the subpolar North Atlantic, the Arctic Ocean, and the Antarctic's Weddell Sea, and drives this very slow global oceanic circulation.

In the subtropics and tropics, steady trade winds, which blow from the east toward the Equator, produce upwellings of cold water from the ocean depths. Along the western coasts of the continents, cold currents therefore flow toward the Equator, strongly affecting weather in these areas, making it dry and relatively cool. However, every few years off the west coast of South America, the trade winds are replaced by westerly ones, which bring unusually warm water to the coast and cause heavy rains. At that time, much of Australia experiences drought. Such events are called El Niño and occur every two to seven years. The opposite phenomenon, called La Niña, often follows.

Tropical cyclones, typhoons, and hurricanes only develop over sea surfaces that have temperatures of at least 80°F (26.7°C) and also contribute to the transport of extra energy out of the tropics.

°F	°C
41.0	5
39.2	4
37.4	3
35.6	2
33.8	1
32	0

+1.8°F (+1.0°C)
+0.36°F (+0.2°C)

Simulating climate This computer model predicts oceanic and atmospheric temperature changes due to warming caused by greenhouse gases in the latter half of the 21st century. It depicts the region over North and South America.

One world ocean Though commonly recognized as separate bodies, the Pacific, Atlantic, Indian, Southern, and Arctic oceans—in descending order of size—comprise one global, interconnected mass of salt water, known as the world ocean.

Creating clouds Côte d'Ivoire is dry for most of the year, but a wet season develops between mid-May and mid-July, when the Intertropical Convergence Zone joins forces with large-scale sea breezes to form clouds along the West African coast.

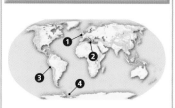

1. Cornwall, England Cornwall, at a latitude of 50°N, has an unusually mild climate. This is caused by the Gulf Stream, the large ocean current that transports warm water from the Gulf

of Mexico, along the East Coast of North America, and across the Atlantic Ocean toward northwest Europe.

Cornwall, England

2. Mediterranean Sea The Strait of Gibraltar connects the Mediterranean Sea to the rest of the world ocean. Being adjacent to the Sahara desert,

the sea receives a lot of solar energy and little rainfall. Its salinity is higher than average, mainly due to evaporation.

Mediterranean Sea

3. Peru Along the coast of Peru, cold and nutrient-rich water wells up from the deep. This makes the coastal area extremely dry, but the sea is full of life. Every few years, however, El Niño

causes warming of the water surface, leading to torrential rainfall, mudslides, and a decline in fish populations.

Peru, South America

4. Weddell Sea In the Weddell Sea, cold temperatures and strong winds lead to huge sea ice formations, which are carried into the Southern Ocean.

The salt expelled from the freezing water increases the density of the upper layer of seawater, which then sinks to great depths.

Weddell Sea, Antarctica

Ocean Climate

The oceans' ability to store heat has a strong impact on climate, and leads to differences between maritime and continental climates. Areas farther away from the ocean have bigger annual temperature ranges and drier climates than coastal regions. Sea surface temperatures, determined by ocean currents, lead to climatic differences along the same latitude.

Sea temperatures (below) Even though solar energy is distributed equally along latitude bands, sea surface temperatures can vary. Satellite images of sea surface temperatures show movements in currents and are used in the study of climate change.

Sea surface temperature, February 2009

28°F (−2°C) ◼◼◼◼◼◼◼◼ 95°F (35°C)

Food chain Life is abundant even in the cold waters around Antarctica, where this leopard seal catches a penguin chick. In most ecosystems, plants absorb solar energy and a food chain begins. Plankton is eaten by small crustaceans called krill; krill are consumed by fish; fish and birds are devoured by seals; and seals are eaten by whales, the top consumers.

Marine forests (above) Kelp forests are highly productive and complex ecosystems found in temperate and polar oceans. They require high concentrations of nitrogen, phosphorus, and light to thrive, provide shelter for a variety of sea life, and protect coastlines.

Endangered species (below) Although coral can grow in cooler waters, the optimum temperature for the majority of species is 78.8–80.6°F (26–27°C). Most shallow-water reefs are found in the tropics. Coral cannot survive in water that is polluted or too warm.

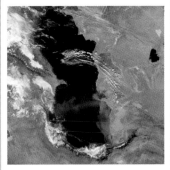
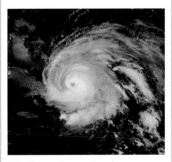

Gyres Ocean vortices, or gyres, are usually generated by surface winds, and form at latitudes around 30° in both hemispheres. These current systems are strongest along their western boundaries, and often spawn warm- and cold-core eddies.

Wind direction

→ Force from above
→ Effective direction of current flow
→ Coriolis effect

Ekman spiral The Coriolis effect causes the direction of water flow to change with increasing depth, until eventually the water is moving in the opposite direction to the wind.

Gyre formation Winds blow clockwise around the subtropics of the Northern Hemisphere, and counterclockwise in the Southern Hemisphere. Gyres form in the main ocean basins.

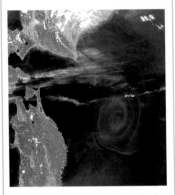

Lasting signature A gyre may also be caused by an intense pressure system. Its signature can be observed long after the driving force has vanished.

Currents

Ocean currents are the weather phenomena of the seas. They occur on the ocean surface and in deep water, and are driven by atmospheric winds, oceanic temperature, and salinity distribution. Ocean currents distribute heat and water mass worldwide and contribute to the exchange of energy across the latitudes, helping to keep the climate system in balance.

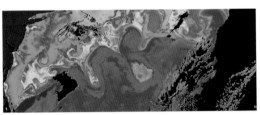

Twisting current The Gulf Stream is not a smooth flow of warm water from south to north. As it makes its way along the North American coast, it starts to become turbulent, creating meanders and eddies that transport heat to the north and cooler water to the south.

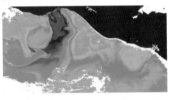

Upwelling Winds are funneled by the mountain gap in the narrowest point between the Gulf of Mexico and Pacific Ocean. This creates upwelling of a narrow band of cold water, leading to large temperature differences at the sea surface.

Mixing currents The cold Malvinas Current springs off the Antarctic Circumpolar Current and stretches along the east coast of South America. It meets and mixes with the warm Brazil Current approximately where the Rio de la Plata enters the ocean, creating nutrient-rich and nutrient-poor cells.

Ocean conveyor (below) The great ocean conveyor belt is a slow-moving band of water encircling the oceans. It is also known as thermohaline circulation. Driven by density differences, it connects deep ocean currents with surface currents.

Global currents Salty, subtropical water cools and sinks in the North Atlantic. Forming a steady stream of cold, deep water, it flows south then east, rises in the Indian and North Pacific oceans, and is driven back into the Atlantic as surface currents. The water can take 1,000 years to circulate.

Meeting currents At the southern tip of South Africa, three large currents meet: the warm Agulhas, the cold Benguela, and the powerful Antarctic Circumpolar current. The interaction of these currents with the bathygraphy, or topography of the sea floor, results in dramatic features, such as this jet-like plume of chlorophyll-rich water, shown in red.

Blooming seas Late summer monsoon winds blow parallel to the coasts of the Arabian Sea, forcing nutrient-rich deep water to the surface. This creates blooms of phytoplankton, shown in red.

Spring blooming In spring, solar heating of the ocean in the Gulf of Alaska can create lenses of warm water in the upper layers. Phytoplankton breed around the edges of these lenses, where the nutrient supply is optimal.

Vertical motion The Ekman spiral causes water currents (blue arrow) to deviate from wind direction (white arrow) by 90 degrees: to the right in the Northern and to the left in the Southern hemispheres. This is called the Ekman transport.

Cold upwelling Northerly winds in the Northern Hemisphere cause surface water to move away from western continental coasts, creating upwelling.

Warm downwelling Southerly winds on a western coastline in the Northern Hemisphere force surface water toward the coast. This brings warm but nutrient-poor water to the shore.

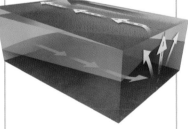

California Current The cold California Current, driven by trade winds of the North Pacific, flows south. Surface water is driven offshore and replenished by nutrient-rich, deep water. The coastline creates eddies filled with phytoplankton.

Cold tongue The Ekman transport is not restricted to coastal areas. Trade winds cause surface water to diverge at the Equator, leading to upwelling and the creation of the Pacific cold tongue.

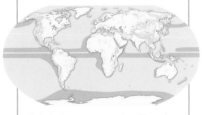

Global phenomenon Upwelling of cold water is a global phenomenon. It occurs along all western continental coasts and in the open ocean.

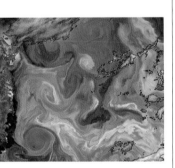

Satellite art Substances in the water and unusual atmospheric conditions may impede the capturing of an accurate satellite image of water temperature. Without correction, this image off the east coast of Tasmania, Australia, would have limited scientific value.

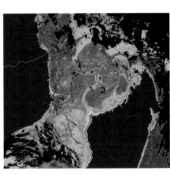

Channel flow A warm current flows south through the Mozambique Channel and forms the Agulhas Current off the coast of southeast Africa. Images of chlorophyll concentrations can help in the study of channel turbulence.

Waves

Winds blowing across ocean waters produce waves. Frictional forces transfer wind energy to the water molecules, deforming the surface and creating waves that range from little ripples to rogue, hurricane-strength waves. The distance between one wave crest and the next is called the wavelength. The time it takes successive crests to pass the same point is the wave period.

Beaufort scale (right) Introduced in 1805 by British Admiral Sir Francis Beaufort to ease communication between sailors, this scale measures wind strength. It distinguishes 12 different classes of wind, depending on the observable state of the sea.

Wave formation (below) The stronger the wind, the higher the waves. Other factors such as the distance the wind blows uninterrupted (fetch), wind speed, duration, and water depth determine the shape of waves.

Force 0: Calm No wind or waves. The sea surface remains flat and calm.

Force 4: Moderate breeze The wind blows from 13 to 18 mph (21 to 29 km/h), creating small waves, including some waves with a white, foamy crest.

Force 8: Gale Wind speeds of 39 to 46 mph (63 to 74 km/h) result in moderately high waves of extended length. The edges of the wave crests break into streaks of foam, which are blown along the wind direction.

Force 12: Hurricane The wind blows at 75 mph (120 km/h) or more. The air is filled with foam and spray, affecting visibility. Depending on circumstances, waves may be 50 feet (15 m) or higher.

Forward motion The speed at which waves travel is equal to the wavelength divided by the period. A wind-generated wave curves up to a narrow crest, then curls down to a low trough.

Inside a wave Water particles inside a wave perform a circular motion: up, forward, down, and back. Waves transport huge amounts of energy, but matter is moved only by surface drag.

Break point When wave height, the vertical distance from a crest to a trough, is close to water depth, the wave becomes too steep and breaks.

Crest

Trough

Closer and higher As waves approach shore, the water continues to arrive at the same speed and the wind force remains constant. As a result, the wavelength shortens and the waves become steeper.

Reduced speed Waves slow down as they travel over shallow water because friction along the seafloor acts like a brake.

Breaking When waves come ashore, they break. Friction slows the base of the wave and the wave crest breaks over the base. Wave energy is released and the water carries sand, pebbles, and other matter back and forth.

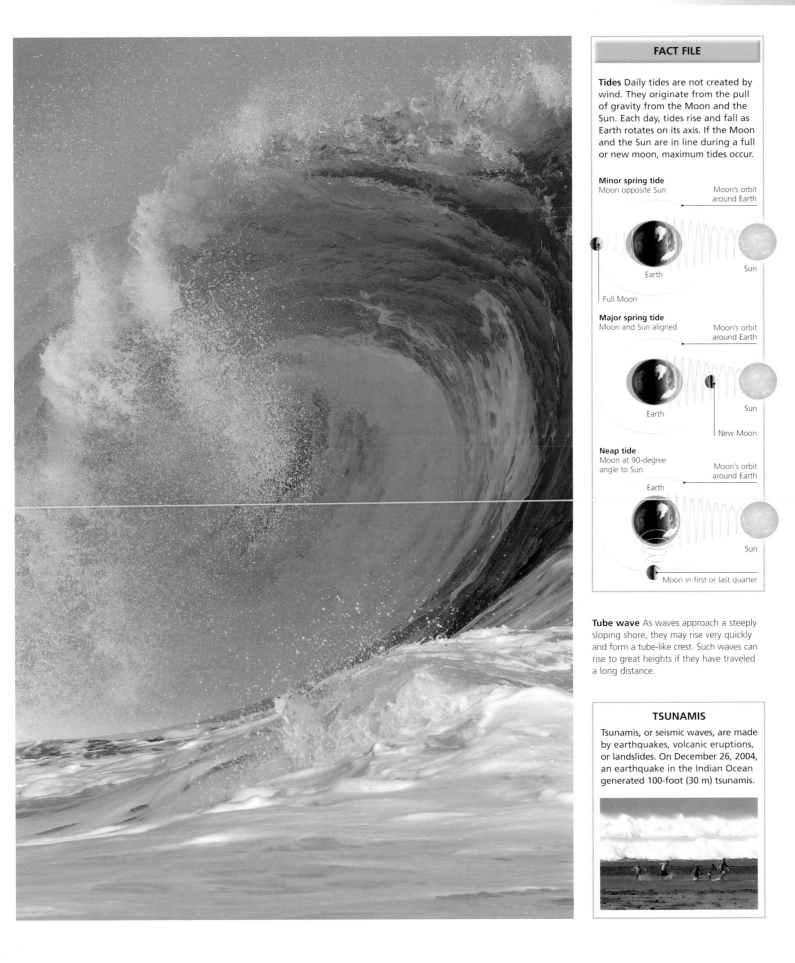

Tides Daily tides are not created by wind. They originate from the pull of gravity from the Moon and the Sun. Each day, tides rise and fall as Earth rotates on its axis. If the Moon and the Sun are in line during a full or new moon, maximum tides occur.

Minor spring tide
Moon opposite Sun

Moon's orbit around Earth

Earth

Sun

Full Moon

Major spring tide
Moon and Sun aligned

Moon's orbit around Earth

Earth

Sun

New Moon

Neap tide
Moon at 90-degree angle to Sun

Moon's orbit around Earth

Earth

Sun

Moon in first or last quarter

Tube wave As waves approach a steeply sloping shore, they may rise very quickly and form a tube-like crest. Such waves can rise to great heights if they have traveled a long distance.

TSUNAMIS

Tsunamis, or seismic waves, are made by earthquakes, volcanic eruptions, or landslides. On December 26, 2004, an earthquake in the Indian Ocean generated 100-foot (30 m) tsunamis.

ACTION

ACTION

WATER

Earth is often called the Blue Planet for it is the presence of liquid water that sets our home apart in the solar system. Water is a relatively simple but changeable substance. One molecule of water consists of two atoms of hydrogen and one atom of oxygen. This basic molecular structure gives water some unique chemical and physical properties that assure it an extremely important role in maintaining Earth's habitable environments. Water exists in three quite different forms—solid, liquid, and gas—depending on the temperature and air pressure to which it is exposed.

THE WATER SUBSTANCE

Molecules in liquid water group together in a fast-changing structure. But the structure changes less rapidly when water is cooled toward the freezing point. On freezing, the molecules lock into a hexagonal (six-sided) lattice. This is the origin of the six-sided symmetry seen in snow crystals. The molecular structure of solid ice is more open than liquid water, hexagonal "holes" extend right through the lattice, so the density of ice is less than that of liquid water, and ice floats on water.

The density of liquid water near room temperature is 62.3 pounds per cubic foot (1 g/cm^3 or 1,000 kg/m^3). Water has a maximum density near 39°F (4°C). Colder water will settle above water at this critical temperature. For this reason, lakes do not freeze from the bottom up, allowing life on the lake floor to overwinter.

Liquid water has a high thermal capacity, meaning it can hold a lot of heat for a given mass. The oceans heat up and cool down much more slowly than do landmasses. This means that ocean

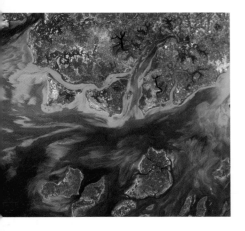

Where river meets ocean Deltas form where a large river deposits sediment as it meets the sea. This satellite photo shows the Beba and other rivers in Guinea-Bissau in West Africa.

Jacksonville, Florida Human settlements often start near a plentiful supply of fresh water. Successful villages grow and eventually develop into large cities.

Rich growth This satellite image reveals the concentrations of chlorophyll on land and in the oceans. The dark blue areas are the ocean's least biologically productive regions. Here, there are few currents to bring deep ocean nutrients to the surface.

currents can carry warm or cold water great distances around the globe. These currents change sea surface temperatures, which, in turn, influence climate, particularly in coastal areas.

When water vapor condenses or evaporates, turns to ice, or melts, a heat exchange occurs involving the transfer of energy to or from the atmosphere. These temperature differences cause air masses to lift and drift, often transporting water great distances.

Approximately 97.5 percent of water on Earth is liquid found in the oceans. These vast reservoirs contain many dissolved salts, mainly sodium chloride, and are also able to hold a certain amount of dissolved carbon dioxide and other gases. The comparatively small amount of fresh water on the planet is mostly locked up in surface ice or is underground.

Water vapor in the atmosphere only constitutes about 0.001 percent of the total, but has a key role in driving the weather. The water vapor content varies from near zero over dry deserts to about four percent in air saturated with vapor.

A continuous interchange of moisture between the oceans, the land, plants, and clouds fuels much of our weather. This process is known as the water cycle.

The water cycle is driven by the Sun which causes water to leave the oceans as a result of evaporation. It recondenses to form small cloud droplets or ice crystals that grow and fall out of clouds, giving us rain and snow. Rivers or groundwater channels carry the water back to the oceans and the cycle continues.

Coral reef Coral reefs are found in shallow waters where sunlight enables coral polyps to thrive. Some reefs are large enough for human habitation, but they tend to be very low-lying and are susceptible to severe storms and rising sea levels.

Crater lake A massive eruption can leave a deep crater in the volcano's center. Sometimes these fill with water to create a lake. In some cases, rainfall can fill the crater to a level where the overspill is sufficient to give rise to a river.

FACT FILE

Water structure A water molecule consists of two hydrogen atoms bound to one oxygen atom. The side of the molecule with the hydrogen atoms has a positive charge, while the other side is negatively charged. This difference provides an electrical attraction between molecules.

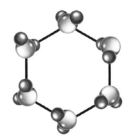

Ice Molecules of water ice form a rigid lattice with a hexagonal structure. Ice has a more open structure than water, so it has a lower density.

Liquid Water consists of flickering clusters of molecules that stay together for only an instant.

Vapor Fast-moving molecules of water vapor bounce off each other and do not become attached.

SALTY AND FRESH WATER

Earth's fresh water

Salt water in oceans

Only 2.5 percent of Earth's water is fresh water. It is distributed among ice caps, groundwater, lakes, vapor, rivers, and inside living organisms.

The Water Cycle

The water cycle is a continuous exchange of water between the oceans, atmosphere, and land. As it moves, water can change states from liquid to vapor to ice. Much of what we call weather is the water cycle at work: water on the move and water changing states. This endless process maintains a balanced system that is critical to the health of planet Earth.

Evaporation of water to form clouds

Growing food (left) Crops are chosen to match local climate and available water supply. In places where rainfall is unreliable, irrigation can make agriculture possible.

The desert environment (below) Deserts receive less than 10 inches (250 mm) of precipitation per year. Deserts can be very hot (Sahara) or very cold (Antarctica).

Rain falls from cloud | Rivers drain into the ocean | Absorption of water by soil and plants | Inland water storages are filled

Subterranean water returns to ocean

Water cycle Water circulates at a very changeable rate, depending on where it is in the cycle. On average, a water molecule will spend about 10 days in the atmosphere but 10,000 years as deep groundwater. Residence in the ocean is longer still, at about 37,000 years.

Rain forest Rain forests cover six percent of Earth's land surface. They occur in the tropics and temperate regions and are always associated with areas of high rainfall. They are home to more than half of Earth's plant and animal species.

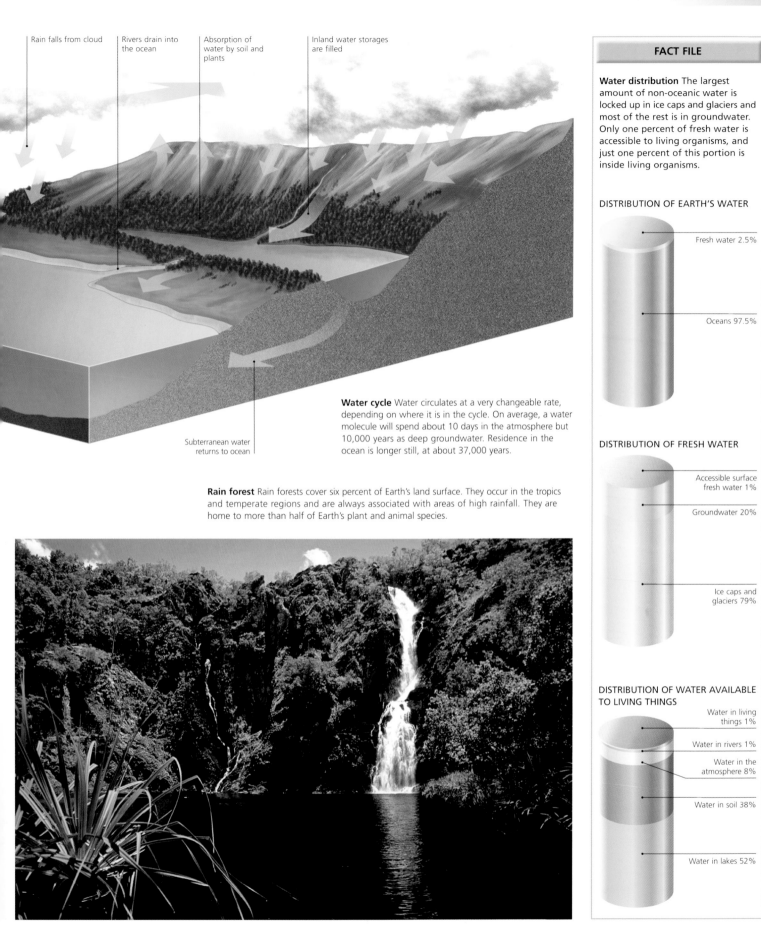

FACT FILE

Water distribution The largest amount of non-oceanic water is locked up in ice caps and glaciers and most of the rest is in groundwater. Only one percent of fresh water is accessible to living organisms, and just one percent of this portion is inside living organisms.

DISTRIBUTION OF EARTH'S WATER

Fresh water 2.5%

Oceans 97.5%

DISTRIBUTION OF FRESH WATER

Accessible surface fresh water 1%

Groundwater 20%

Ice caps and glaciers 79%

DISTRIBUTION OF WATER AVAILABLE TO LIVING THINGS

Water in living things 1%

Water in rivers 1%

Water in the atmosphere 8%

Water in soil 38%

Water in lakes 52%

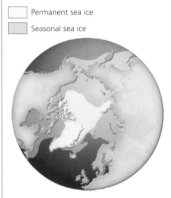

North Pole A permanent ice shelf floating on the sea extends to the north of Greenland. The shelf grows in winter to form a much larger expanse of sea ice.

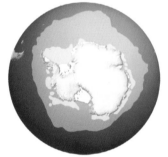

South Pole The Antarctic continent is covered in a thick ice sheet that extends over the Southern Ocean in places. Much thinner sea ice changes extent with the seasons.

LIVING ON THE ICE

Only scientists and support staff live in Antarctica, staying for short periods, but many indigenous people live permanently in the Arctic. Adept at ice fishing and hunting, they have become successful inhabitants of this inhospitable region.

Ice

Ice covers the coldest parts of Earth's surface. There is permanent ice cover over the Antarctic, the polar seas, Greenland, and several other Arctic islands, glaciers, and high mountain ranges. In colder months, sea ice is much greater in extent. Ice in the polar regions reflects the Sun's energy which has a stabilizing effect on Earth's climate.

Glacier Glaciers are slow-moving rivers of ice made up of snow laid down many years previously. Where glaciers meet the sea, the vertical ice wall breaks up to form icebergs and floating ice debris.

Spitsbergen, Arctic Ocean Communities that live on the ice have developed novel forms of transport. Snowshoes, skis, dog-sleds, and skidoos were all invented to give freedom of movement on ice and snow.

Polar wildlife (below) Polar bears are the top predators in the Arctic. They make the most of the long summer days to build up their fat reserves for winter hibernation in dens beneath the snow. The cubs are born in the spring.

Ice cores Ice core samples drilled from ice caps and glaciers give a history of climate stretching back up to 750,000 years. Analysis of trapped air shows the atmospheric concentrations of gases and other particles at the time the snow fell.

Ice core sampling In 1959 hand drills were used to obtain ice core samples. Modern drills have provided samples to a depth of 11,886 feet (3,623 m).

ICE FORMATION

Much of the snow that falls in cold climates does not melt. As time goes by, layers of new snow bury the old, compacting it and turning it to ice.

Falling snow

Fresh snowfall

Small granules of ice

Firn (compacting ice)

Solid ice

Glaciers

In mountainous and polar regions, so much winter snowfall can build up that the snow remains frozen even after the following summer's sunshine. Under the weight of snow, glacial ice slides downhill over wet ground beneath. Deep water-filled inlets in coastlines (fjords) are created by the scouring action of a valley glacier on its way to the sea.

LOUIS AGASSIZ

Swiss–American scientist Louis Agassiz is known as the father of glaciology. Although others had studied glaciers before him, he was the first to see evidence of an ice age, when Earth had been largely covered by a thick ice sheet.

Antarctic dry valleys (below) The volume of the Antarctic Ice Sheet has fluctuated greatly over geological time. The ice-free valleys in this photo were scoured by glaciers millions of years ago. They are kept dry by the action of constant wind.

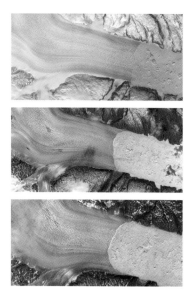

Glacial retreat (above) Greenland's Helheim Glacier was photographed by NASA in 2005 (top), 2003, and 2001. The break up of the front into icebergs has accelerated in recent years. It is now retreating about 6 miles (10 km) per year.

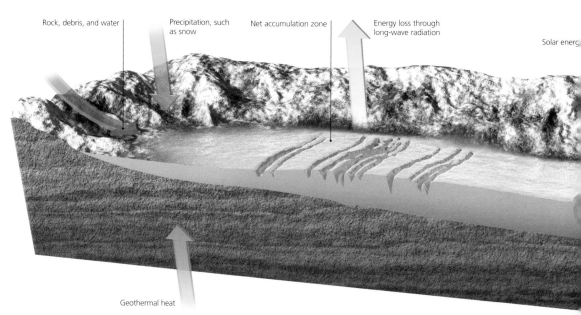

Rock, debris, and water

Precipitation, such as snow

Net accumulation zone

Energy loss through long-wave radiation

Solar energy

Geothermal heat

How a glacier flows A valley glacier is fed from above with ice, water, rock, and other debris. Geothermal heat warms the base of the glacier, causing it to flow downhill. At the upper levels, snow continues to accumulate in the net accumulation zone, but water is also lost by evaporation, sublimation, and wind (deflation). In the net ablation zone, below the equilibrium line, the glacier loses mass faster than it receives it.

Glacier breakup (above) This satellite image of Greenland's west coast shows glaciers flowing around mountaintops and into Baffin Bay. The Greenland Ice Cap is the birthplace of countless icebergs that then drift out to sea on ocean currents.

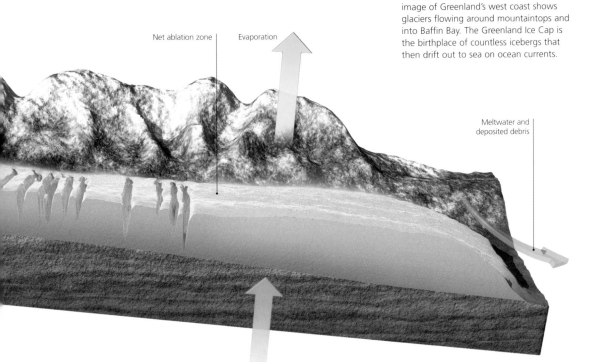

Net ablation zone

Evaporation

Meltwater and deposited debris

Geothermal heat

FACT FILE

Shaping the land As they flow, glaciers carve the underlying rocks into shapes an expert eye can recognize thousands of years later.

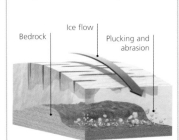

Bedrock

Ice flow

Plucking and abrasion

Glacial plucking As a glacier moves over uneven ground, it breaks the bedrock into pieces that are transported downhill until they are deposited as a terminal moraine.

Before glaciation Between ice ages, the climate is warm and mountains are covered by vegetation. Valleys are V-shaped.

During glaciation During an ice age, the climate is too cold for plants and they disappear. Glaciers flow through the valleys.

After glaciation When the ice retreats, the valleys that remain are U-shaped, with finger lakes in deeper places.

Iceberg classification There are two basic types of iceberg: tabular and non-tabular. Tabular icebergs have sheer sides and a flat top like a table-top. There are five sub-types of non-tabular icebergs.

Tabular A solid mass of ice with vertical sides and a flat top. These originally formed part of an extensive ice sheet.

Dome These icebergs have rounded tops, usually with a smooth surface.

Pinnacle A few spires of ice protrude from the bulk of the submerged ice.

Wedge Like a wedge of cheese, these icebergs have a steep flat face on one side and a smooth gentle slope on the top side.

Drydock These icebergs have two or more tall columns of ice, with a deep U-shaped slot in between that reaches down to near water level.

Blocky Like tabular icebergs, blocky icebergs have steep, vertical sides and a flat top, but are proportionally taller and resemble huge ice-cubes.

Icebergs

Icebergs are created when ice breaks away from glaciers that extend to the sea or from the breakup of floating ice sheets and ice shelves. They are composed of pure water that fell as snow thousands of years previously. Winds and tides cause icebergs to drift great distances and hundreds of them enter shipping lanes every year.

Penguins on ice Penguins use small icebergs and ice floes as temporary bases. They provide safety from predators but eventually the penguins must return to the sea to feed.

An iceberg is born (right) Upon reaching the ocean, a glacier breaks apart when melting of its lower layers overcomes the structural integrity of the ice mass. Floating icebergs break up into smaller pieces under the stresses of ocean waves and subsurface currents and tides.

Melting iceberg (below) An iceberg in water only a few degrees above freezing can take months to melt completely. It may also lose mass when eroded by heavy waves.

Iceberg size classification
The International Ice Patrol classifies icebergs by their observed dimensions above water. Typically, nine-tenths of the volume of an iceberg is below the surface.

ICEBERG SIZE CLASSIFICATION		
Size category	Height	Length
Growler	less than 3 feet (<1 m)	less than 16 feet (<5 m)
Bergy bit	3–13 feet (1–4 m)	15–46 feet (5–14 m)
Small	14–50 feet (5–15 m)	47–200 feet (15–60 m)
Medium	51–150 feet (16–45 m)	201–400 feet (61–122 m)
Large	151–240 feet (46–75 m)	401–670 feet (123–213)
Very large	over 240 feet (>75 m)	over 670 feet (>213)

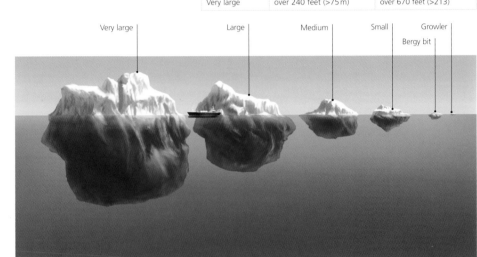

Very large Large Medium Small Growler Bergy bit

FACT FILE

Glacier meets ocean The Drygalski Ice Tongue is a floating mass of ice that reaches more than 50 miles (80 km) into the Southern Ocean. It is an extension of Antarctica's David Glacier which is fed by ice from over 180 miles (300 km) inland.

Drygalski Ice Tongue David Glacier is so strong and thick that, unlike most glaciers, it does not quickly break up on reaching the ocean.

Ice strike In March 2006 an eight by seven mile (13 × 11 km) section of the Drygalski Ice Tongue was broken off when it was struck by a giant iceberg.

HIDDEN DANGER

Approximately 80–90 percent of an iceberg lies beneath the surface of the water. The more dense an iceberg is, or the less air trapped inside it, the greater the amount of iceberg found underwater.

SHIPS AND ICE

Travel by ship requires an open passage across the seas, but in winter some waters can freeze over, making this impossible. Sea ice, both seasonal and permanent, frustrated the efforts of polar explorers for centuries. Even in regions where the sea remains unfrozen, icebergs that have broken off the polar ice shelves may pose hazards. In areas where the transport of goods by sea is economically important, icebreakers can maintain a passage through the ice for some time. Even so, many ports are closed in the heart of winter.

ICE HAZARDS

Ships crossing the oceans in polar regions have to avoid icebergs, which are mostly hidden underwater. The RMS *Titanic* was just one of many ships lost after striking an iceberg. The *Titanic* was designed with watertight compartments to keep the ship afloat after a collision. However, the accident caused so many of these compartments to fill with water that the ship sank. The supply of lifeboats on board met the safety standards of the time but was insufficient for the large number of passengers and crew, and the loss of life was very high.

Nowadays, cruise ships take tourists to the Arctic and Antarctic oceans. Even with all the advantages of modern technology, icebergs are still a serious danger. The *Explorer*, cruising from South America to Antarctica, was seriously damaged and sank in November 2007 after striking what was believed to be ice. The *Explorer* was the first cruise ship to be custom-built for polar waters. All passengers and crew were rescued.

There have been numerous expeditions to the polar regions, involving ships as an integral part of the exercise or simply to transport the voyagers. British explorer Captain James Cook became one of the first to cross the Antarctic Circle, during his voyage of 1772–75. He sailed almost all the way around Antarctica but did not see land. In 1820, explorer Fabian Gottlieb von Bellingshausen, leading a Russian expedition, became the first person to see the Antarctic continent.

In the 19th century, a number of expeditions were mounted to explore the Arctic Ocean, reach the North Pole or seek out the fabled Northwest Passage. Many of these exploration ships were fated to be crushed by ice. Notable voyages include John Franklin's

Headline news (left) Designed to be unsinkable, the *Titanic* was holed by an iceberg while on her maiden voyage from Southampton, England, to New York, U.S.A. Of the 2,223 people on board, only 706 survived.

Maritime disaster (below) On April 14, 1912 at 11.40 pm, RMS *Titanic* struck an iceberg and sank in less than three hours. The forward watertight compartments flooded and dragged the ship down. The vessel broke in two as it sank.

"lost expedition" commanding the HMS *Erebus* and HMS *Terror* in 1845; U.S. Arctic explorer Charles Francis Hall, aboard the USS *Polaris* in 1871–73; British explorer Albert Markham, HMS *Alert*, 1875–76; Adolphus Greely, USS *Proteus*, 1881–84; and Norwegian Fridtjof Nansen, *Fram*, 1893–96.

Explorers soon turned their attention to Antarctica. In 1907, British explorer Ernest Shackleton's expedition, delivered by the *Nimrod,* came as close as 112 miles (180 km) from the South Pole. In 1914 he was determined to reach the South Pole and cross Antarctica by dogsled, but his ship, HMS *Endurance* was crushed by ice in the Weddell Sea before it reached land. The crew over-wintered on the ice on Elephant Island, while Shackleton and several others rowed for 15 days to reach South Georgia for help. The whole crew was eventually rescued.

More recently, submarines have been able to travel under the Arctic ice floes. The first submarine to reach the North Pole was the nuclear-powered USS *Nautilus*, in 1958. In August 2007, traveling in two mini-submarines, Russian explorers planted a titanium flag on the North Pole seabed to lay a since disputed claim to Arctic resources.

Track of the *Fram* in open water

Track of the *Fram* in polar ice

Track of Nansen and Johansen's sled journey

Track of Nansen and Johansen home in the *Windward*

The voyage of the *Fram* The specially-strengthened *Fram* was locked and drifting in the ice for three years from September 1893. Fridtjof Nansen and Hjalmar Johansen left the rest of the crew in March 1895 in a thwarted attempt to reach the Pole by dogsled. Fifteen months later, they happened across a British expedition in Franz Joseph Land and returned home with them.

Nansen and *Fram* (above) Norwegian explorer Fridtjof Nansen hoped to be carried to the North Pole aboard the intentionally icebound ship *Fram*. When it became apparent that the currents were not in his favor, he abandoned *Fram* in the ice and tried to reach the Pole on foot.

Shackleton and *Endurance* (right) During Sir Ernest Shackleton's 1914 expedition to Antarctica, his ship, *Endurance*, became stuck in sea ice and sank.

Advance guard (below) Icebreakers clear channels for shipping in high latitudes. Their strengthened hulls are designed to ride up onto the ice and the weight of the ship breaks it.

Humidity

Humidity refers to the amount of water vapor in the atmosphere. Vapor levels depend upon temperature and vary considerably across the planet. The amount of water vapor in air is usually measured by its relative humidity: the water vapor content as a percentage of the saturation level at the ambient temperature.

Global water vapor This false color satellite image reveals the atmospheric water vapor content on a late January day. Humidity varies from high values at the Equator to low values near the poles and deserts. Black areas indicate cloud cover.

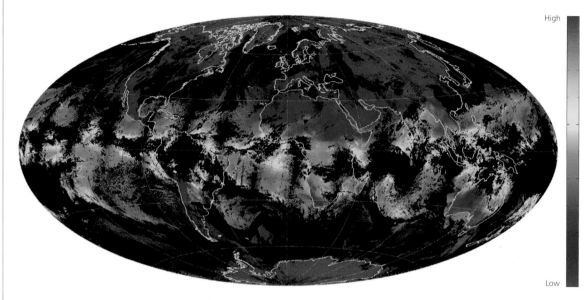

High

Low

FACT FILE

1. Beijing, China A combination of relative humidities over 70 percent and high temperatures make life in Beijing very uncomfortable in summer. The humidity aggravates pollution and reduces visibility. Winter temperatures fall below 14°F (−10°C) with relative humidities of 50 percent.

Beijing, China

2. Brisbane, Australia Brisbane, on the east coast of Australia, has a subtropical climate like that of Florida in the U.S.A. Maximum relative humidity exceeds 70 percent in late summer, with maximum temperatures of 82°F (28°C).

Brisbane, Queensland, Australia

3. Phoenix, Arizona In summer, Phoenix is among the hottest places in North America, with an average high temperature in July of 106°F (41°C). Fortunately for its residents, the punishing heat is moderated somewhat by a low average summer humidity of just 12 percent.

Phoenix, Arizona, U.S.A.

4. London, England London's maximum humidity occurs in winter, with an average of 87 percent, but the average winter temperature is only 43°F (6°C). In summer, with average highs of 72°F (22°C), the humidity is around 70 percent.

London, England

5. Kolkata, India Built on reclaimed wetlands near the Bay of Bengal, Kolkata has a subtropical climate that leads to an uncomfortable 88 percent relative humidity in August. The maximum temperature in summer is 106°F (41°C) and the minimum in winter is 50°F (10°C).

Kolkata, West Bengal, India

Life in the desert In hot, dry desert conditions, the body perspires copiously and sweat evaporates rapidly. In the Middle East and North Africa, people wear flowing robes that cover most of the body and help to reduce water loss from perspiration.

Measuring humidity The amount of water that an air mass can hold gradually increases with temperature. Air incapable of holding more water vapor is said to be saturated. The temperature at which saturation occurs is known as the dew point.

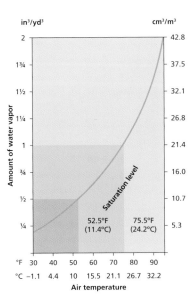

Global Humidity Index The Global Humidity Index is a measure of mean annual potential moisture availability. It is based on the ratio of annual precipitation and annual potential evapotranspiration (the sum of evaporation from lakes, streams, soil, canopy, and plant transpiration). Low values indicate deserts—high values indicate high humidity regions.

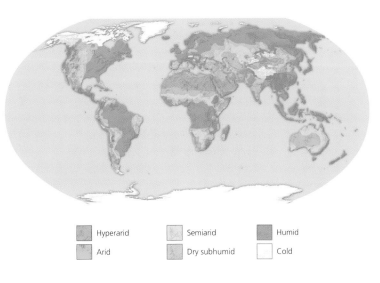

Hyperarid
Arid
Semiarid
Dry subhumid
Humid
Cold

Life in the tropics Keeping cool can be a struggle when both humidity and temperature are high. Near-nudity is favored by many inhabitants of the tropics, including this hunter from the Mentawai Islands, Indonesia.

Humidity and comfort In hot weather, the body's primary strategy for keeping cool is to sweat. However, high humidity makes this much more difficult.

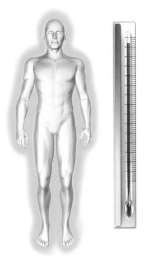

Heat is on In hot weather, the heart pumps blood away from the body's core to the skin.

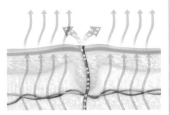

Keeping cool When the air is dry, sweat can easily evaporate, taking some of the body's heat with it.

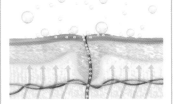

Hot and bothered Water vapor in humid air slows the evaporation, decreasing the cooling effect.

Shifting humidity Rising global temperatures may alter the global distribution of rainfall. Moisture in the air is less likely to condense to produce rain in higher temperatures, so deserts may spread. Warmer oceans lead to higher humidities over the sea, which may favor more severe tropical storms.

Dew and Frost

On a cold, clear night, the ground is cooled as it radiates energy into space. The air near the ground also cools and if the dew point is reached, water vapor will condense on to surfaces near the ground. If the temperature is above freezing, dew is formed; if it is below freezing, the vapor will form frost.

Frost on trees (below) In areas where temperatures fall below freezing at night, frost can deposit on tree leaves, needles, and branches. Frost-tolerant local varieties are preferred in managed woodlands.

Frost on grass (above) This is a magnified view of frost on a blade of grass. Frost may form as fine feathery growth or denser structures, depending on temperatures.

Dew and frost formation (below) Dewdrops begin when water vapor molecules deposit on a cold surface. Once a liquid film has formed, surface tension draws the molecules together to make dewdrops that grow over time. At temperatures below freezing, water molecules build preferentially on embryonic ice crystals to form frost crystals.

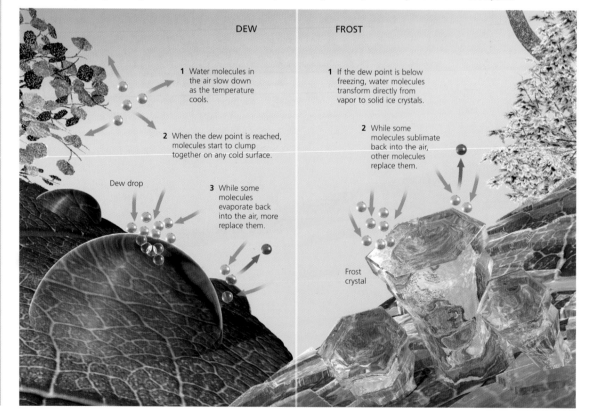

DEW

1 Water molecules in the air slow down as the temperature cools.

2 When the dew point is reached, molecules start to clump together on any cold surface.

3 While some molecules evaporate back into the air, more replace them.

Dew drop

FROST

1 If the dew point is below freezing, water molecules transform directly from vapor to solid ice crystals.

2 While some molecules sublimate back into the air, other molecules replace them.

Frost crystal

Dewdrop (below) An array of small dewdrops has formed on a blade of grass. When a pair of droplets has grown large enough to reach a neighbor, it coalesces to form a larger dewdrop.

Frosted leaves (above) Leaf edges and other surfaces with fine irregularities are favored sites for frost crystal growth.

FACT FILE

Frost and crops Frost damage to crops can be severe. If water in a plant's cells freezes and expands, the cells will be damaged and the plant may die. Protective measures include kerosene-fuelled heaters and removable covers.

Vineyards Frost damage can ruin a grape harvest. But frosting is essential in grapes used for some dessert wines, to concentrate the sugars.

Cover up This vegetable crop is protected from frosts by plastic sheeting. Such covers allow for an extended growing season.

ADAPTED TO ICE

The caribou is in its element in cold and frosty conditions. Sharp-edged hooves secure traction on ice and crusted snow.

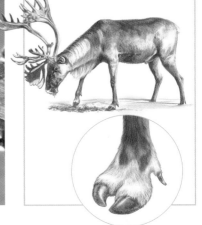

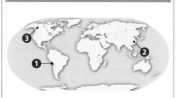
1. Camanchaca fog This fog forms when the cold oceanic Humboldt Current cools moist marine air. The fog can extend several miles inland over the Atacama Desert. Plants and insects in this virtually rainless region depend entirely on this fog for water.

Atacama Desert, Chile

2. Chongqing This city, located at the confluence of the Changjiang and Jialing rivers, is known locally as "Foggy City." Water vapor from the two rivers feeds a fog that shrouds the city for more than 100 days each year, mostly during winter. Visibility is often reduced to 30 feet (9 m) or less.

Chongqing, Sichuan, China

3. Fairbanks Surface temperatures in Fairbanks can fall to −45°F (−43°C). Even colder air higher up traps a stable layer of air near the surface, where a thick fog of ice crystals often forms. Moisture for ice crystals comes from local sources such as automobile exhaust and power plant cooling towers.

Fairbanks, Alaska, U.S.A.

WATER FROM AIR

In parts of southern Africa, fog is common but rain is not. Inspired by the fog-basking beetle *Onymacris unguiculari*, some communities harvest precious water using fog-catching nets.

Mist and Fog

Fog is really cloud that forms near the ground, and, like cloud, develops as a result of condensation. Fog usually consists of water droplets that adhere to atmospheric particles such as dust specks. In extremely cold conditions, fog can form out of ice crystals. If visibility is between half a mile and one and one-quarter miles (1–2 km), the fog is known as mist.

Grounded Fog is a major aviation hazard because it can hide an airstrip from a pilot's view. It can also have economic impacts when flights are grounded or diverted. Some pilots and aircraft are certified to land in even the thickest fog, using automated landing technology installed in the aircraft and on the ground.

Mediterranean Sea fog This satellite image shows early-spring advection sea fog hugging the shoreline around the Gulf of Taranto, in southern Italy.

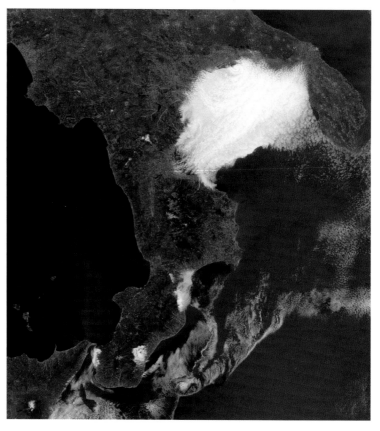

Types of fog Fog usually forms when humid air is cooled to its dew point, causing water vapor to condense into tiny drops. Less often, evaporation of water into air can play a role. This illustration shows four common ways that fog can form.

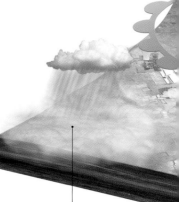

Frontal fog Frontal, or precipitation, fog occurs when some falling rain or snow evaporates into the air below the cloud. The water vapor both cools the air and increases its moisture content. Fog forms when the dew point of the air is reached.

Upslope fog This fog occurs when moist air is blown up a hill or mountain, where it cools.

Radiation fog This fog usually forms on clear nights when the ground cools rapidly, chilling the air that is in contact with it. Valley radiation fog can last for several days in calm conditions.

Advection fog This fog occurs when warm, moist air flows over cold ground or water.

San Francisco fog
Summer is peak time for fog in San Francisco, when warm, moist air is chilled as it crosses cold currents just offshore in the Pacific Ocean. The fog is then carried by morning breezes into San Francisco Bay and over the city.

FACT FILE

The Great Smog In December 1952, London's characteristic fog combined with pollutants from low-grade heating coal to create a deadly smog. It lasted for five days in the stable winter air, resulting in the deaths of more than 12,000 people.

Pea souper Visibility was greatly reduced during the Great Smog. At times, driving and even walking safely became impossible.

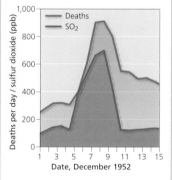

Death rate The number of deaths recorded in London spiked along with the pollution levels. Many more died in the weeks that followed.

STEAM FOG

When very cold air moves across a body of water that is at a warmer temperature, vapor evaporating from the surface is cooled and forms fog. Steam fog is often seen over lakes and streams during cold winter conditions.

FOG SIGNALS

When fog obscures lighthouses and other maritime navigational aids, foghorns are used to warn of shorelines, shoals, and other dangers. Foghorns aboard boats are used to signal the size of the vessel and its position.

Mist and Fog continued

Extensive areas of fog may obscure large areas of terrain, creating significant hazards for mariners and pilots. This can cause problems for the motorist as well, particularly in mountainous areas. To those on foot, fog can be a disorientating and sinister presence. At other times, it can be one of the most beautiful and tranquil manifestations of weather.

Slovenian Alps (right) Two types of fog are shown in this scene: at the top of the mountain ridge, fog resulting from orographic lift is pouring over from the far side, while on the near side, rising air is cooling to produce fog.

Jura, France (above) Stratus fog has developed overnight in this valley. A shallow layer of fog indicates a steep temperature change from the surface to the warmer air only a short distance above. In the background, the fog is lifting to form low-lying cloud.

Iceland (below) The Westfjords peninsula in northwestern Iceland is locally renowned for a thick, fast-setting fog. It occurs when air warmed by a branch of the Gulf Stream meets much colder air over the Greenland Sea.

Huangshan Mountains, China (below) Fog is almost a permanent fixture in the steep valleys of the Huangshan Mountains in southeast China. The fog is fed by high rainfall and humidity, and the frequent arrival of warm air masses from the Pacific.

Morning fog On a clear, still night, cold ground can chill humid air to its dew point. The droplets grow until a stable layer of fog blankets the ground. In the morning, the action of the sun can form a layer of fog stratus.

Sunrise The sun's rays first heat the ground near the edges of the fog, causing the perimeter to dissipate. The sun also heats the air, and convection currents overturn the fog layers, bringing them up into the sunlight.

Early morning The fog is thinning fast under continued solar heating and the stirring effect of convection. The sun's rays can now penetrate the deck. The heat from the ground then begins to evaporate the fog at low levels.

Mid-morning The fog has eroded from the edges towards the center and from the ground up, resulting in a layer of fog some distance off the ground.

CLOUDS

Never the same twice, clouds are among the most dynamic features of our natural world. They can vary from benign, fluffy white masses drifting in a summer sky to towering storm clouds that lash the ground with rain and hail and illuminate the sky with vivid flashes of lightning. No matter what shape they take, every cloud is a visible mass of tiny water droplets or ice crystals. They are essential in the production of life-giving rainfall. With rare exceptions, clouds exist entirely within the troposphere, the lowest layer of the atmosphere.

THE NATURE OF CLOUDS

Clouds are formed when moist air is cooled sufficiently for its water vapor content to reach saturation, followed either by condensation to form water droplets, or by deposition to produce ice crystals. Usually, the cooling occurs when relatively warm air from near the ground is carried up and expands into colder, lower-pressure levels higher in the atmosphere. The formation of droplets or crystals normally requires the presence of tiny airborne nuclei in the atmosphere, which serve as sites for condensation or deposition to occur.

When conditions are above freezing, water droplets will form. Over time, the droplets grow in size as more vapor molecules are drawn toward them and are captured. This process continues until the droplets are about 0.001 inch (0.02 mm) in diameter, then the growth rate slows and the cloud is made up of a stable population of droplets.

Precipitation occurs when some of the droplets or ice crystals that constitute a cloud grow large enough to fall to Earth under the influence of gravity. Two processes can cause this to happen, and they can occur both individually and in combination.

The first process, known as coalescence, is most likely to occur in very moist cumuliform clouds and when air temperatures are above freezing. The water droplets in clouds are generally so small that they are kept aloft by air resistance and rising air currents, despite the effects of gravity. However, if turbulence within the cloud causes these droplets to collide, they may merge with one another to form larger droplets, and eventually these droplets will be heavy enough to fall from the cloud. As they fall, they collide with more

droplets, continuing to grow in size until they reach the ground as rain.

The second process, called the Bergeron process, requires ice crystals to be present in the cloud. It is most likely to occur in thick clouds at middle or high latitudes, where supercooled water droplets (water cooled below freezing but not frozen) and ice crystals coexist.

Ice crystals and supercooled water droplets have different

saturation points, so when they coexist in a cloud, water molecules will move from the water droplets to the ice crystals. Under these conditions, the ice crystals will grow quickly at the expense of the water droplets until they are large enough to fall. When falling, they will tend to grow further through the process of accretion, and may reach the ground as ice pellets, or may melt on the way down to fall as rain.

South Georgia (above) Clouds are often created or modified through interaction with landforms. In this satellite image, cumulus clouds are being deflected around the island of South Georgia while thicker marine stratocumulus is forming from air lifted over the island's highest peaks.

Clouds over ocean (above) In the late afternoon, these broken offshore altocumulus clouds allow crepuscular rays of light to cast a spotlight on the ocean's surface.

Noctilucent clouds These rarely seen blue-gray clouds form high in the mesosphere, on the edge of space, and are only visible when the sun is over the horizon. They are observable in the early summer from latitudes between 50 and 60 degrees in both hemispheres. They may consist of ice-coated micrometeorite fragments.

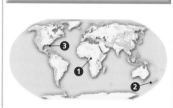

1. Tropical Africa Towering cumulonimbus clouds producing intense precipitation are regular visitors to tropical African skies. They

are the result of powerful convection cells that rise from rain-soaked ground heated by solar radiation.

Tropical Africa

2. Cardrona This ski resort at an altitude of 5,500 feet (1,700 m) is subject to fast-changing weather caused by orographic uplift. Moist air

from the west is lifted up by the mountains, where it condenses to form rain and reliable snowfall in winter.

Cardrona, New Zealand

3. Florida Air masses moving east from the Gulf of Mexico and air masses moving west from the Atlantic Ocean regularly encounter each other over the Florida peninsula. This forces some air upward, where it condenses to form

clouds. Any warm air overlying the landmass is likely to be pushed up by the denser, cooler oceanic air.

Florida, U.S.A.

LAKE IN THE SKY

Clouds float because they are less dense than the air beneath them. But they do have weight. The water contained within a modest cumulus cloud measuring about 0.6 miles (1 km) wide and high weighs more than a 747 jetliner fully loaded with passengers and fuel.

Cloud Formation

Almost all clouds form from the movement of warm, moist air upward into cooler, less dense levels of the atmosphere, where the water vapor condenses. There are various mechanisms that drive this movement and each gives rise to a distinctive set of cloud types. The type and strength of the uplift can result in anything from benign cumulus or cirrus clouds to violent cumulonimbus storm clouds.

Convection Convection occurs when an air mass, heated by contact with warm ground, becomes buoyant and moves upward. This produces dome-shaped cumulus or stormy cumulonimbus if the convection is sufficiently powerful.

Orographic lifting When an air mass encounters a mountain range, the landmass forces the air upward, often lifting it to condensation level. This usually produces layers of stratus-type clouds.

Lifting mechanisms The rising motion of air to produce a cloud can be associated with convection, orographic, or dynamic uplift. Convection occurs when warm air becomes more buoyant than its surroundings and moves upward. Orographic uplift happens when air moves over mountain barriers. Dynamic uplift is associated with large-scale air movement into surface low-pressure systems or along frontal surfaces.

Frontal formation When two air masses of different temperatures meet along a front, the warmer air mass is forced to rise. If the warm air is sufficiently moist, cloud will then form.

Convergence When two or more air masses collide, some air is forced upward. In the midlatitudes these convergences can seed extratropical cyclones that bring stormy weather to large parts of Earth.

Cloud maker Not all clouds are produced by natural mechanisms. Warm moist air rises in the cooling towers of this nuclear power station then cools in the air above to condense and form clouds.

FACT FILE

Air stability A rising mass of unsaturated air cools at a rate of 5.4°F per 1,000 feet (9.8°C per km). As long as the air mass remains warmer than the surrounding air, it will continue to rise.

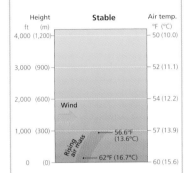

Height		Stable	Air temp.
ft	(m)		°F (°C)
4,000	(1,200)		50 (10.0)
3,000	(900)		52 (11.1)
2,000	(600)	Wind	54 (12.2)
1,000	(300)	Rising air mass ← 56.6°F (13.6°C)	57 (13.9)
0	(0)	← 62°F (16.7°C)	60 (15.6)

Stable If an air mass quickly reaches the temperature of the surrounding air (and therefore stops rising), conditions are said to be stable.

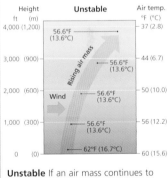

Height		Unstable	Air temp.
ft	(m)		°F (°C)
4,000	(1,200)	56.6°F (13.6°C) →	37 (2.8)
3,000	(900)	Rising air mass → 56.6°F (13.6°C)	44 (6.7)
2,000	(600)	Wind → 56.6°F (13.6°C)	50 (10.0)
1,000	(300)	→ 56.6°F (13.6°C)	56 (12.2)
0	(0)	← 62°F (16.7°C)	60 (15.6)

Unstable If an air mass continues to rise, conditions are said to be unstable.

DROP SIZES

A typical raindrop is about one-tenth of an inch in diameter (2.5 mm). Cloud droplets may measure up to one-hundredth of an inch (0.025 mm). All droplets start with water molecules collecting on a cloud condensation nucleus, which is 100 times smaller than a cloud droplet.

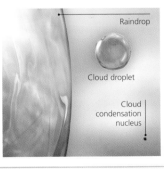

Raindrop

Cloud droplet

Cloud condensation nucleus

Vortices and Waves

Moving air can behave just like a fluid, swirling around in a fast-flowing river. A vortex develops when fluid or airflow rotates, and can range in scale from weather frontal systems such as cyclones and anticyclones, and airflow over mountains, down to the eddies that form at the tips of aircraft wings.

Vortex waves over the Canary Islands
In this satellite view, the Canary Islands off the west coast of Africa are obstructing the airflow blowing from left to right. This produces turbulent wake patterns in the clouds downstream of the islands.

FACT FILE

1. Rocky Mountains, U.S.A. The Colorado Rockies are a frequent source of multi-stacked lenticular wave clouds located over and downwind of the mountains. The smooth edges of the clouds emulate the smooth curve of the wavelike airflow across the range.

Rocky Mountains, U.S.A.

2. Pennines, England Airflow across the Pennines from the east intensifies as it drops down the steep western slopes of the range. Approximately 6 miles (10 km) further downwind, a stationary rotor cloud often forms parallel to the hills. This roll cloud, called The Helm Bar, can last for days.

Pennines, England

3. Gulf of Carpentaria, Australia The Morning Glory is a spectacular roll cloud that sweeps across Australia's Gulf of Carpentaria region in spring. The cloud can be up to 600 miles (1,000 km) long and a mile (1.6 km) high. It attracts glider pilots who "surf" the updrafts at the leading edge of the cloud.

Gulf of Carpentaria, Australia

WAVE CLOUDS OFF AFRICA

When cool, dry air from North Africa meets a stable layer of moist, warm air over the Atlantic Ocean it pushes up the moist air. The air cools and condenses at the crest then falls, to be pushed up again and form a slightly lower cloud. This results in a series of decreasing wave clouds.

Von Kármán vortex (right) Alexander Selkirk Island off the Chilean coast is over a mile (1.6 km) high. The extreme steepness of the island produces an elegant vortex pattern as the air is swept around it.

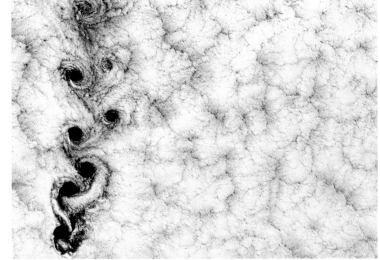

Lenticular clouds over Mount Cook Lenticular waves are a common sight over New Zealand's Southern Alps. They are formed when moisture-rich air from the Tasman Sea is lifted up by the range.

Moist air

Lenticular cloud

Formation of lenticular clouds Air rising over the crest of a mountain range condenses to form smooth clouds. The airflow continues its wavelike pattern and creates a series of lens-shaped clouds at the wave peaks, downwind of the mountains.

Von Kármán vortices A repeating pattern of swirling vortices called Von Kármán vortex streets, is caused when large bodies disturb steady airflow. They are named after the engineer and fluid dynamicist Theodore von Kármán.

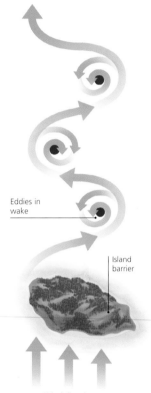

Eddies in wake

Island barrier

Wind direction

Island vortex Where an island extends through the temperature inversion that caps low-level vertical motion, air is forced around the barrier rather than over it.

Theodore von Kármán This Hungarian–American scientist (1881–1963) was a leading figure in the field of aeronautics and astronautics. He established a laboratory at Caltech that later became the NASA Jet Propulsion Laboratory.

Classifying Clouds

Many different types of clouds are observed in nature and each type is identified by its special name. In 1803, Englishman Luke Howard first used Latin words to describe four main types: *cirrus* "wisp", *cumulus* "heap", *stratus* "layer", and *nimbus* "bringing rain". With the addition of the word *alto* for mid levels, a combination of these names is used to describe all main clouds.

Living Earth (below) This image is made up of data from a number of satellites to represent the global environment in which we live. Cloud type, depth, and extent was recorded using infrared detectors on four geostationary satellites.

Clear sky, 0 oktas There are no or very few clouds in the sky.

Scattered clouds, 2 oktas Approximately one-quarter of the sky is covered.

Broken clouds, 6 oktas Approximately three-quarters of the sky is covered.

Overcast, 8 oktas A layer of cloud completely covers the sky.

Obscured This symbol is used if the sky is obscured by fog, heavy snowfall, or smog.

Clouds low and high In this cloudscape, high-altitude cirrus cloud consisting of ice crystals overlies a lower region of altocumulus, made up of water droplets. The region between is too dry for clouds to form.

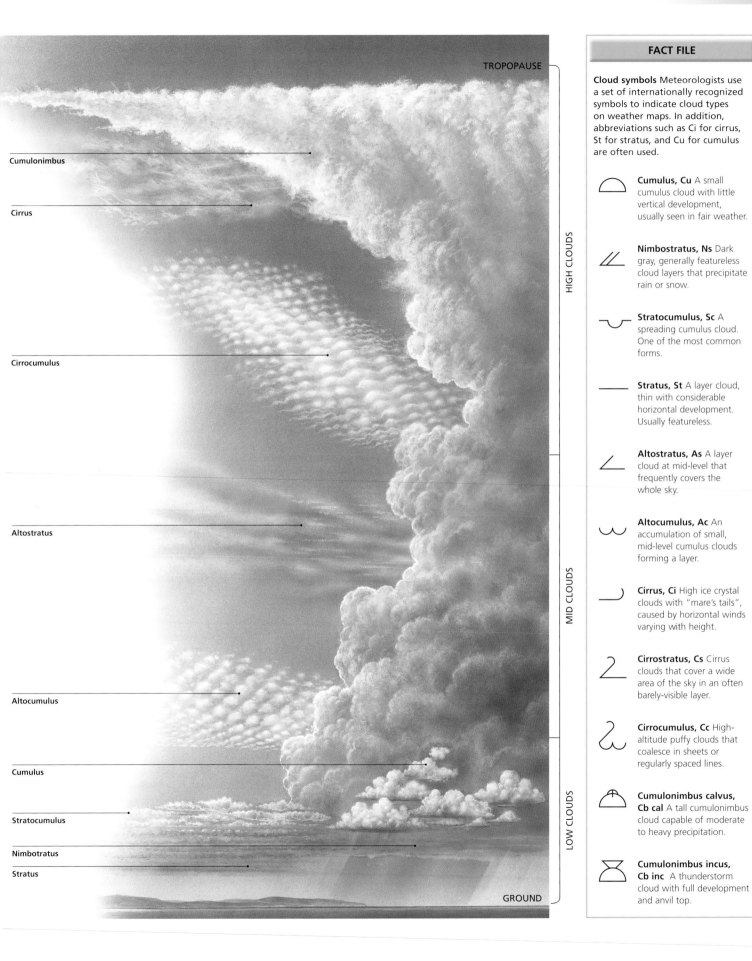

TROPOPAUSE

HIGH CLOUDS

Cumulonimbus

Cirrus

Cirrocumulus

MID CLOUDS

Altostratus

Altocumulus

LOW CLOUDS

Cumulus

Stratocumulus

Nimbotratus

Stratus

GROUND

FACT FILE

Cloud symbols Meteorologists use a set of internationally recognized symbols to indicate cloud types on weather maps. In addition, abbreviations such as Ci for cirrus, St for stratus, and Cu for cumulus are often used.

Cumulus, Cu A small cumulus cloud with little vertical development, usually seen in fair weather.

Nimbostratus, Ns Dark gray, generally featureless cloud layers that precipitate rain or snow.

Stratocumulus, Sc A spreading cumulus cloud. One of the most common forms.

Stratus, St A layer cloud, thin with considerable horizontal development. Usually featureless.

Altostratus, As A layer cloud at mid-level that frequently covers the whole sky.

Altocumulus, Ac An accumulation of small, mid-level cumulus clouds forming a layer.

Cirrus, Ci High ice crystal clouds with "mare's tails", caused by horizontal winds varying with height.

Cirrostratus, Cs Cirrus clouds that cover a wide area of the sky in an often barely-visible layer.

Cirrocumulus, Cc High-altitude puffy clouds that coalesce in sheets or regularly spaced lines.

Cumulonimbus calvus, Cb cal A tall cumulonimbus cloud capable of moderate to heavy precipitation.

Cumulonimbus incus, Cb inc A thunderstorm cloud with full development and anvil top.

Low Clouds

Low clouds have cloud bases below 6,500 feet (2,000 m) so their various shapes and characteristics are clearly evident to us on the ground. They are somewhat more varied and quick to change form than higher clouds, subject as they are to fluctuating ground heating effects and the vagaries of even quite modest topography. Low types include stratus, stratocumulus, nimbostratus, and fog, which is cloud at the ground. Low clouds are usually made up of water droplets because their development is in the warmer layers close to the ground, but they can extend to heights where freezing starts, producing ice crystals and snow.

Rain clouds (right) This thick layer of rain-heavy cloud allows very little light through.

Low-cloud chart (below) This chart shows the four main types of low cloud in the altitude range expected in the midlatitudes.

LOW CLOUD TYPES

These are the lowest types of clouds to affect our weather. They can be cumulus (Cs), nimbostratus (Ns), stratocumulus (Sc), or stratus (St). Fog close to the ground can also be included in this category. These clouds generally consist of water droplets. These droplets are initiated by water vapor diffusion toward cloud condensation nuclei. Once formed, the droplets continue to grow from available water vapor. Growth is accelerated when the larger droplets start to collect smaller ones while moving through the cloud. This collision and coalescence process can eventually lead to rainfall, when the raindrops are too heavy to stay suspended in the cloud.

Sometimes there is enough available energy in the cloud to lift it up through the freezing level. The droplets remain as liquid until an ice-nucleating material causes them to freeze. The ice particles continue to grow by collecting water vapor and by colliding with other droplets that freeze on contact. The larger ice particles will eventually fall and may reach the ground, although usually they will melt to produce rain. At altitudes above the freezing level, ice crystals can be initiated and grow into very small snow crystals.

The character of these clouds can be seen from the ground. Stratus is a uniform gray layer. Low-level nimbostratus clouds are often dark and produce moderate precipitation.

Precipitation from relatively high nimbostratus and cumulonimbus clouds can provide moisture for regions of stratus clouds below.

Fog often forms during the night near the ground, when moist air is cooled by the cold surface beneath. With a little heating in the morning, the fog can lift to produce a low-lying cloud bank that eventually burns off during the day or may form a layer of stratus. Fog formed over the sea often drifts to the shore and forms low-level stratus.

If the cloud layer is thin, corona rings can often be seen circling the Sun or Moon. These rings are sometimes colored by a diffraction effect if the droplets are of uniform size. If ice crystals have formed, refraction may cause a much larger 22-degree halo to be visible around the Sun or Moon. These rings often indicate that a new weather system is on its way.

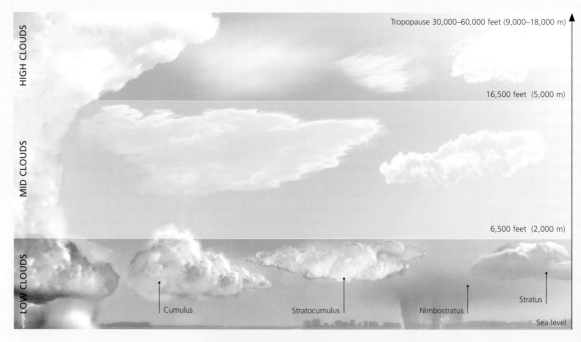

Tropopause 30,000–60,000 feet (9,000–18,000 m)

16,500 feet (5,000 m)

6,500 feet (2,000 m)

HIGH CLOUDS

MID CLOUDS

LOW CLOUDS

Cumulus

Stratocumulus

Nimbostratus

Stratus

Sea level

LOW CLOUD ALTITUDES

Type	Polar regions	Temperate regions	Tropical regions
Fog			
Cumulus			
Nimbostratus	surface to 6,500 feet (0–2,000 m)	surface to 6,500 feet (0–2,000 m)	surface to 6,500 feet (0–2,000 m)
Stratocumulus			
Stratus			

Lake cloud (left)
A combination of moisture from the lake and the arrival of cold air drifting down the valley from above has created a low-level bank of cloud. These clouds usually form at night. With solar heating, the airflow is reversed and the cloud dissipates.

Clouds over London (right) These scattered cumulus clouds over London are illuminated by the evening sun. Such early-evening clouds may produce light rain showers.

Mountain clouds These Spanish mountaintops are shrouded by cumulus. The clouds will eventually dissipate into the clear skies above.

Cumulus

Cumulus clouds are usually isolated clouds with a sharp outline and a flattish base. Driven by solar heating of the ground, which causes convection and transport of moist air aloft, they develop during the day, with a peak in growth after midday. Within the cloud, water droplets condense and grow larger. Given enough time, cumulus clouds may produce precipitation.

Summer clouds "Fair-weather" cumulus clouds are formed on warm days when moist air heated near the ground rises, then cools and condenses into fluffy clouds. Each cloud has a fairly short lifetime so does not produce rain.

Cumulus humilis   Cumulus mediocris   Cumulus congestus

Cumulus species Three principal species of cumulus cloud are recognized: humilis, mediocris, and congestus. They are distinguished by their level of vertical development, which indicates the strength of the convection currents that create them.

Jacob's Ladder In this French painting from c.1490, Jacob rests while angels climb to heaven, situated among summer cumulus clouds. Clouds are often used in art to illustrate the heavenly realm, with angels, saints, and God appearing in billowing clouds.

Turbulence Small cumulus clouds present no problems for aircraft. But cumulus mediocris and especially cumulus congestus are warning signs of possible severe turbulence, so aircraft avoid these clouds whenever possible.

FACT FILE

NIMBOSTRATUS

Appearance	dark gray, featureless, sometimes accompanied by pannus clouds
Species	none
Meaning of name	rain layer
Precipitation	prolonged moderate to heavy rain

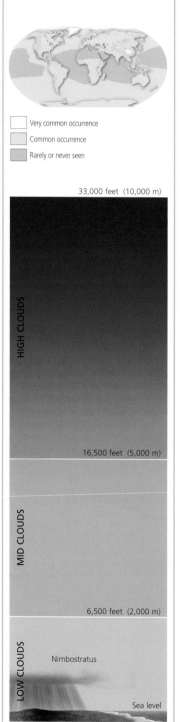

☐ Very common occurrence

☐ Common occurrence

■ Rarely or never seen

33,000 feet (10,000 m)

HIGH CLOUDS

16,500 feet (5,000 m)

MID CLOUDS

6,500 feet (2,000 m)

LOW CLOUDS

Nimbostratus

Sea level

Nimbostratus

These clouds completely mask the sun and produce precipitation at the ground but they are not associated with hail, thunder or lightning. Nimbostratus present as a thick layer that can extend from close to the ground to as high as 18,000 feet (5.5 km) and are composed of suspended water droplets that may be supercooled, with raindrops and snow.

Rainy day, Vancouver Island, Canada
Although they are found worldwide, nimbostratus clouds are more often seen in the midlatitudes, especially in places where the prevailing winds come off the ocean.

Boar Lane, Leeds, John
Atkinson Grimshaw, 1881
Many artists of the late 19th
century were fascinated by
the colors and lighting effects
associated with rainfall,
particularly in urban settings.
This painting shows the
darkening sky veiled by
nimbostratus clouds.

Snowfield, Japan Nimbostratus clouds can extend to heights where it is cold enough for
ice to form and grow into snow crystals. These crystals aggregate as they fall to make large
snowflakes that can cover the ground.

Nimbostratus landscape Over a windswept landscape in Iceland, nimbostratus clouds
darken the sky and produce continuous rainfall that soaks the ground, runs into gullies and
eventually into rivers that carry the water to the sea.

Stratocumulus

One of the most common clouds worldwide, stratocumulus is a good indicator of moisture in the lower levels of the atmosphere. These clouds are a product of weak convection or form when a frontal system lifts a large, moist air mass. Optical ring effects around the Sun or Moon are common when stratocumulus clouds are thin.

Stratocumulus from above In this aerial view, ascending moist air that forms the cloud has reached a height where stable air slows the ascent, forcing the cloud to spread out. There is a region of more vigorous air movement at the back.

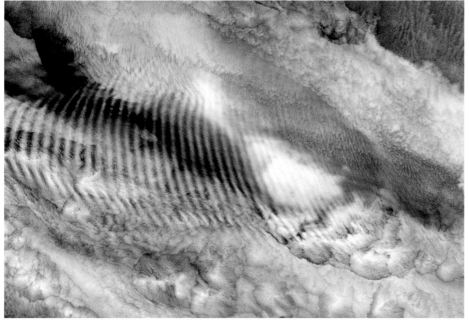

Stratocumulus gravity waves (above) These ripples form when a relatively stable cloud layer is disturbed by a vertical trigger, such as a thunderstorm updraft or an underlying mountain or island.

Sunset over wheat field (below) The underside of a sheet of broken stratocumulus reflects light from the setting sun. Stratocumulus clouds are often arranged in orderly groups.

Land and sea As Earth rotates, northerly winds deflected by the Coriolis effect push warm surface away from the northwest coast of the U.S.A., to be replaced with upwelling colder water. This leads to marine stratocumulus cloud formation. The warmer land is cloud-free.

La Mer Calme, **Gustave Courbet, 1869** This painting depicts a broken layer of stratocumulus over a calm sea. These marine stratocumulus clouds are made up of water droplets that reflect sunlight back to space and thereby help to cool the planet.

Stratus

Stratus clouds form in sheets or layers and occur when relatively large areas of moist air rise gently in a stable atmosphere to a level where condensation occurs. A layer of stratus clouds may cover hundreds of square miles. Orographic stratus clouds are only distantly related, being stationary clouds associated with elevated landforms.

Gray sky (below) Stratus are gray to nearly white and usually featureless, although they can sometimes have a ragged base. Stratus vary in thickness from a semi-transparent sheet a few feet thick to a deck of around 1,500 feet (460 m).

Banner cloud off the Matterhorn This cloud is continually being created at the mountain peak as it evaporates downwind.

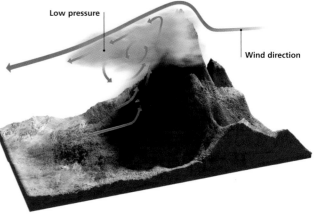

Low pressure

Wind direction

Banner cloud formation Banner clouds occur when strong wind blowing around an isolated mountain peak lowers the atmospheric pressure on the lee (downwind) side. This draws moist air from below to create a cloud.

Philadelphia dawn Stratus cloud can sometimes occur when a layer of fog that has developed at ground level starts to rise as it is warmed by the sun. This formation is known as fog stratus.

Orographic cap Orographic clouds are most common near coasts, where there is a plentiful supply of low-altitude moisture. This example has formed over a mountain on the Arctic island of Spitsbergen.

Wind direction

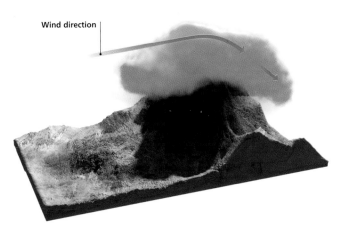

Orographic cloud formation These stratus clouds occur where the prevailing winds carry moist air over an elevated landform, such as a mountain range, to a level where condensation takes place.

Mid Clouds

Found between about 6,500 feet (2,000 m) and 16,500 feet (5,000 m), these clouds are intermediate in height between low and high clouds. Typical clouds found at this altitude are altostratus and altocumulus. The temperatures in this mid-level are below freezing, so any water droplets in the clouds are supercooled, which allows ice crystals and ice pellets to form and grow. Turbulence in the winds circulating in the middle levels of the atmosphere leads to the appearance of wavelike patterns in some forms of these clouds. Precipitation at the ground can then be a mix of rain, melted ice particles, and snow.

Mid-cloud chart (below) This chart shows the two main types of mid cloud in the altitude range expected in the midlatitudes. Towering cumulonimbus clouds can extend right through this layer.

MID CLOUD TYPES

There are two main types of mid-level cloud: altostratus and altocumulus. Although the prefix "alto" is derived from the Latin word *altus*, meaning high, these clouds are, in fact, found below cirrus clouds but well above low-level clouds.

Altostratus clouds are most common in midlatitudes but are found all over the globe. They are typically featureless sheets of cloud and always gray or pale blue in color. They never appear white but can sometimes be thin enough to let the sun shine through. They are the result of the lifting and condensation of a large air mass, usually by an incoming frontal system. This can result in an extensive deck of cloud, which may extend over thousands of square miles. If sufficiently thick, altostratus can produce rain or snow over a wide area. Sometimes the precipitation evaporates before reaching the ground, when the rain or snow shaft below the cloud is known as virga.

The composition of an altostratus cloud can change with altitude. It may have a lower region made up of supercooled water droplets, a central region consisting of supercooled water droplets together with ice crystals and snowflakes, and an upper region that is all ice crystals.

For pilots, the layers of supercooled droplets can be a cause for concern because ice may build up on the aircraft's wings.

Altocumulus clouds are mid-level cumulus clouds that, like altostratus, normally occur when a large air mass is lifted by a frontal system. The principal difference between the two formations is that altocumulus are affected by instability in the surrounding atmosphere. This gives rise to their distinctive cumuliform shape.

They usually accumulate in clusters or parallel bands, but can also be shapeless in an extensive layer.

If the extent of altocumulus formation appears to be increasing during the course of a day, this may be a sign of an approaching frontal system and the possibility of impending thunderstorms and heavy rain.

When altocumulus clouds are formed by the lifting of moist air over landforms, they can produce spectacular lenticular clouds (altocumulus lenticularis) that have been reported as flying saucers.

Evening sun over the Goshute Mountains, Nevada, U.S.A. (right) In this scene, high layers of cirrostratus overlie a mid-layer of scattered altostratus and a patch of altocumulus. In the distance, the clouds have smoother contours characteristic of mountain airflow.

HIGH CLOUDS

MID CLOUDS

LOW CLOUDS

Tropopause 30,000–60,000 feet (9,000–18,000 m)

16,500 feet (5,000 m)

Altostratus

Altocumulus

6,500 feet (2,000 m)

Sea level

MID CLOUD ALTITUDES			
Type	**Polar regions**	**Temperate regions**	**Tropical regions**
Altostratus	6,500 to 13,000 feet (2,000–4,000 m)	6,500 to 23,000 feet (2,000–7,000 m)	6,500 to 26,000 feet (2,000–8,000 m)
Altocumulus			

Sunset over London (above) A broken layer of altocumulus over the Palace of Westminster is brightly illuminated from beneath by the setting sun.

Altocumulus over wheat fields, Washington State, U.S.A. (below) A broken field of altocumulus clouds form over land that has been heated on a hot summer's day.

Altostratus

Altostratus result from the lifting of a large moist air mass, usually by an incoming frontal system. This can result in an extensive cloud deck that may extend over thousands of square miles. Altostratus are typically featureless, and can range from a thin, white veil of cloud through to a dense, gray mantle that may completely block out the sun.

Altostratus at sunset During the day, altostratus is usually a uniform gray sheet. However, the low angle of the rising or setting sun can reveal variations in cloud density and fine textures in the base.

Thin altostratus (right) This region of altostratus is illuminated from above by the sun. The variegated pattern of gray and white indicates instability in the atmosphere at the height of the cloud deck.

Frosted sun When seen through altostratus, the sun has a diffuse, blurred appearance. This is in contrast to lower-altitude stratus, where the sun's outline is usually apparent.

Altocumulus

While altostratus are often flat and featureless, altocumulus usually create interesting and varied skies. In some cases, many thousands of small altocumulus clouds will be strung together across the sky in spectacular formations. These clouds may consist of supercooled water droplets, or ice crystals, or both, depending on the temperature and cloud history.

Altocumulus undulatus (below and below left) These parallel roll clouds form as a result of wind shear, which occurs when one layer of air slides over another layer moving at a different speed or in a different direction (or both).

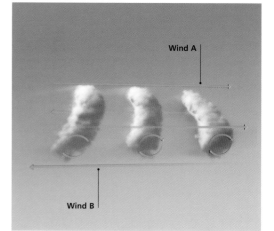

Altocumulus over dunes (below) Breezes from the Atlantic Ocean bring clouds and fog to the Namib Desert but very rarely rain. Less than 0.4 inches (10 mm) of rain falls here each year.

Lenticular cloud over Mount Fuji This cap cloud is the product of moist air rising up the slopes of the volcano, where it cools. The cloud is continually forming on the upwind side and evaporating downwind. The soft appearance of the cloud edges indicates that it is composed of ice crystals.

Standing wave clouds (left and below) Airflow over a sharp mountain ridge can set up a wave pattern downwind of the mountain. Moist air cresting a wave is cooled and condenses to form a cloud. The air then descends to a warmer layer, where the cloud dissipates, only to reform at the next wave crest. Closest to the mountain, the wave intensity can be so great as to cause a rotor, or circulatory flow.

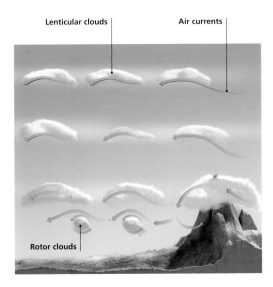

Lenticular clouds

Air currents

Rotor clouds

High Clouds

These clouds are found at levels above about 15,500 feet (5,000 m) up to the tropopause, where the temperature of the atmosphere starts to increase with altitude and so puts a temperature inversion cap on normal cloud height. High cloud types, increasing in altitude, include cumulonimbus, cirrus, cirrocumulus, nacreous, and noctilucent clouds. The latter two exist above the tropopause. Cloud particles in this region grow in a low temperature regime that leads to the formation of ice particles, hail, ice crystals, and snow, but supercooled water droplets are also present at temperatures warmer than −40°F (−40°C).

High-cloud chart (below) This chart shows the main types of cloud to be found above 16,500 feet (5,000 m). Note that the top of the chart does not indicate the usual height of the tropopause.

HIGH CLOUD TYPES

The most common clouds in this highest region of the troposphere are cirrus, cirrostratus, and cirrocumulus clouds. Towering cumulonimbus clouds also reach into these altitudes. Above the tropopause, other forms of ice crystal cloud are sometimes observed—nacreous and noctilucent clouds.

Cirrus clouds consist of ice crystals that form in temperatures below −22°F (−30°C). When supersaturation occurs at these temperatures, water vapor deposits directly onto ice nuclei to initiate ice crystal formation. Small supercooled droplets freeze and also act as ice crystal embryos. The crystals grow by diffusion into pristine plates and columns that can create optical effects such as halos and sun dogs.

From the ground, cirrus clouds look white, thin, and wispy. Wind shear at these altitudes can move crystals falling from the cloud at a different velocity to the cloud itself, giving them a distinctive shape known as "mare's tails." The crystals evaporate before reaching the ground.

When a thunderstorm dissipates at the end of its life, it can leave behind an anvil-shaped region of ice crystals that merge into any cirrus layer present. Hurricanes can also leave a cirrus residue. Aircraft at these levels produce condensation trails of ice crystals when water in their hot exhausts freezes.

Cirrostratus clouds form a widespread cirrus layer when a large area of moist air is lifted. Such layers are almost entirely translucent, though sometimes the ice crystals create a halo of light round the Sun or Moon. Like cirrus clouds, these extensive cirrus layers are found worldwide.

Cirrocumulus clouds are composed of individual cloud elements and often aggregate to form spectacular patterns known as "herringbone" or "mackerel sky." They tend to be transient, developing from and returning to cirrus and cirrostratus forms. Cirrocumulus clouds can produce precipitation that shows up as virga.

Cumulonimbus with great vertical extent can reach from the condensation level at the cloud base up to the tropopause, as high as 60,000 feet (18 km). The largest of these clouds occur in the tropical regions of Earth that provide both high heat energy and a plentiful supply of moisture. They can also form along a front when cold air pushes warm, moist air aloft.

HIGH CLOUDS — Cumulonimbus — Cirrostratus — Cirrus — Tropopause 30,000 to 60,000 feet (9,000–18,000 m) — Cirrocumulus — 16,500 ft (5,000 m)

MID CLOUDS — 6,500 ft (2,000 m)

LOW CLOUDS — Sea level

Arctic cirrus (right) Trails of cirrus clouds form above a Canadian mountain range north of the Arctic Circle. The clouds are drawn out in the direction of the prevailing winds at high altitude.

Cirrus mare's tails (below) Ice crystals in a layer of cirrus clouds fall out of the cloud and form wisps when they experience a change in horizontal wind speed. This change in speed is called "shear." These are a good indication of an approaching weather system.

HIGH CLOUDS ALTITUDES			
Type	Polar regions	Temperate regions	Tropical regions
Cirrus			
Cirrostratus	10,000 to 26,000 feet (3,000–8,000 m)	16,000 to 45,000 feet (5,000–14,000 m)	21,000 to 60,000 feet (6,000–18,000 m)
Cirrocumulus			
Cumulonimbus	0 to 26,000 feet (0–8,000 m)	0 to 45,000 feet (0–14,000 m)	0 to 60,000 feet (0–18,000 m)

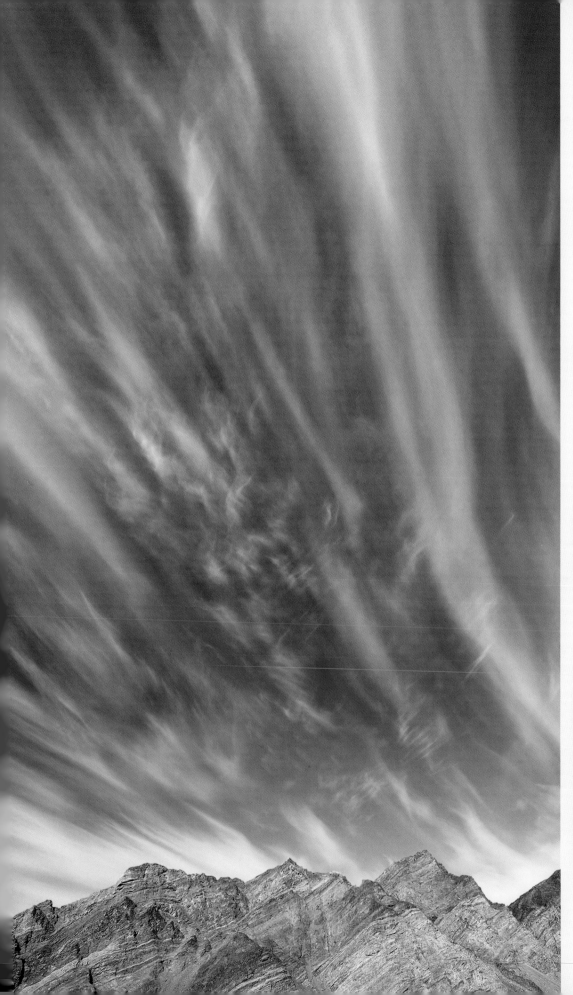

An evening mix In this photo, an overlying layer of cirrostratus cloud is illuminated by the setting sun while small fragments of cumulus are evident in the foreground. A thunderstorm is developing to the right.

Cirrocumulus over Japan These cirrocumulus clouds have lined up in a pattern known colloquially as "herringbone" or "mackerel sky." This pattern indicates some wind shear and turbulence aloft.

View from above In this satellite image, bright areas indicate the presence of noctilucent, or "night-shining" clouds. The black area above the North Pole is where no data could be acquired.

Cirrus

When warm, moist air encounters cold air, it overruns the cold air, creating a front between two air masses with a gradient of about 50 degrees. The warm air is pushed up the slope, where it cools and develops cloud. At the top of the slope, where the air is coldest, cirrus cloud forms. Cirrus cloud is therefore the first sign of an approaching warm front.

Cirrus in the evening A layer of cirrostratus is mirrored in a placid lake. The cloud cover does not extend to the horizon, so the low sun is able to illuminate the underside of the cirrus layer.

Kelvin-Helmholtz waves (below) Wind shear normally creates a series of gently undulating cloud formations along the tops of the waves. In the case of the Kelvin–Helmholtz formation, the eddies are more powerful, and carry the cloud over the peak and down the other side, so that the waves "break" in the same manner as ocean waves approaching the shore.

Cirrus over Olvera (above) Cirrus clouds have been swept into sinuous curves by high-altitude winds above this hillside town in Andalucia, southern Spain.

Study of Cirrus Clouds, John Constable, c.1822 (below) English painter John Constable was a keen observer of clouds. Here, cirrus streaks are depicted, together with small cumulus clouds.

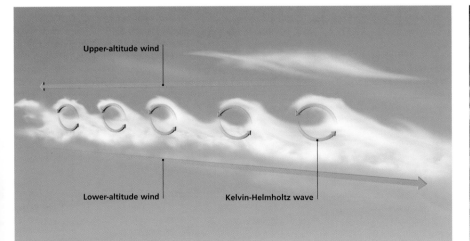

Upper-altitude wind

Lower-altitude wind

Kelvin-Helmholtz wave

Cirrostratus

Extensive layers of cirrus clouds are called cirrostratus. These clouds consist of ice crystals formed when moist air is gently lifted over a wide area. There are two common types: fibratus, consisting of long, thin filaments that spread across a wide area of the sky; and nebulosus, a thin, featureless layer that can be difficult to discern.

Cirrostratus fibratus The evening sun shows up the long, thin filaments known as striations that are a feature of this cloud type. The even texture results from ice crystals being blown by strong, steady, high-level winds.

Solar halo (right) This halo around the Sun is a product of hexagonal ice crystals that act as transparent prisms. As a light beam enters and leaves a crystal it is deflected by 22 degrees. This happens for crystals seen all around the Sun and so produces the halo.

Lunar halo (below right) At night, when moonlight shines through a cirrus layer, the light beams are refracted in exactly the same way as for the halo around the Sun. These halo effects are not to be confused with much smaller rings around the Sun and Moon caused by a thin cloud of water droplets.

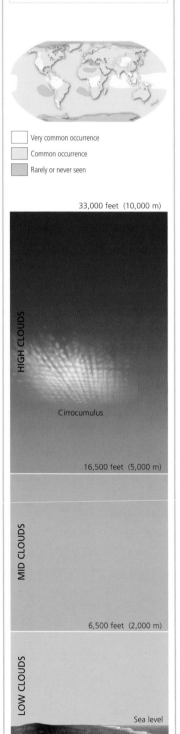

☐ Very common occurrence

☐ Common occurrence

☐ Rarely or never seen

33,000 feet (10,000 m)

HIGH CLOUDS

Cirrocumulus

16,500 feet (5,000 m)

MID CLOUDS

6,500 feet (2,000 m)

LOW CLOUDS

Sea level

Cirrocumulus

Cirrocumulus, like cirrostratus, occur when a large area of moist air at a high level of the atmosphere reaches saturation and forms ice crystals. What differentiates cirrocumulus from cirrostratus is the presence of instability at cloud level. This gives the cloud its cumuliform appearance. Cirrocumulus is one of the most attractive of all clouds, often forming spectacular patterns that may stretch for hundreds of miles across the sky.

Cirrocumulus sky (right) A field of small, isolated cirrocumulus clouds drifts high over a frost-covered forest. Their great height makes the individual cloud elements appear very small from the ground.

Isolated cirrocumulus (above) A well defined area of cirrus indicates convective lifting of a localized vapor source. Around the edges of the main cloud mass, smaller elements of cirrocumulus are visible.

Arctic cirrocumulus (below) These clouds have formed in low temperatures over the ocean. Limited vapor availability in the Arctic means that the cloud cover is thin, featuring small, isolated cloudlets.

Very common occurrence

Common occurrence

Rarely or never seen

33,000 feet (10,000 m)

Cumulonimbus

HIGH CLOUDS

16,500 feet (5,000 m)

MID CLOUDS

6,500 feet (2,000 m)

LOW CLOUDS

Cumulonimbus

Sea level

Cumulonimbus

These giants of the cloud world extend over the whole height of the troposphere and can generate thunderstorms that release as much energy as an atom bomb. Cumulonimbus clouds develop in unstable air that continues rising because it is continually warmed by the latent heat of condensation.

Storm cloud over Galapagos (right) The flat base of this towering cumulonimbus indicates a uniform region of rising air carrying moisture into the cloud. A dense shaft of rain is falling from the center of the cloud.

Thunderstorm from space (above) This view from the International Space Station clearly shows the vast area covered by the ice-crystal anvil at the top of this storm. New convective turrets are rising in cloud fields beyond the shadow of the anvil.

Anvil formation (below) The lower parts of this cloud formation show the rounded forms and sharp edges associated with the presence of water droplets, while the anvil top is clearly glaciated with ice crystals.

Cumulonimbus continued

Cumulonimbus clouds penetrate the atmosphere to great heights, and their rapid updraft speeds and turbulence are best avoided by aircraft, but it is the effects closer to the ground that are most evident to us. The downdraft associated with the fall of precipitation can be intense, producing a strong, cold storm outflow near the ground.

Arcus cloud (right) These menacing-looking horizontal roll clouds form at the leading edge of a severe storm's outflow.

Wall cloud These accessory clouds extend from the base of a cumulonimbus cloud and indicate the lower portion of a very strong updraft, usually associated with a supercell or severe multicell storm.

Mammatus and pileus These unusual clouds are associated with the powerful dynamics at work in a storm. Mammatus look like hanging pouches. Pileus clouds are formed over the tops of rising cumulonimbus.

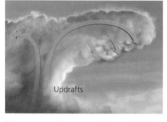

Updrafts

Mammatus These pendulous globules of cloud are formed on the underside of an anvil top when strong downdrafts push parcels of warm moist air into an area of cooler air.

Pileus

Pileus These layers of cirrus cloud form above cumulonimbus when rising air pushes up a cold, moist layer sufficiently fast for it to cool further, saturate, and grow ice crystals.

Flanking line (above) Powerful thunderstorms often feature a line of towering cumulus clouds leading up to the main storm complex. These "flanking line" clouds are fed by an outflow of cold air from aloft and may eventually merge into the parent storm.

Fractus cloud (below) These shreds of ragged cloud are often seen below cumulonimbus. They are accessory clouds, having been torn off a larger cloud mass by strong winds. Generally short-lived, they may develop, merge with other clouds, or dissipate.

Occasional occurrence

Never seen

Noctilucent | 53 miles (85 km)

MESOSPHERE

STRATOSPHERE

Nacreous
6.2 miles (10 km)

TROPOSPHERE

Nacreous and Noctilucent

Nacreous clouds are located in the stratosphere at altitudes between 9 and 16 miles (15–25 km) at temperatures below –108°F (–78°C). They are also known as polar stratospheric clouds, or mother-of-pearl clouds for their vivid colors, as seen in oyster shells. Noctilucent clouds form on the edge of space, at altitudes of about 50 miles (80 km) and a temperature of –148°F (–100°C).

High fliers (below) This photograph taken aboard the International Space Station shows noctilucent clouds between the orange-colored troposphere and a crescent Moon.

Estonian sky (above) Resembling cirrus clouds, these thin layers of blue, silver, red, or orange-colored clouds are seen only at around 60 degrees south or north of the Equator during summer. Their composition is unknown.

Nacreous clouds over South Orkney Islands (left) The beautiful iridescent appearance of nacreous clouds is the product of optical effects in the small ice crystals that compose these clouds.

Out of the shadows (right) Noctilucent clouds can only be observed when the Sun is in the region of 6–12 degrees below the horizon and lower-altitude clouds are in Earth's shadow.

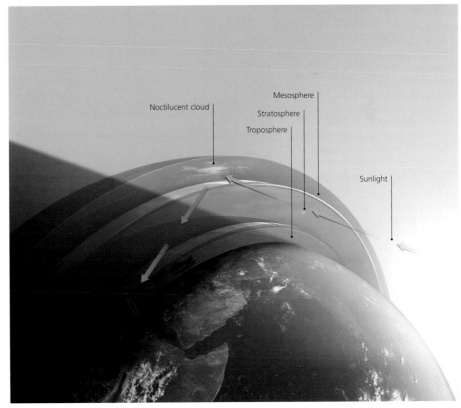

Noctilucent cloud

Mesosphere

Stratosphere

Troposphere

Sunlight

MANIPULATING CLOUDS

For thousands of years humans have attempted to exercise some control over the impact that weather has on their lives. The earliest measures took the form of prayers and ceremonies. With advances in the understanding of cloud physics during the 20th century, scientists began intentionally seeding clouds to stimulate precipitation, clear fog, reduce hail size, or weaken the force of hurricanes. While some of these cloud-seeding efforts have been moderately successful, they remain controversial since their efficacy is difficult to assess.

Project Stormfury This US government project was intended to reduce the strength of hurricanes by disrupting their structure. Between 1962 and 1971 hurricanes were seeded with silver iodide, but with no great success. However, a great deal was learned about hurricanes.

Sioux rain dance (above) Native American rain dances were performed to inform the gods that rainfall was needed. Similar rites have been carried out all over the world.

CLOUD SEEDING

Natural clouds are composed of small water droplets and ice crystals that grow on nuclei. More droplets and ice crystals can be created by introducing artificial nuclei into a region supersaturated with water vapor. If this "cloud seeding" is done in the right place and in the right atmospheric conditions, it is possible to modify clouds and increase the likelihood of rain.

Smoke from forest fires is a natural form of cloud seeding; Native American shaman may have used fire in their rain-making ceremonies. Cities are also a source of particulates that can initiate clouds, or such output can impede ice nucleus activity. This is inadvertent weather modification.

Deliberate attempts to modify clouds by cloud seeding followed the discovery by American scientist Vincent Schaefer that seeding clouds with tiny crystals of dry ice could cause precipitation. Dry ice has a very low temperature (–109°F/ –78.3°C) and the tiny crystals rapidly attract water droplets until they are large enough to fall as

snow or rain. Schaefer's colleague Bernard Vonnegut later found that silver iodide crystals formed ideal freezing nuclei and so could also be used to seed clouds. Both techniques have been used in many cloud-seeding experiments and projects since their discoveries in 1946.

The Soviet Union famously seeded clouds before they reached Moscow to try to prevent rain falling on military parades. Cloud-seeding experiments have been conducted in Israel since 1960 and are among the few that have claimed statistical evidence of rain enhancement. In the 1970s, cloud-seeding experiments in the San Juan Mountains, Colorado, showed that it was possible to increase snowfall on the western side of the mountains, thereby increasing river flow towards California. In 2006, a new program of seeding to increase water input to the river was initiated in Colorado, Wyoming, and Utah.

Clouds have also been seeded in order to prevent the development of large hailstones that can destroy crops. Countries where this has been tried include Australia, Italy, Bulgaria, and Russia.

Vincent Schaefer Schaefer created an artificial cloud of supercooled water droplets at General Electric Research Laboratory in New York State.

Bernard Vonnegut Vonnegut demonstrated that silver iodide is an effective ice nucleus. He also researched the properties of lightning.

Water wings (above) Cloud-seeding aircraft carry tanks of propane containing silver iodide. By burning off the propane, the silver iodide is released into the atmosphere, where it can initiate ice crystals.

Keeping the rain away (right and below) Before key events of the 2008 Olympic Games, Chinese authorities fired silver iodide-loaded rockets into clouds to encourage them to produce rain before they reached the Olympic Stadium. The Chinese have many years' experience using these rockets in regions where rainfall is needed.

Unusual and Artificial Clouds

Not all clouds are formed by the usual engines of weather. Clouds can be born from wildfires, the heat of a volcano, and a variety of human activities. Jet aircraft draw long contrails across the sky. Some industries pump clouds into the air. Even the mushroom cloud of a nuclear bomb is a product of the condensation of water vapor.

FACT FILE

Mapping contrails The most important determinant of contrail occurrence is the density of air traffic. However contrails will form at lower altitudes in winter and last longer when humidity in the upper atmosphere is high.

Very common occurrence

Common occurrence

Occasional occurrence

Rarely or never seen

33,000 feet (10,000 m)

Contrails

Volcanic pyrocumulus

16,500 feet (5,000 m)

Pyrocumulus

6,500 feet (2,000 m)

Volcanic pyrocumulus, Montserrat (below) Pyrocumulus clouds are a regular feature of volcanic eruptions. They occur when water vapor present in volcanic gases and the surrounding air is lifted by convection.

Contrails Where the air temperature is below –40°F (–40°C), water vapor produced in the combustion of jet fuel rapidly freezes to create "condensation trails" or contrails. The trail persists until mixing with dry air causes sublimation of the ice crystals.

Wildfire pyrocumulus (above) The heat from this forest fire in Australia has fuelled strong convection currents that, in turn, have created a vigorous pyrocumulus cloud. Such clouds have been known to produce lightning and even rain that extinguishes the blaze below.

Cloud factory (above) Warm, moist air rises in these power station cooling towers and cools sufficiently to condense into water droplet clouds.

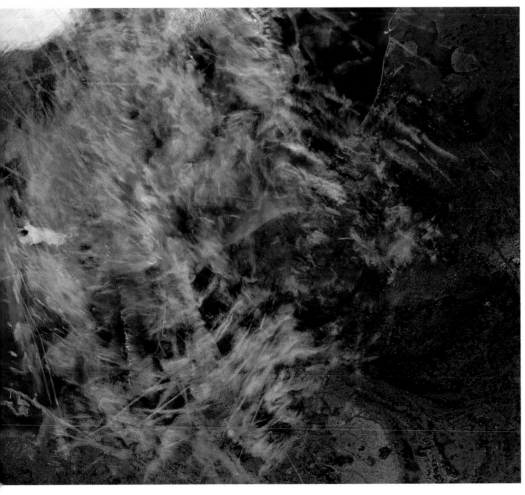

Vapor cone (above) An aircraft nearing supersonic speed causes a rapid increase in air pressure due to compression, followed by a sudden expansion that cools the air enough to cause condensation and cloud formation. The air quickly returns to atmospheric pressure and the cloud evaporates.

Contrails over the English Channel (left) If the surrounding air mass is close to saturation, contrails will spread out and may persist for hours. They can cover 50 percent of the sky in regions with high air traffic.

PRECIPITATION

Any of the various forms of water that fall from the atmosphere and reach the ground are called precipitation. It includes both liquid and solid particles and comprises drizzle, rain, ice crystals and snow in all its forms, soft hail pellets (graupel), and hailstones. Having reached the ground, the amount of precipitation can be measured using a rain gauge, which gives the depth of liquid falling over any area in a given time. The depth of water collected—including melted ice particles—is quoted in inches or millimeters of precipitation.

WATER IN THE AIR

The atmospheric water cycle begins with the evaporation of surface water. This vapor is transported aloft in rising air parcels, where it cools. The air becomes saturated and begins to condense, and becomes liquid. The temperature at which this happens is known as the dew point. When water vapor condenses onto tiny specks of dust called condensation nuclei, it forms water droplets. When the condensation occurs onto ice nuclei, it produces ice crystals. The newly formed particles grow by colliding and merging with others, until they are large enough to fall, collecting smaller particles on the way to the ground. This process, called coalescence, produces rain-sized particles. The aggregation of ice crystals creates large snowflakes, and when these falling ice crystals collect supercooled droplets they produces hail pellets. On the ground, ice precipitation may build ice caps and glaciers, while rainfall

Hail (above) Hailstones fall fast enough and are heavy enough to do damage to people and property. Layers of ice are built up, sometimes trapping air bubbles, as the hailstone passes through areas of differing temperature and humidity.

and melted snow irrigates the soil or runs off into lakes, rivers, and the sea to complete the cycle.

The total volume of water on Earth is about 332,500,000 cubic miles (1,386,000,000 km³). Over 97 percent of this is in the oceans. The other portion, the fresh water, is in the ice caps, lakes and rivers, and a tiny fraction in the atmosphere. If all the atmospheric water vapor suddenly fell as precipitation, it would produce worldwide rainfall of 1.2 inches (30 mm). On average, water molecules remain in the atmosphere for around 11 days before falling as precipitation.

Average rainfall varies markedly across the globe. In equatorial or tropical regions there can be over 120 inches (3,000 mm) of rainfall per year, while deserts receive less than 10 inches (250 mm) per year.

The type of precipitation reaching the ground depends on conditions within a cloud, and the temperatures in and below it. Very cold air within a cloud may produce rain, freezing rain, or snow, depending on the temperature of the air beneath the cloud. Rain may become supercooled and form freezing rain by passing through a layer of freezing air. Ice crystals may melt and become rain; or remain frozen, grow, and reach the ground as snow. Rain or ice crystals may pass through a layer of warm, dry air and evaporate completely.

Hail is possibly the most destructive form of precipitation. These lumps of ice require the strong updrafts of thunderstorms to keep them in the air long enough to grow, layer by layer. When they fall to the ground, they can cause considerable damage. Hailstones have been classified according to size: from pea and grape, through walnut and golf ball, to grapefruit and melon sizes. In June 2003, the largest hailstone ever recorded—seven inches (18 cm) in diameter—hit Aurora, Nebraska, in the U.S.A. This is within an area known as Hail Alley.

Rain (above) Shafts of rain below a dense cumulonimbus show the locations of active storm cells within the cloud. When water droplets fall from a cloud but evaporate in mid-air this creates an effect called virga, which resembles a dark fringe hanging from the cloud.

Snow (right) The snow-covered forests of Yosemite Valley, California, create a glorious winter landscape. The weight of snow on the trees can be considerable, sometimes enough to break branches. On the ground, the surface layer of snow crystals melt and recrystallize to form a crunchy layer of snow.

Global precipitation distribution
The most severe storms and heaviest precipitation occur in areas of both high moisture and high temperature. Low rainfall occurs in arid and polar regions, where there is a meagre supply of moisture.

0	10	20	40	80	120	in
0	250	500	1,000	2,000	3,000	mm

Average annual precipitation

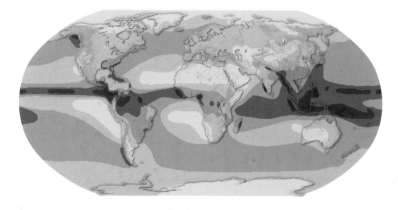

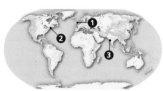

1. Borrowdale Prevailing moisture-laden westerly winds from the Atlantic Ocean bring high rainfall to the hilly Lake District, in the northwest of England. Borrowdale Valley holds the

record for the wettest inhabited place in England, where the mean annual rainfall exceeds 118 inches (3 m).

Borrowdale, Cumbria, England

2. Boston Located on the east coast of the U.S.A., Boston experiences a humid continental climate. Annual rainfall is around 42.5 inches (1.1 m). The driest

month is July, with 3 inches (76 mm), and the wettest month is November, with 4 inches (100 mm) of rainfall.

Boston, Massachusetts, U.S.A.

3. Mumbai India's most populous city experiences a tropical, humid climate. Even in winter the average daily high temperature is 88°F (31°C). Almost all of the annual rainfall of 94 inches

(2.4 m) falls during monsoon from June to September. It does not rain at all between December and April.

Mumbai, Maharashtra, India

SHAPE OF A RAINDROP

Small raindrops are almost spherical, while large raindrops attain an oblate spheroid shape, with a flattened or indented bottom due to air resistance. Larger still and they will open up like a parachute before bursting.

0.04 inch (1 mm) spherical

0.12 inch (3 mm) indented

>0.18 inch (>4.5 mm) parachute

burst

0.08 inch (2 mm) flattened

Rain

After a certain period of growth, the drops in a cloud become too large to be supported by the cloud updraft. As they fall, they continue to collect smaller droplets until they reach the cloud base. They fall out of the cloud to the ground as rain. Precipitation from low stratus clouds often consists of small raindrops, called drizzle, which are usually smaller than 0.02 inches (0.5 mm) in size.

Precipitation, evaporation and runoff (below) Much of the water that falls as rain evaporates before humans have a chance to use it. This map reveals surface water availability by continent. Total precipitation is given in cubic miles. Runoff is the movement of land water to the oceans.

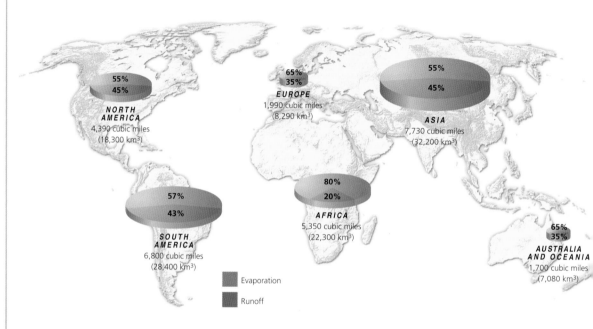

55%
45%

NORTH AMERICA
4,390 cubic miles
(18,300 km³)

65%
35%

EUROPE
1,990 cubic miles
(8,290 km³)

55%
45%

ASIA
7,730 cubic miles
(32,200 km³)

57%
43%

SOUTH AMERICA
6,800 cubic miles
(28,400 km³)

80%
20%

AFRICA
5,350 cubic miles
(22,300 km³)

65%
35%

AUSTRALIA AND OCEANIA
1,700 cubic miles
(7,080 km³)

Evaporation

Runoff

London drizzle London experiences a temperate marine climate. Its average annual rainfall of 23 inches (585 mm) falls mostly as drizzle or light rain, and is fairly evenly distributed throughout the year.

Amritsar downpour A continuous deluge of rain is typical of the monsoon regions of Asia. The monsoon season brings moisture from the Indian Ocean and the Arabian Sea and provides 80 percent of total rainfall to the region.

Intense downpour A localized storm cloud releases its precipitation in a clearly marked rainshaft. The defined nature of the cloud is revealed by its sharp shadow. The cloud is just one of a series that develops rapidly, produces its rainfall, and dissipates just as fast.

Rain forest tree
Rain forest trees have plenty of water and can grow to be among the largest of all living organisms.

Pachypodium **tree**
Trees of this genus survive droughts and dry seasons by using water stored in their enlarged trunks.

Desert cactus
Cacti store water in their fleshy stems. Sharp spines lose less water than regular leaves and deter animals.

Global vegetation Rain brings life and is the most important determinant of local ecosystem type. This global map of biome type correlates closely to the global precipitation map on page 126.

Tropical forest	Mediterranean forest and scrub
Seasonal tropical forest	Midlatitude grassland
Desert	Midlatitude forest
Tropical grassland and savanna	Boreal forest

Tundra
Ice sheet
Mountain vegetation

Rainfall intensities Definitions of rain intensity vary among meteorological authorities but the rates below are common to many. Drizzle is characterised by both low intensity and small drop size.

Drizzle
Drop size < 0.02 inch (0.5 mm)
Intensity < 0.05 in/hour (1.3 mm/h)

Light rain
Drop size > 0.02 inch (0.5 mm)
Intensity < 0.1 in/hour (0.25 mm/h)

Moderate rain
Drop size > 0.02 inch (0.5 mm)
Intensity 0.1 to 0.3 in/hour (2.5–7.6 mm/h)

Heavy rain
Drop size > 0.02 inch (0.5 mm)
Intensity 0.3 to 0.6 in/hour (7.6–15 mm/h)

Downpour
Drop size > 0.02 inch (0.5 mm)
Intensity > 0.6 in/hour (>15 mm/h)

Hail and Freezing Rain

Solid ice precipitation forms in various ways. Graupel, or soft hail pellets, grow in clouds when ice crystals collect freezing supercooled droplets. The tumbling crystals continue to accrete while they are carried along in the storm, and may eventually grow into hailstones. Ice pellets are raindrops that freeze while on their way to the ground.

Ice pellets Unlike hailstones, which only form in a thundercloud, ice pellets may fall from any cloud that can produce rain.

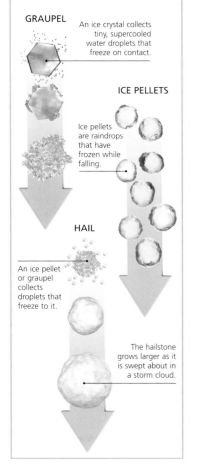

GRAUPEL

An ice crystal collects tiny, supercooled water droplets that freeze on contact.

ICE PELLETS

Ice pellets are raindrops that have frozen while falling.

HAIL

An ice pellet or graupel collects droplets that freeze to it.

The hailstone grows larger as it is swept about in a storm cloud.

Hailstones These small hailstones were gathered after a sudden downburst from a passing storm. Most hailstones are about the size of a pea but some are much larger.

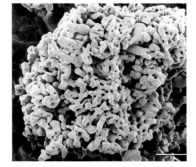

Graupel This electron micrograph image of graupel reveals a low-density structure of frozen droplets that have completely smothered the initial ice crystal.

Rime ice When supercooled water droplets from fog or low clouds freeze on a surface, they form a white ice deposit known as rime.

HAILSTONES

The onion-ring structure inside a large hailstone shows the history of its growth. Alternate layers of frozen droplets and ice crystals are collected while the hailstone is carried around inside the cloud.

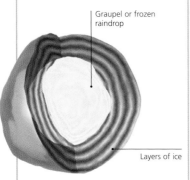

Graupel or frozen raindrop

Layers of ice

WARM AIR

Freezing rain When snow falls into a thin band of warmer air, it melts to form raindrops. If the air below is sufficiently cold, the raindrops will cool below freezing, but will not freeze until they hit objects on the ground.

Ice pellets Ice pellets are produced when rain encounters a deep layer of subfreezing air that turns them solid.

Ice storm During an intense fall of freezing rain, trees can be coated with a thick, heavy layer of ice. The extra rigidity and weight is often more than the branches can bear, so trees can be badly damaged during ice storms.

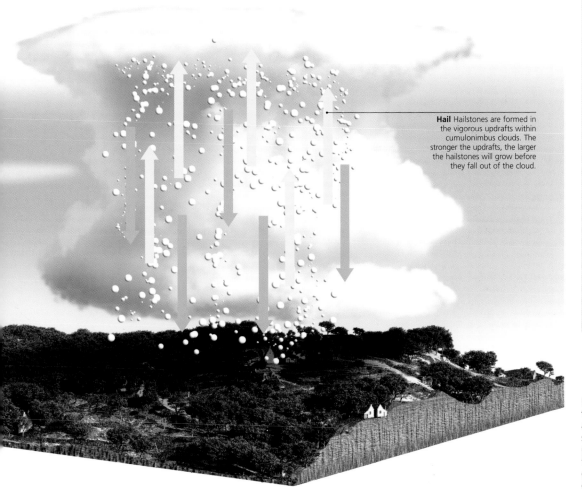

Hail Hailstones are formed in the vigorous updrafts within cumulonimbus clouds. The stronger the updrafts, the larger the hailstones will grow before they fall out of the cloud.

FACT FILE

1. Sydney The costliest hailstorm on record occurred over Sydney, Australia in April 1999, when hailstones over 3.5 inches (90 mm) in size fell. An estimated 500,000 metric tonnes of hail fell during the storm. The insured loss totaled over A$1.7 billion (US$1 billion).

Sydney, New South Wales, Australia

2. Gopalganj The heaviest hailstone on record fell in Gopalganj, central Bangladesh, on April 14, 1986. The hailstone, which fell during a storm that killed 92 people, weighed 2.25 pounds (1.02 kg). Bangladesh is susceptible to severe hailstorms that cause death and crop destruction.

Gopalganj, Bangladesh

3. Munich On July 12, 1984, Munich, Germany was pummeled by a hailstorm that injured more than 400 people and damaged 700,000 homes and 200,000 cars. Large hail, with a diameter up to 4 inches (10 cm) fell along a 155-mile (250 km) path.

Munich, Bavaria, Germany

4. Saint Lawrence Valley In January 1998, freezing rain covered parts of Quebec, Ontario, and New England with up to 5 inches (12.5 cm) of ice. The weight of the ice tore off branches and pulled down power lines. Car accidents and hypothermia were responsible for 35 deaths.

Saint Lawrence Valley, Canada and U.S.A.

Freezing rain, ice pellet, and hail formation The formation of freezing rain and ice pellets requires layers of warm and subfreezing air in the atmosphere. Ice pellets can fall almost anywhere, while freezing rain requires freezing conditions at ground level. Hail is only produced inside cumulonimbus storm clouds.

Snow

Snow is formed in clouds where ice crystals grow. Like raindrops, snowflakes fall when they grow too heavy to be kept suspended by air currents. In warm regions, snow melts before reaching the ground, but in colder regions snow arrives in flurries, snowstorms, blizzards, and in steady snowfalls that blanket the ground in white.

FACT FILE

Snow Snowflakes take many different forms but all are variations of a basic hexagonal crystal shape. The shape of the flake depends on the environment within the cloud in which it forms.

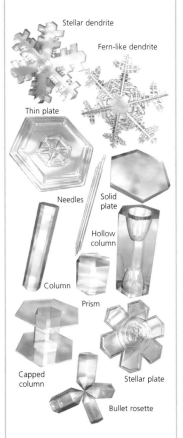

Stellar dendrite

Fern-like dendrite

Thin plate

Needles

Solid plate

Hollow column

Column

Prism

Capped column

Stellar plate

Bullet rosette

Crystal types The shape that an ice crystal takes is determined by the temperature at which it forms and, to a lesser extent, the humidity level in the air. Higher humidity, above water saturation, leads to more branched growth on the crystals.

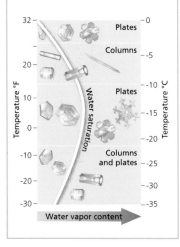

Snow clouds (below) Brooding snow clouds overhead suggest that recent falls may not be the last. The ice crystals in the cloud reflect the sunlight upwards, so the snow clouds appear dark when viewed from the ground.

Living with snow (left) Even a modern city like New York can almost grind to a halt following heavy snowfalls. Going underground is one way to escape. A blizzard in 1888 was the impetus for the construction of New York's subway system.

Fall speeds (below) Raindrops and ice particles falling from a cloud reach terminal velocity when the force of gravity is counteracted by air resistance. Below are the terminal velocities of three examples of precipitation and their fall times from a height of 10,000 feet (3,000 m).

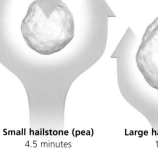

3.1 miles per hour (5 km/h)

14.5 miles per hour (23 km/h)

24 miles per hour (39 km/h)

64 miles per hour (103 km/h)

Snowflake
36 minutes

Raindrop
8 minutes

Small hailstone (pea)
4.5 minutes

Large hailstone (golf ball)
1.75 minutes

FACT FILE

1. Sapporo Japan's fifth-largest city, on the island of Hokkaido, is the snowiest metropolis in the world. Its annual average snowfall is 248 inches (630 cm). Each February, over two million people attend the Sapporo Snow Festival to admire hundreds of snow and ice sculptures.

Sapporo, Japan

2. London In February 2009, the worst snowstorm in 18 years led to traffic chaos, the cancellation of all buses, many trains, and more than 650 flights from Heathrow Airport. Because heavy snowfall is so rare, the city was not well prepared for such a severe event.

London, U.K.

3. Mount Cayambe Snow is most common near the Poles and increasingly rare toward the Equator. However, snow is found on some high mountaintops in equatorial regions. In fact, the Equator passes right through the permanent snow cover around the peak of this 18,996-foot (5,790 m) volcano.

Mount Cayambe, Ecuador

Upper layer The temperature here can be as low as –40°F (–40°C) and the clouds, which are thinly dispersed, are formed mainly of ice crystals.

Middle layer Warm updrafts and cold downdrafts meet about halfway down the cloud, where there is a mixture of ice crystals, snowflakes, and liquid water droplets.

Lower layer This lower layer is close to freezing. A rising air parcel brings water vapor that condenses to form cloud.

Snow Snow forms if the freezing level is below a height of 1,000 feet (300 m) above the ground and the ice crystals do not have time to melt before they reach the surface.

Rain When water droplets grow and become too heavy to be held up in the cloud they fall to the ground as rain.

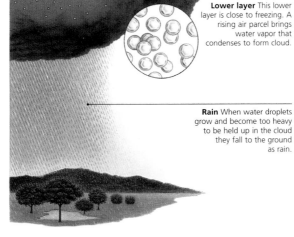

Snowmaker (above) A large cumulonimbus cloud can produce rain, hail, or snow. As ice crystals are swirled up and down by air currents inside the cloud, they collect droplets to form hailstones. Ice crystals join up to form snowflakes.

Mountain snow (below) When air ascends a mountain slope it cools with altitude and its capacity to hold water decreases, causing an increase in rain and snow. Earth's highest mountains are permanently snow-covered.

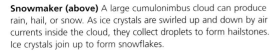

WILSON A. BENTLEY

This farmer from Vermont, U.S.A., took the first detailed photograph of a snowflake, in 1885, and would take thousands more over his lifetime. His book *Snow Crystals* has become a classic and the phrase "no two snowflakes are alike" can be credited to him.

Snow continued

Winter snowfall presents increased dangers to drivers and older people with reduced mobility. In regions with regular heavy snowfall, people have become used to living with snow and enjoy the opportunities it presents. In other places, snow is unusual, and people have greater difficulty dealing with the problems it brings.

FACT FILE

Snow shelters The polar regions are the coldest on Earth. To survive there, humans must build shelters to protect them from the snow, cold, wind and, at times, the wildlife. The most famous polar shelter, the igloo, is built from snow.

Inuit igloo Snow is a good insulator, so it can be quite warm inside an igloo. Stoves can cause the blocks to melt a little, but the moisture freezes overnight.

Sami tent A traditional Sami tent, or "lavvu," has wooden poles and a reindeer-hide covering. Modern versions use aluminum poles and light fabrics.

Hydraulic home The Amundsen–Scott South Pole Station sits on huge hydraulic jacks, which can raise it to stay above the accumulating snow.

WEATHER WATCH

Snow and climate change Fresh water from melting ice shelves threatens to dilute salty seawater, which, in turn, may alter ocean currents and global weather. In addition, a reduction in snow cover will decrease reflected solar energy, leading to more global warming and reduced snowfall.

Cold, dry air is dense and heavy. Strong mountain airflow may lift it, but even if ice crystals form, they are likely to sublimate in the dry air as they fall.

Saturated air at 0°F (−18°C) contains about one-quarter of the water vapor as at 32°F (0°C).

0°F (−18°C)   32°F (0°C)

Too cold to snow? While theoretically it can never be too cold to snow, it does become less likely at temperatures well below freezing. Very cold air cannot hold much water vapor. Cold air is also dense and heavy, and therefore less likely to be lifted to expand and condense to form snow.

Polar bears (above) Snow is home to polar bears. They are born in a den under the snow in spring and when strong enough, they emerge with their mother to learn how to catch food for themselves.

Igloo (right) The traditional Inuit igloo is a dome built from blocks of hard, wind-compacted snow. The blocks are placed in a spiral pattern, each one tilted slightly inwards. This makes a very stable structure.

Snow reflects most of the light and heat from the Sun.

Snow chills air near the surface during the day.

Skiing (above) Skis were first used by inhabitants of Scandinavia and northern Russia to travel about during the snowy winters. In the 20th century, skiing became a popular recreation and skis were redesigned for speed and control.

At night, snow radiates energy at infrared wavelengths and more heat is lost.

On a clear night, the loss of energy from snow-covered ground further chills the air.

Dark surfaces absorb energy, warming the air and melting nearby snow.

Snow compounds cold (left) Under clear skies, radiated heat from snow and the ground is not trapped by clouds, so it escapes to space, thereby cooling the snow. Cold snow also cools the air layer above the surface.

MAKING SNOW

In many ski resorts there is a shortage of snow, particularly at the beginning or end of the season. The solution is to make artificial snow. This is done by spraying a very fine mist of water droplets into the air, often with the addition of a nucleating agent.

OPTICAL EFFECTS

As light travels through Earth's atmosphere, air molecules and particles scatter the rays in different directions, producing optical effects. The type of event depends mainly on the size and shape of the particles, and the wavelength of the light. The effects can range from the everyday scattering of sunlight to make the sky appear blue, to the rare incidence of iridescence. These atmospheric phenomena can help us to interpret and predict the weather.

COLORS IN THE SKY

The visible part of the solar spectrum covers the wavelength range from red to violet—the colors of the rainbow. Water droplets can refract, or bend, the light from the Sun. This happens when light enters the droplet and again when it leaves it. Different colors of the spectrum refract at slightly different angles. Viewed from the ground, with the Sun behind and a rain shower ahead, all the droplets act together to produce the arc of a rainbow. Sometimes, a second arc is visible outside the primary arc, and the order of the colors is reversed because of an extra reflection inside each water droplet.

Clouds look white because they scatter all the colors of the spectrum equally. Large cloud particles, such as hailstones or snowflakes, can make clouds look dark because they are opaque and block the path of light. The surface of an ice crystal can act as a mirror; for example, ice crystals in a thunderstorm anvil cloud have been observed to give a flash of reflected sunlight during a lightning strike.

Ice crystals can also act as small prisms, refracting light as it enters and leaves. Most ice crystals are hexagonal and the most common refraction is 22 degrees. Billions of ice crystals act together to deviate sunlight and produce a 22-degree halo, a white or faintly colored ring around the Sun or Moon. Halos of differing angles of refraction are also possible. Sun dogs, bright spots on either side of the evening Sun, are produced under the same conditions as the 22-degree halo but occur only when the ice crystals are oriented horizontally as they fall.

Small water droplets can also give rise to a bright ring around the Sun or Moon, called a corona. This is caused by the diffraction of light as it passes through the cloud. Smaller droplets create larger coronas, but they are always much smaller than the ice-crystal halos. The corona is most visible when it occurs around the Moon as the Sun's brightness tends to blind the viewer to this subtle effect.

When aerosols—small particles suspended in the air—attract a coating of water, they become haze. These particles are too small to grow into cloud droplets. The sky appears white around the Sun due to the haze particles scattering the light forward, away from the Sun and toward the observer.

Air molecules, which are even smaller particles, also scatter light. The so-called Rayleigh scattering is affected by wavelength, so that short waves are scattered more than long waves. This results in light at the blue end of the spectrum being dispersed more than red light, and explains why the sky appears blue. In the evening, when sunlight has a long path through the atmosphere, so much blue light is scattered out that the evening Sun appears red.

Alpine glow (below) The rays of the setting Sun, shining on mountains to the east, create a series of colors from yellow, through pink, red, and violet. The color order is reversed in the morning when looking toward mountains in the west.

Sunset (above and right) During the day, the sky appears blue because air molecules and small dust particles scatter the blue rays in all directions more effectively than the other colors. At sunset, the Sun's rays travel further through the atmosphere. Air molecules scatter out much of the short-wave blue light, leaving the longer wavelength red end of the spectrum to reflect from low clouds.

Iridescence (left) When the Sun, or moon, shines through mid-level clouds, such as altostratus or altocumulus, this effect appears as diffuse patches of color. A cloud composed of a large number of uniformly-sized small droplets produces the best spectacle. Iridescence is not a common phenomenon but can appear in any part of the world.

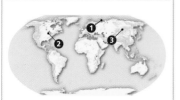

Optical Effects continued

Rainbows that form in clouds of water droplets are very familiar to us; on a sunny day they can even be seen in a spray from a garden hose. But other optical phenomena are rare because they need the presence of ice crystals or unusual conditions of temperature inversion.

Rainbow (right) In this scene, raindrops directly ahead are refracting and returning some sunlight in the form of a rainbow. The colors range from blue on the inside to red on the outside.

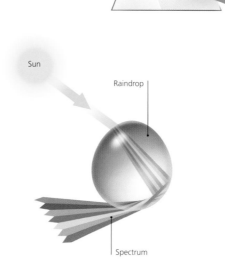

Sun
Prism
Spectrum

Sun
Raindrop
Spectrum

Light spectrum White light passing through a glass prism or raindrop is refracted into its component colors (left). Sunlight passing into and out of a spherical raindrop is bent and dispersed twice into its component colors (below left). If the sun is within approximately 42 degrees of the horizon and there is nearby falling rain, an observer facing away from the sun will see a rainbow (below).

Rain shower
Sun
Observer
Rainbow

Mirage (below) Air over a hot desert has low density near the ground and increasing density away from the surface. Light passing near the ground from distant objects is bent downward by refraction as it crosses these layers of changing density, producing a false image below the horizon.

FACT FILE

Sun dogs These bright spots appear on either side of the sun and are often observed in ice crystal clouds in polar regions. They can also be seen elsewhere when the sun is low and cirrus cloud is forming in the sky.

Sun dogs over Stockholm This painting depicts multiple sun dogs and arcs seen over the Swedish capital in April 1535. The occurrence was thought to be a bad omen or a warning from God.

Sun dogs over Nova Scotia Sun dogs are most commonly seen in winter, when the sun is low in the sky.

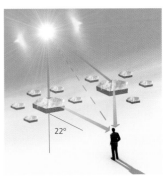

Refracted vision Sun dogs appear on one or both sides of a low sun when its rays are turned through 22 degrees by refraction in hexagonal, horizontally aligned plate ice crystals.

22°

Sun pillar (above and left) Plate ice crystals tend to fall with their large sides parallel to the ground. Sunlight passing through a cloud of these crystals can be reflected off their undersides to create a column of light.

EXTREMES

EXTREMES

THUNDERSTORMS

At any given time, about 2,000 thunderstorms rage across Earth. All thunderstorms have lightning but beyond that storm properties vary widely. Most thunderstorms bring brief, intense rainfall, accompanied by a few strokes of lightning. A small proportion of thunderstorms can be classified as producing extreme weather. Such severe storms most commonly have strong, horizontal winds. Less common features are flooding rains, rarer still is heavy hail, and rarest of all are damaging tornadoes. Thunderstorms can occur anywhere in the world, but most frequently in equatorial regions.

Global distribution (below) This map indicates the average number of thunderstorm days per year across Earth. Thunderstorm formation over ocean is less frequent because ocean water does not warm as rapidly as land.

LIGHT AND SOUND

A thunderstorm is essentially a cloud that produces lightning: a gigantic spark created when an enormous imbalance of positive and negative charges occurs. The lightning greatly heats the air, causing it to expand violently, resulting in the crashing noise known as thunder.

Three main ingredients are needed for thunderstorms to form: moisture, instability, and lift. Thunderstorms develop by convection. One way to initiate convection is by heating the bottom of the atmosphere, for example when sunlight is absorbed by the ground and the heat is conducted into the air. The heated air becomes buoyant and rises, similar to the air in a hot-air balloon. Provided the atmosphere remains unstable—the air parcel is warmer and more buoyant than the surroundings—the air will continue to rise and form cloud.

Multicell thunderstorms consist of several individual storms and can be classified into clusters, lines, or rings. An individual cell usually lasts less than an hour, but organized groups can persist for hours if the circulations from earlier storms initiate new thunderstorms on their flanks.

Thunderstorms are formed in several ways. Large-scale atmospheric circulations, such as Hadley cells and summer monsoons, create low-level converging winds over the oceans, especially near the Equator. This convergence forces warmer, moist air to rise. Variations in Earth's surface, such as islands or mountains, can influence the development of thunderstorms. Islands absorb sunlight and heat up more rapidly than adjacent waters. Also, winds blowing off the ocean are slowed by the greater friction over land and this enhances the convergence there. Mountains magnify the effects of solar heating and friction, causing thunderstorms to form first over ridgelines. Over the oceans, thunderstorms develop

above regions of warmer surface water. Thunderstorms are ubiquitous in the tropics where they are usually concentrated over land areas.

Severe thunderstorms are most likely to occur in the midlatitudes, where the lower atmosphere is warmer and more humid than the air above. This provides the energy for a storm to develop, creating saturated and extremely buoyant air, and leading to vigorous updrafts. Wind shear—layers of air moving in different directions or at different speeds—is often present. The wind shear interacts with the strong updrafts to create a complex circulation pattern that allows storms to persist and provides the seed for a tornado.

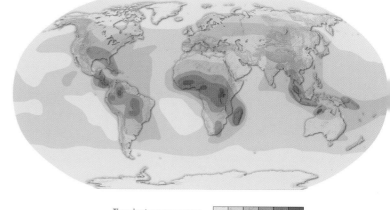

Thunderstorm occurrence

0	5	20	60	100	180

Days per year

Stormy skies (below) Lightning snakes across the sky above Tucson, Arizona, U.S.A. Intense rain falls from a thunderstorm cell on the left of the photograph. Accompanied by a strong downdraft, the rain spreads horizontally where the downdraft strikes the ground. Such downdrafts are hazardous to aviation.

Ominous signs (above) A menacing cloud crosses the Minnesota, U.S.A., prairie ahead of a thunderstorm. The turbulent base of the cloud is a roll cloud, riding the top of the downdraft gust front. The smoother cloud above is a shelf cloud, where inflowing air first condenses.

Towering cumulus Condensing water to make the cloud heats the air. Updrafts prevail within the cloud and the cloud rapidly grows in height.

Mature The updraft spreads out and forms the anvil when striking the stable stratosphere. A downdraft forms where air is cooled, lowering the freezing level. Precipitation begins falling.

Dissipating Eventually, downdrafts predominate, shut off the supply of warm air, and the cloud dissipates.

Thunderstorms continued

The power in a thunderstorm is immense. Energy is released when water vapor condenses and rain falls out. The amount of energy generated during an average thunderstorm is about 10 times the energy produced by the world's largest hydroelectric dam. Most of this energy heats the air, but a small fraction drives the strong winds of the thunderstorm.

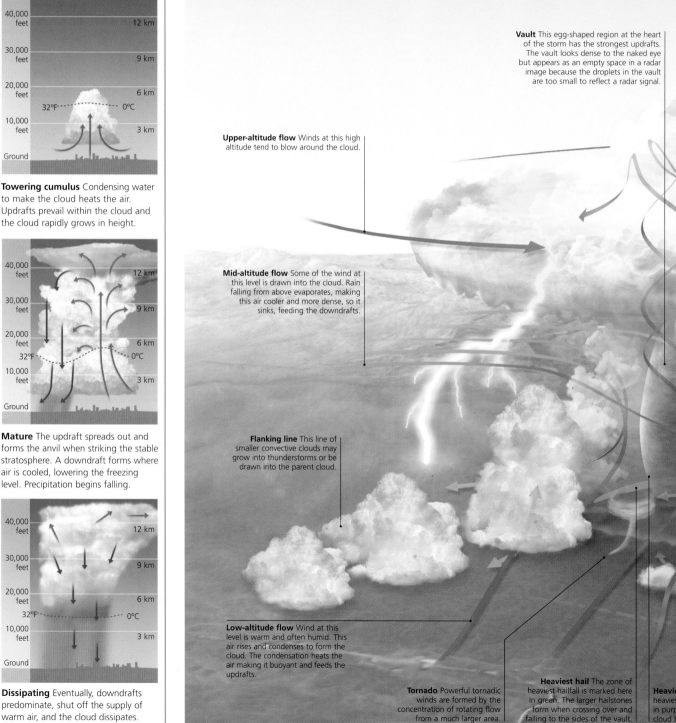

Overshooting top The strongest updrafts have caused this part of the cloud to punch through the tropopause and into the stratosphere.

Vault This egg-shaped region at the heart of the storm has the strongest updrafts. The vault looks dense to the naked eye but appears as an empty space in a radar image because the droplets in the vault are too small to reflect a radar signal.

Upper-altitude flow Winds at this high altitude tend to blow around the cloud.

Mid-altitude flow Some of the wind at this level is drawn into the cloud. Rain falling from above evaporates, making this air cooler and more dense, so it sinks, feeding the downdrafts.

Flanking line This line of smaller convective clouds may grow into thunderstorms or be drawn into the parent cloud.

Low-altitude flow Wind at this level is warm and often humid. This air rises and condenses to form the cloud. The condensation heats the air making it buoyant and feeds the updrafts.

Tornado Powerful tornadic winds are formed by the concentration of rotating flow from a much larger area.

Heaviest hail The zone of heaviest hailfall is marked here in green. The larger hailstones form when crossing over and falling to the sides of the vault.

Heaviest rain The zone of heaviest rain is marked here in purple at the level of the cloud base.

Inside a thunderstorm This illustration of a supercell thunderstorm shows all the major components. All thunderstorms have updrafts and downdrafts. Severe storms have a complex structure, where downdrafts reinforce updrafts. Some storms occur in an environment where the wind direction changes with height, initiating horizontally rotating winds within the storm.

Anvil top Rising air in cloud tends to spread out when it reaches the tropopause, forming a flat top.

Mammatus These pendulous globules of cloud hang from the underside of the anvil. They form when a very high concentration of condensed water is present and are associated with severe storms. These clouds are the only clouds that grow downward.

Shelf cloud This horizontally smooth cloud skirt, near the bottom of the thunderstorm, indicates inflowing air.

Gust front The gust front is the leading edge of the outward-spreading air created when downdraft reaches the ground.

Roll cloud These horizontal clouds sometimes form at the leading edge of a storm's outflow.

Lightning Thunderstorms are defined by the presence of lightning. Most lightning occurs within clouds. Lightning that strikes the ground can start fires and kill people unlucky enough to be caught in the open.

FACT FILE

Thunderstorm organization Flow rising up a mountain favors a single storm. Sometimes a group of storms persist when the spreading downdrafts initiate new convection. At other times, a weather front organizes storms into a line.

Single storm This is an isolated thunderstorm. The spreading anvil has cast a shadow onto lower clouds to the left. Smaller convection clouds flank the sides of the thunderstorm.

Storm cluster Here, three overshooting tops from a cluster of thunderstorms cast shadows on the anvil cloud. In the tropics, such clusters can be the seed for a tropical cyclone.

Squall line A cold front has created a line of thunderstorms, with developing storms on the right. Dissipating storms are on the left, where anvil tops have merged to form a single broad deck of high cloud.

Types of lightning When opposite electrical charges within a storm cloud develop to a sufficiently high level, lightning occurs between them. The type of lightning depends on the locations of the charges in and around the cloud.

Intracloud lightning The most common type of lightning is an arc between upper and lower parts of a cloud that have an opposite charge.

Intercloud lightning A less common discharge is between adjacent clouds, when their areas of opposite charge are located close to each other.

Cloud-to-ground negative polarity This lightning strike connects the negative charge at the bottom of the thunderstorm and the positive charge attracted in the ground beneath.

Cloud-to-ground positive polarity This lightning strike connects the positive charge high in the thunderstorm and a negative charge attracted in the ground beneath.

Lightning

Lightning is a massive spark between accumulations of negative and positive charge in a thunderstorm. Air is a poor conductor, so a huge voltage difference must develop before a path is activated along which electrons flow. When the spark does occur, the air is instantly heated to many thousands of degrees and expands explosively to create thunder.

Three types of lightning (below) On the lower left are two cloud-to-ground negative polarity bolts. The brighter flash on the right is a positive polarity strike, extending from the anvil to the surface. The highest, faint stroke is a cloud-to-cloud bolt.

Charge separation (right) The separation of charge in a cumulonimbus cloud that results in lightning is believed to be the product of countless contacts between snowflakes, heavier hailstones, and graupel pellets within the cloud.

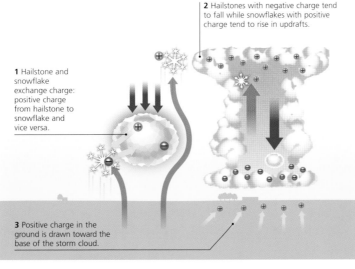

2 Hailstones with negative charge tend to fall while snowflakes with positive charge tend to rise in updrafts.

1 Hailstone and snowflake exchange charge: positive charge from hailstone to snowflake and vice versa.

3 Positive charge in the ground is drawn toward the base of the storm cloud.

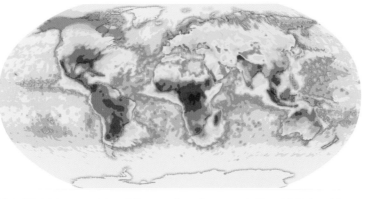

Global lightning map (above) The map shows the concentration of lightning strikes across the globe as recorded by a satellite. Lightning is much more common over land than over ocean. The highest frequency of strikes occurs over central Africa.

Lightning inferno Smoke from over 2,000 fires obscures most of northern California in this satellite view taken on June 25, 2008. The fires were caused by lightning associated with a rare outbreak of summer thunderstorms a few days before.

Lightning stages A flash of lightning consists of a "stepped leader" stroke from cloud to ground and one or many "return strokes." The return stroke produces more than 99 percent of the luminosity we see as lightning.

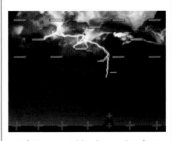

1 A faint stepped leader reaches from inside the cloud base toward the ground in a series of steps, each about 160 feet (50 m) in length.

2 A positively charged ground leader reaches up from the ground in the direction of the descending stepped leader.

3 The stepped leader and ground leader connect. Positive charge streams up the pathway in a bright "return stroke."

4 Charge in other parts of the cloud can be drained along the established pathway, causing the flickering often seen in a lightning flash.

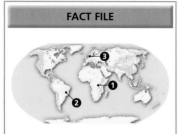

Lightning continued

Lightning is dangerous because of its power and its unpredictability. To zip across several miles of atmosphere requires an enormous current. A typical bolt has about 1,000 times the current that an electricity provider delivers to the average house. Thunder travels more slowly, taking five seconds to traverse a mile or three seconds to cover a kilometer.

Spider lightning (below) Lightning can spread across the underside of a thunderstorm cloud, along multiple paths at the same time. Each path discharges successive patches, with the whole progression taking several seconds.

Thunder (below) The temperature of a lightning bolt exceeds 55,000°F (30,000°C). When a bolt forms, the air around it is superheated, which causes the air to expand then contract rapidly. This creates sound waves we hear as thunder. The nature of the sound we hear is determined by our distance from the lightning bolt.

Thunderclap Thunder starts with low and high sound waves. When both are present, we hear a bang!

Rumble Short waves are rapidly absorbed by the air, so an observer a mile away only hears the low rumble.

Silent storm Sound waves are refracted from warmer toward cooler air. Lightning is rarely heard beyond 10 miles (16 km).

miles

kilometers 100

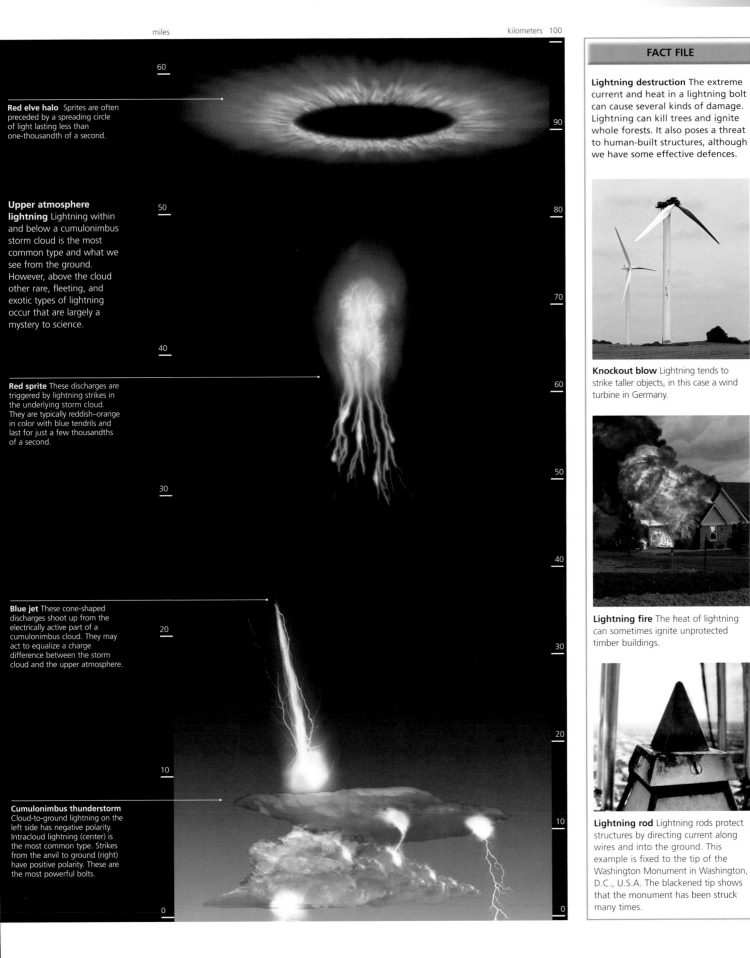

60

90

Red elve halo Sprites are often preceded by a spreading circle of light lasting less than one-thousandth of a second.

Upper atmosphere lightning Lightning within and below a cumulonimbus storm cloud is the most common type and what we see from the ground. However, above the cloud other rare, fleeting, and exotic types of lightning occur that are largely a mystery to science.

50

80

70

40

Red sprite These discharges are triggered by lightning strikes in the underlying storm cloud. They are typically reddish–orange in color with blue tendrils and last for just a few thousandths of a second.

60

50

30

40

Blue jet These cone-shaped discharges shoot up from the electrically active part of a cumulonimbus cloud. They may act to equalize a charge difference between the storm cloud and the upper atmosphere.

20

30

10

20

10

Cumulonimbus thunderstorm Cloud-to-ground lightning on the left side has negative polarity. Intracloud lightning (center) is the most common type. Strikes from the anvil to ground (right) have positive polarity. These are the most powerful bolts.

0

10

0

FACT FILE

Lightning destruction The extreme current and heat in a lightning bolt can cause several kinds of damage. Lightning can kill trees and ignite whole forests. It also poses a threat to human-built structures, although we have some effective defences.

Knockout blow Lightning tends to strike taller objects, in this case a wind turbine in Germany.

Lightning fire The heat of lightning can sometimes ignite unprotected timber buildings.

Lightning rod Lightning rods protect structures by directing current along wires and into the ground. This example is fixed to the tip of the Washington Monument in Washington, D.C., U.S.A. The blackened tip shows that the monument has been struck many times.

Hailstorms

Severe thunderstorms generate wild, gusty winds and damaging hailstones. Hailstorms are most common in the midlatitudes, particularly in spring and summer. While most hailstones are the size of a pea, some grow as large as golf balls or even oranges. Lobes and asymmetric shapes can occur in rare, extremely large hailstones.

Splashdown (right) A large hailstone splashes into a puddle near the bottom of this photo. Other hailstones accumulate in the street and on the grass of this U.S. city. Some of the larger stones exceeded three inches (7.6 cm) in diameter.

Skylight A hailstorm in Stavropol, in southern Russia, left this woman's roof peppered with holes. Hundreds of homes were damaged in the city.

Crop damage A villager surveys a destroyed corn crop in central China. A single severe hailstorm can reduce a farmer's annual income to zero.

Broken nose This aircraft was unlucky to get caught in a hailstorm en route from Geneva to London. Aircraft are designed to withstand such punishment.

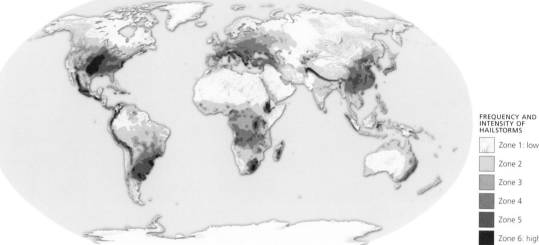

FREQUENCY AND INTENSITY OF HAILSTORMS

Zone 1: low
Zone 2
Zone 3
Zone 4
Zone 5
Zone 6: high

Global hailstorm map (above) Hailstorms tend to be most frequent and damaging where both thunderstorms are frequent and the topography is elevated. Hailstones falling toward high ground have less time in the air to melt completely or shrink to a less damaging size.

FACT FILE

Updrafts Hailstones grow inside a thunderstorm until they are too heavy to be supported by the updrafts. This graphic indicates the updraft speeds required to support a variety of hailstone sizes.

Pea 24 mph (39 km/h)

Marble 35 mph (56 km/h)

Pinball 49 mph (79 km/h)

Golf ball 64 mph (103 km/h)

Tennis ball 77 mph (124 km/h)

Grapefruit 98 mph (158 km/h)

Icebound (left) On November 3, 2007, the strongest hailstorm ever seen in Bogatá, Colombia blanketed the city with ice. Heavy rain flushed the hail in to this underpass, burying dozens of cars in up to 59 inches (1.5 m) of ice and freezing water.

Hail growth (right) Hail follows a looping path in an updraft. In the colder upper cloud, the stone grows a cloudy layer through contact with snow and ice. In the warmer lower cloud, commonly around and above the vault, liquid water spreads over the stone and freezes into a clear ice layer.

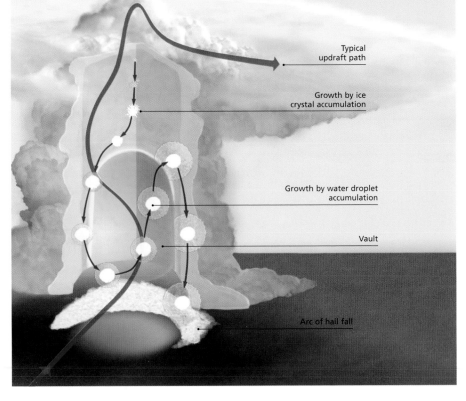

Typical updraft path

Growth by ice crystal accumulation

Growth by water droplet accumulation

Vault

Arc of hail fall

RECORD HAILSTONE

The largest hailstone ever recorded (pictured here) fell on the town of Aurora, Nebraska, U.S.A. on June 22, 2003. Its circumference measured 18.75 inches (47.62 cm).

HAZARDS TO AVIATION

Extreme weather and transportation is a dangerous mix. Mariners and motorists must make allowances for the elements but it is pilots who must be most wary. Unlike vehicles on land and water, aircraft are enveloped by weather, travel at great speeds, and fly at high altitudes. Weather was a greater threat in the early days of aviation, when meteorological information was minimal and planes could not fly above the weather. Now the risks are well known and a great deal of technology has been developed to make air travel one of the safest means to traverse long distances.

Runway conditions (left) On February 6, 1958, a plane carrying the Manchester United football team crashed in Munich, Germany. A build-up of slush at the end of the runway prevented the aircraft from successfully taking off.

never catastrophic. The lightning travels safely through the metal skin of the plane, exiting at another point. The plane's fuel systems are isolated to minimize the risk that a stray spark will ignite any fumes, but lightning can occasionally damage onboard computers or flight instruments.

Finally, atmospheric conditions such as fog and low clouds, smoke, dust storms, and volcanic ash can hamper visibility. However, aircraft and airports are equipped with instruments that allow planes to fly and land safely, even when visibility is poor.

Wind shear (below) The hazard of wind shear was first identified during an investigation into this August 2, 1985 crash in Dallas, Texas, U.S.A. While attempting to land, the plane encountered a powerful downdraft, lost lift, and crashed.

DANGER IN THE AIR

An airplane gains lift from air flowing over its wings. If this airflow is disrupted, disaster can result. Wind shear is a term that refers to an abrupt change of wind direction. It is particularly hazardous during takeoff and landing, when the wind shifts from a headwind, which enhances lift, to a tailwind, which reduces lift. Signs of wind shear are subtle but not invisible. Doppler radars and anemometers have been installed at major airports to help detect wind shear and direct pilots to avoid it.

Turbulence is familiar to every regular flier but pilots can anticipate turbulence and it rarely poses a danger. However, one form of turbulence—clear air turbulence (or CAT)—comes out of the blue and can violently throw an aircraft about, injuring people inside. However, forecasters are getting better at predicting where CAT may be encountered and pilots who encounter turbulence report their experience so that aircraft following can avoid it.

Icing occurs when frozen water accumulates on the airplane. It can happen on the ground or when a plane's surface is well below freezing as it flies through supercooled water droplets. The coating of ice can become quite heavy, disrupting airflow over the plane, increasing drag, and even jamming some control surfaces, such as wing flaps. Areas where icing conditions are likely to be encountered can be forecast from temperature and moisture data. Modern aircraft have various systems and procedures to inhibit and shed ice buildup.

Another form of ice—hail—is found only in thunderstorms. Cockpit windows are designed to resist hail impacts but, given the speed of an airplane, hail can severely damage the leading edges of wings, the tail, engines, and nose. Thunderstorms and hail can generally be identified using onboard radar and avoided.

Thunderstorms also produce lightning; strikes on aircraft are relatively common and almost

De-icing a plane (above) Similar to car antifreeze, de-icing fluid lowers the freezing point and prevents ice from accumulating on the aircraft prior to takeoff.

Wind speed (below) Doppler radar can detect when strong winds are moving in opposite directions. This pinpoints an area of dangerous wind shear and pilots can be alerted.

Icy wings (right) This aircraft crashed on January 13, 1982 shortly after takeoff. The plane had been de-iced, but sat waiting on the runway. An hour later—with ice on the wings and engine—the plane attempted to take off and crashed into the Potomac River, Washington D.C.

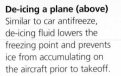

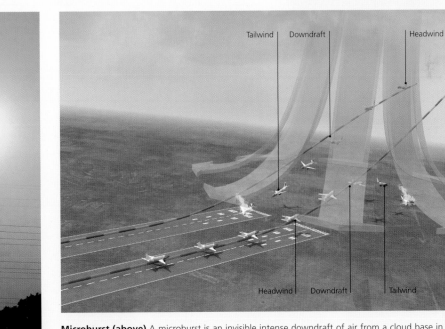

Tailwind | Downdraft | Headwind

Headwind | Downdraft | Tailwind

Microburst (above) A microburst is an invisible intense downdraft of air from a cloud base in stormy conditions. When it reaches the ground, the air blows out in all directions, creating dangerous wind shear conditions. A pilot attempting to land or take off first encounters a strong headwind, creating lift, is then driven downward inside the downdraft, and finally encounters a tailwind, which further robs the aircraft of lift.

TORNADOES

The tornado is the most extreme of extreme weather. No wind is stronger than that which occurs within a suction vortex of an EF-5 tornado. Its occurrence is unpredictable, its path sometimes erratic, and the damage is as random as it is devastating. The tornado is the most destructive member of a family of vortices that includes waterspouts, landspouts, and benign dust devils. The savage winds, often carrying projectiles, prevent direct measurement, but modern instruments can measure the tornado at a safe distance and have revealed hidden structures within.

POWER AND MYSTERY

Tornadoes are just a small part of a thunderstorm, yet they fascinate us for many reasons: their unharnessed power, their capriciousness, their rarity, and their mysteries.

A typical thunderstorm may be 5–10 miles (8–16 km) across at its base and tower more than 40,000 feet (12 km) into the air. From the base of a thunderstorm, a wall cloud a mile (1.6 km) wide may extend down near the rear of the storm. A smaller tornado may appear from beneath that wall cloud. The damage swathe of a tornado can range from 33 feet (10 m) to two-thirds of a mile (1 km) across, but is typically about 275 yards (250 m).

The tornado base moves slightly slower than the upper part of the tornado cloud; so the tornado stretches over time, often resembling a rope in its final stages. Its path of destruction can vary from about 100 feet (30 m) to more than 60 miles (96 km).

Tornado wind speed varies considerably and is a basis for rating tornado intensity. Size does not reliably relate to intensity. Doppler radars, able to operate at a safe distance from the storm, have measured winds at nearly 300 miles per hour (480 km/h). At such speeds, cars become four-wheeled wrecking balls.

The interplay of shear and updraft cause a large portion of the storm to spin. This rotation may become concentrated. As the winds increase, the pressure falls inside the whirlwind. The tornado cloud forms when air pressure at the center of the tornado becomes so low that condensation occurs. A tornado cloud appears to grow downward from a thunderstorm as the pressure falls beneath the wall cloud. An especially dangerous situation develops when the air near the ground is too dry to condense, even with the low pressures accompanying the tornadic winds; in that case, no tornado cloud is visible to warn of the danger.

To develop, tornadoes need intense updrafts in combination with wind shear—layers of air moving in different directions or at different speeds. These conditions occur in many places around the world, but especially over the U.S. central plains, where air that has warmed and dried out from precipitating over the Rocky Mountains blows east above hot, humid air drawn north from the Gulf of Mexico.

Conditions similar to those that lead to tornadoes also cause less violent vortices like waterspouts, landspouts—their counterpart over land—and even dust devils. Strong thunderstorms occur in the tropics, but they rarely produce tornadoes because wind shear is not present.

Baffling discoveries following tornadoes are legendary. Sightings of featherless chickens, tales of frogs and fish raining down, and all manner of delicate objects found embedded in stronger ones attest to the extreme winds and destructive power of tornadoes.

Record weather A tornado slammed a thin plastic phonograph record into a power pole before blowing the pole over. A record is more brittle than a CD or DVD, and much weaker than the pole it is rammed into, yet the record is unbroken.

Out of the blue (above) Although often associated with the U.S. central plains, tornadoes occur in many parts of the world. Three people perished along the 6-mile (10 km) path of this August 4, 2008 tornado in the town of Hautmont, in northern France.

Small but strong (right) A tornado's size is not necessarily related to its intensity. This small tornado, which occurred on June 8, 1995 in Pampa, Texas, U.S.A., was strong enough to pick up six cars at the same time while crossing a parking lot.

No match for a tornado (below) This large pickup truck is barely recognizable after the intense winds of an EF-5 tornado stripped off the metal skin and bent the frame around a pole. The tornado was one of 22 that occurred around Oklahoma City, U.S.A., on May 3, 1999.

Tornado Structure

A supercell thunderstorm is a severe storm characterised by a strong, rotating updraft. Under the right conditions, this system extends downward to become more compact, causing it to rotate faster, finally reaching the ground as a tornado. Within the body of the tornado, incredibly intense mini tornadoes, called suction vortices, sometimes form.

Inside a tornado Warm moist air entering the updraft (**1**) rises over the cool air of the downdraft (**2**). Part of the downdraft spirals inward, and rises on the outside of the tornado (**3**) while the whole column of air rotates (**4**). Inside the tornado the air pressure is very low. This creates the visible

Building rotation Wind speed increasing with height, a gust front diving under the inflow, or a wall cloud downdraft meeting an updraft, are three ways wind shear is created. When an updraft raises air on part of that line of shear, the updraft begins to rotate about a vertical axis.

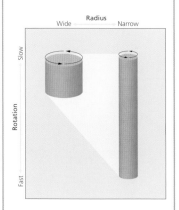

Vortex stretching A column of air the width of the wall cloud is rotating. The top of the column rises as the updraft builds, but the bottom is defined by ground. The column is stretched which makes it longer and narrower.

Angular momentum conservation Stretching the column draws air closer in. Air spiraling inward gains speed at a rate inversely proportional to the distance from the rotation axis. An ice skater does this when drawing her arms closer to her body.

condensation funnel and causes air in the core to sink (**5**). Because the winds around each suction vortex (**6**) are much faster than the winds around the tornado (**4**), each suction vortex carves its own looping path into the ground (**7**) as it circulates the perimeter of the tornado.

Moving music Tornado suction vortices are astonishingly powerful. This grand piano was picked up and hurtled one quarter of a mile (400 m).

Rope tornado This tornado swayed across an Oklahoma field on May 22, 1981. As the gap between top and base of the tornado grows, the funnel cloud narrows until it breaks apart and dissipates.

FACT FILE

Suction vortices Some large diameter tornadoes develop as many as six suction vortices within them. They form within the chaotic tornado winds by a process known as vortex breakdown.

1 Surface inflow (blue arrows) converge and rise inside the tornado (orange). The net motion is a spiral (green).

2 The air pressure inside drops as the tornado grows stronger. The air in the core starts to sink (yellow).

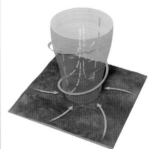

3 The tornado continues to intensify, further lowering the air pressure inside. The air in the core sinks even farther.

4 When the downward moving air reaches the ground and spreads out, suction vortices form from the complex interaction with the inflow.

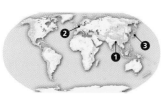
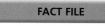

TORNADO SAFETY

When a tornado threatens, the safest place to be is in a shelter designed for the purpose, such as a basement room made of reinforced concrete. Automobiles can be tossed violently through the air and occupants rarely survive. If caught in the open, lying low in a ditch is the best option. Sheltering under a highway overpass is a mistake that has cost some people their lives.

Tornado Climatology

Tornadoes form where thunderstorms grow in the presence of wind shear. Tropical thunderstorms are numerous and powerful but form where there is little shear. In midlatitudes, only certain regions have cool, dry air moving perpendicular to warm, moist air beneath. That combination triggers huge thunderstorms and the right kind of shear for tornadoes.

U.S. tornadoes (below) Over the U.S. central plains, warm, moist air from the Gulf of Mexico collides with cold, dry air from the Rockies. The result is Earth's greatest concentration of tornadoes. Over 50 years of U.S. tornadoes are mapped below.

U.S. tornado paths January 1950–June 2003

Fujita scale (since 2007 the Enhanced Fujita Scale is used)

	F5	Incredible
	F4	Devastating
	F3	Severe
	F2	Significant
	F1	Moderate
	F0	Gale

Tornadoes by the hour (below) Tornadoes can occur at any time of day, but most occur in the afternoon because solar heating often builds up more and more vigorous convection over the course of the day.

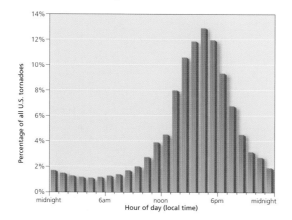

Tornadoes by the month (below) Tornadoes can occur in any season, but are most common in spring, when stronger contrasts develop between cold and warm air masses. Thus, stronger fronts generate more powerful thunderstorms.

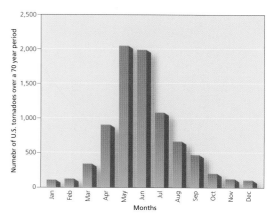

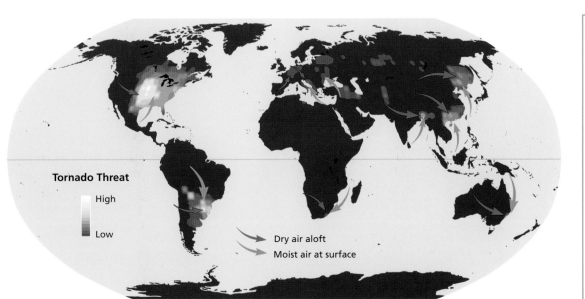

Tornado Threat

High

Low

Dry air aloft

Moist air at surface

Tornadoes around the world (above)
Tornadoes occur where warm, moist air can move at an angle to cool, dry air above. This occurs mostly in the midlatitudes near the eastern sides of continents.

Enhanced Fujita Scale (right)
This scale classifies the intensity of tornadoes by analyzing the nature of the wreckage they leave behind.

Storm chasing
People are fascinated by tornadoes, but few have seen one up close. So experienced, commercial "tornado chasers" offer tours that position their customers close to the action.

ENHANCED FUJITA SCALE

Wind speed, mph (km/h)	Damage
EF0 65–85 (105–137)	Light damage. Some roof covering lost, some glass breakage, tree limb breakage.
EF1 86–110 (138–177)	Moderate damage. Parts of house roof deck and all its covering lost, all windows and glass doors shattered, trees denuded.
EF2 111–135 (178–217)	Considerable damage. House roof supports ripped off, some exterior walls collapsed, shifted off foundation; tree trunks snapped off.
EF3 136–165 (218–266)	Terrible damage. Only interior, reinforced house walls still stand; pre-cast roofs of reinforced concrete structures removed.
EF4 166–200 (267–322)	Severe damage. Home foundation swept clean; reinforced concrete buildings lose exterior walls.
EF5 more than 200 (322)	Devastating damage. Reinforced concrete structures demolished; debris in chaotic piles.

FACT FILE

Tracking tornadoes In the U.S., the conditions that spawn tornadoes are routinely detected by special equipment available to weather forecasters. As a result, the average warning time for a likely tornado strike has improved to 15 minutes.

Watchful eye Meteorologists monitor data captured by advanced instruments. Computer programs highlight features that signal severe weather.

In the field This mobile Doppler radar can be moved into close proximity of a severe thunderstorm.

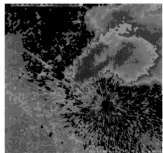

Radar This radar reflectivity image shows heavy rainfall (red) bent in a small hook shape. This "hook echo" prompts a severe weather warning.

10 DEADLIEST TORNADOES

Year	Place	Death toll
1989	Daultipur to Salturia, Bangladesh	1,300
1996	Madarganj to Mrizapur, Bangladesh	700
1925	"Tri-State," U.S.A.	689
1973	Manikganj, Bangladesh	681
1969	Dhaka, Bangladesh	660
1551	Valletta, Malta	600
1964	Magura and Narail, Bangladesh	500
1851	Western Sicily, Italy	500
1977	Madaripur, Bangladesh	500
1984	Ivanovo and Yaroslavl, Russia	400

NOTABLE TORNADOES

Thousands of tornadoes touch down somewhere on Earth every year, however few of these are very destructive and some go unwitnessed. While most tornadoes have been lost to history, some are remembered decades or even centuries later because of their unusual strength or duration, because they strike where they are uncommon, or because they happen to hit a heavily populated area and cause great loss of life and destruction. Just a few of these notorious twisters are discussed here.

TERRIBLE TWISTERS

The earliest recorded tornado in history struck the village of Rosdalla, Ireland on April 30, 1054. Witnesses described it as a "steeple of fire" surrounded by an innumerable flock of dark-colored birds that snatched up animals and uprooted an oak tree.

In subsequent centuries, a number of exceptional tornadoes have struck Europe. In the mid-1550s an armada was wrecked by a waterspout on the harbor at Valletta, Malta with the loss of about 600 lives. On December 28, 1879 two or three waterspouts brought down the Tay Rail Bridge in Scotland. A passenger train plunged into the estuary below, with the loss of 75 lives. The most deadly European tornado outbreak of recent times struck several towns in Russia, reportedly killing 400 people on June 9, 1984.

North America has experienced numerous notorious tornadoes. Any list includes the great Tri-State Tornado of March 18, 1925. Lesser known are twisters that struck Natchez, Mississippi on May 7, 1840 and St. Louis, Missouri on May 27, 1896. These and others each killed hundreds. The largest one-day outbreak was April 3–4

1974: 148 tornadoes touched down in 13 states. On May 3, 1999, 22 tornadoes spawned near Oklahoma City, including an EF5 within city limits that destroyed 2,200 homes. The greatest one-week outbreak was May 4–10, 2003, when 400 tornadoes tore through 19 states; one crossing a path taken in 1999.

The most deadly tornado in history ripped across Manikganj District, Bangladesh on April 26, 1989 killing about 1,300 people and leaving 80,000 homeless along an eight-mile (13-km) path. Bangladesh has witnessed many such devastating tornadoes, including an outbreak that struck south of Dhaka on May 13, 1996, though some of the 700 fatalities were from blows by extremely large hailstones.

Anecdotes abound of things sent flying. A 1915 tornado that trashed Great Bend, Kansas, rained debris 80 miles (130 km) away, carried a flour sack 110 miles (177 km), and deposited a check 305 miles (491 km) downwind. On March 12, 2006 a Missouri teenager earned an unenviable record when he was tossed by a tornado 1,307 feet (398 m) and survived.

A tornado important in the history of meteorological science passed down a Lubbock, Texas street in 1970. It demolished a substantial brick church, yet spared a wooden shack just a few steps away. This counter-intuitive event inspired research that uncovered previously unrecognized structures within some tornadoes: a suction vortex had skipped the shack but hit the church.

St. Louis, 1896 (above) This contemporary illustration depicts a tornado that blasted through the heart of St. Louis, Missouri, and East St. Louis, Illinois on May 27, 1896. Adjusted for inflation, it was the costliest tornado in U.S. history. A half-mile (800 m) wide damage swathe swept structures away or piled debris in drifts. More than 250 people perished.

Among ruins (above) A man searches through his brother's home for salvageable items. The neighborhood was ripped asunder by a powerful tornado, one of the infamous twisters that barreled into Oklahoma City on May 3, 1999.

Bangaldesh (left) A tornado smashed this home and hundreds more north of Dhaka in April 2004. Bangladeshi tornadoes are particularly deadly, due to the high population density and lack of safe shelter. This couple was lucky; they survived and can rebuild their lives. But first, they salvage their rice.

Demolition (below) About three dozen twisters strike the U.K. each year. Among the more powerful examples swept through this London neighborhood on December 8, 2006. It touched down for just a few minutes, but long enough to damage 100 homes.

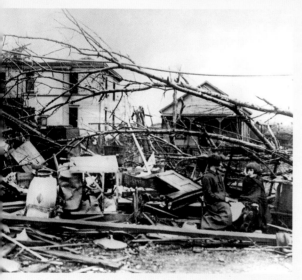

Tri-State Tornado The great Tri-State Tornado (mapped right) was likely a series of two or three tornadoes strung together: as one finished, the next started nearby. The tornado traveled at speeds of up to 73 miles an hour (117 km/h) over a 3.5-hour period, covering 219 miles (352 km). It was the most deadly twister in U.S. history, claiming 689 lives, including 234 in the town of Murphysboro, Illinois (left).

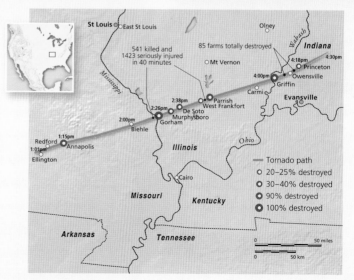

St Louis
East St Louis
Olney
Wabash
Indiana
541 killed and 1423 seriously injured in 40 minutes
85 farms totally destroyed
Mt Vernon
4:18pm
4:30pm
4:00pm
Princeton
Owensville
Mississippi
2:38pm
Parrish
West Frankfort
Carmi
Griffin
Evansville
2:26pm
De Soto
Murphysboro
Gorham
2:00pm
Biehle
Ohio
Redford
1:15pm
1:01pm
Annapolis
Illinois
Ellington
Cairo

Tornado path
○ 20–25% destroyed
◎ 30–40% destroyed
◍ 90% destroyed
● 100% destroyed

Missouri
Kentucky

Arkansas
Tennessee

0   50 miles
0   50 km

RAINING ANIMALS

Animals falling from the sky is a very rare meteorological phenomenon, but one that has been reported throughout history. The creatures most commonly reported are tadpoles, fish, and frogs. Sometimes the animals are alive; at other times they are frozen or in pieces. The phenomena is thought to be caused by waterspouts drawing up the animals from ponds and rivers, and carrying them away.

Related Types

There are other kinds of rotating winds apart from tornadoes. A dust devil is a spiraling, dust-filled vortex of air. They vary in height from only a few feet to over 1,000 feet (300 m). A waterspout can be a tornado that happens to be over water, but many form by a different process that does not require a supercell thunderstorm.

Waterspout (below) A strong waterspout churns up sea spray at the surface. Winds channeled by hills in the background can create convergence zones with wind shear. This, coupled with an updraft, results in a waterspout.

Waterspout dynamics Non-tornadic waterspouts (and landspouts) form where there is an updraft from a growing cloud located over a near-surface shear line along which the wind changes direction abruptly. Unlike a tornado, the vortex grows from the surface up.

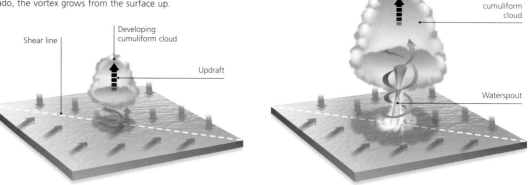

Shear line

Developing cumuliform cloud

Updraft

Mature cumuliform cloud

Waterspout

Fire tornado (right) Whirlwinds of flame often form at the edge of a fire. They are produced when updrafts of heated air encounter a wind shear boundary along the edge of the fire. They can greatly increase the risk of the fire spreading.

Spreading the dust Dust devils occur mainly in desert and semi-arid areas, where the ground is dry and high surface temperatures produce strong thermals. This photo shows a dust devil in the Atacama Desert in Chile. Towering dust devils have been observed on the desert planet Mars.

Evolution of a dust devil These swirling columns of wind, dust, and debris occur when high ground temperatures produce a strong uplift of air, and prevailing winds are deflected to become a vortex.

1 The sun heats the ground which heats air in contact with it. The heated air rises in an environment where wind speed increases with elevation.

Air displaced by rising air, sinks.

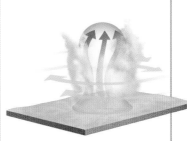

2 The rising air tends to collect into a "bubble" known as a thermal. As it rises, some air above must sink to compensate.

3 Sinking air brings down faster winds on one side of the thermal. Slower winds are raised on the other side. This shear causes rotation.

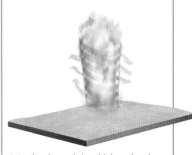

4 As the thermal rises higher, the air spins around a narrower radius which increases the speed. Soon soil and debris are lifted.

HURRICANES

A tornado may produce the strongest winds on Earth, but for destructive power nothing matches a hurricane. These spiraling storm systems can measure up to 500 miles (800 km) in diameter and can produce torrential rain, winds of up to 190 miles per hour (300km/h), and an enormous high tide called a storm surge. "Tropical cyclone" is the generic meteorological name and also the name by which they are known in and Australia and around the Indian Ocean. They are called hurricanes in the Atlantic, Caribbean, and Eastern Pacific, apparently in reference to a god of the indigenous Caribbeans named "Huracan." In the Western Pacific they are called typhoons, from the Cantonese word *tai fung,* meaning "big wind."

SUPER STORMS

All mature hurricanes have certain features in common. In the center is an eye—a clear, almost calm area bordered by a ring of extremely vigorous convection, known as the eye wall. The extreme updrafts within the ring are fed by lines of converging winds that, in turn, form bands of thunderclouds and heavy rain.

The clusters of storms that produce hurricanes occur only where sea temperatures are at least 79°F (26°C). This means that they usually originate in the tropics. To develop its distinctive rotation, a system must be at least five degrees from the Equator, because this is where the Coriolis effect begins to have an influence (see page 39).

The warm ocean powers the hurricane via a process of energy exchange and release. First, the warm ocean heats the air above it, thus increasing the amount of vapor needed before the air reaches saturation. The ocean is essentially an unlimited source of water, so a huge amount of water is evaporated into the air (much more than is possible over land). The heat that caused the water to evaporate is released when the moisture condenses into clouds, particularly rain clouds. If the clouds are in a group, that heat especially warms an area in the center of the group, causing the air to expand and the pressure to drop. The lower the pressure, the faster the wind blows around the group which becomes a hurricane. The stronger the wind blows over the ocean surface, the greater the rate of evaporation, so, as the hurricane spins faster, it grows stronger. This positive feedback process continues for as long as conditions are favorable for hurricane development.

Once spinning, a hurricane tends to move erratically because its wind speeds greatly exceed the weak "steering winds" of its surrounding environment. Most storms move first towards the west, driven by the trade winds. Then they move poleward, driven by winds around a subtropical high. Farther from the Equator the cyclone encounters westerly flow and tends to move to the east.

If a hurricane strikes land, its storm surge and rains may flood large areas and its winds create a wide path of devastation. But when a hurricane leaves the ocean, it also leaves behind its supply of moisture. As it moves inland, it weakens rapidly and soon dies out.

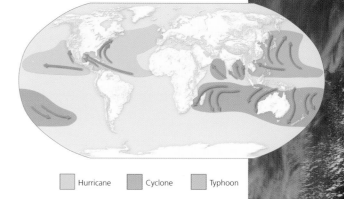

| Hurricane | Cyclone | Typhoon |

Hurricane Rita (right) Hurricane Rita churns across the Gulf of Mexico on September 21, 2005. Rita was one of four rare Category 5 hurricanes to roar across the gulf during the most active hurricane season ever in the Atlantic.

New Orleans after Katrina (right) The flooding in the wake of a hurricane is usually more destructive than the high winds. For days after Hurricane Katrina people were stranded in flooded neighborhoods. Houses were ruined by floodwater and the growth of mold.

Dumped ashore (left) This yacht was swept several miles across a section of the Florida Everglades during Hurricane Andrew in 1992.

FACT FILE

Parts of a Hurricane All hurricanes share several characteristic structures. Many of these are revealed by satellite photography, but high cloud can obscure some features. Radar can unmask many of these complex internal structures.

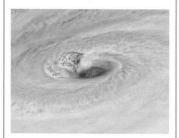

Eye Sinking air in the hurricane center suppresses all but the lowest clouds in the eye and winds are weak. The eye can be 5–50 miles (8–80 km) in diameter.

Eye wall Winds and rain are strongest in the eye wall. Horizontal wind speeds decrease with height, so the lower cloud rotates faster than the upper cloud.

Rainfall bands Bands of intense rain spiral in toward the eye wall. Two bands are often more prominent and can extend for 1,000 miles (1,600 km).

Rainfall revealed A radar mounted on an orbiting satellite shows individual convective cells of intense rainfall as green columns.

Hurricanes continued

As a tropical storm builds up to hurricane force, several characteristic structures appear. As air spirals inward, it accelerates in a way similar to the motion seen in a tornado. This acceleration is revealed by lines of converging air masses that produce spiral bands of convection with varying intensity. In the center of this revolving spiral of clouds is a mostly cloud-free "eye."

1
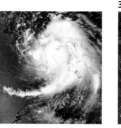

2

3

4
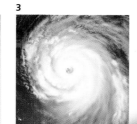

Hurricane Katrina (above) When a tropical disturbance first forms (**1**), the clouds are asymmetric and there is no eye. It intensifies into a tropical depression (**2**), with an irregular, partial ring of cloud. The storm becomes a hurricane (**3**), with deep convection, a prominent eye, symmetric ring, and spiral arms. If it reaches land (**4**), it loses its supply of warm oceanic air and quickly weakens.

Anatomy of a hurricane A mature hurricane consists of bands of thunderclouds (**1**) spiraling about the eye (**2**), a clear, almost calm area at the center of the storm. Each band is fed by updrafts of progressively warmer and wetter air as the band spirals towards the eye (**3**). The air is warmest and circulates fastest at the base of the hurricane (**4**). The air spirals (counterclockwise in the Northern Hemisphere and clockwise in the Southern Hemisphere) into the center, accelerating as it does so. The air then spirals upward in the torrential rain-producing eyewall cloud that surrounds the eye of the storm. In the upper levels the air spirals outward, away from the hurricane's center (**5**). As the hurricane approaches the coast, a storm surge can drive seawater deep inland (**6**).

1

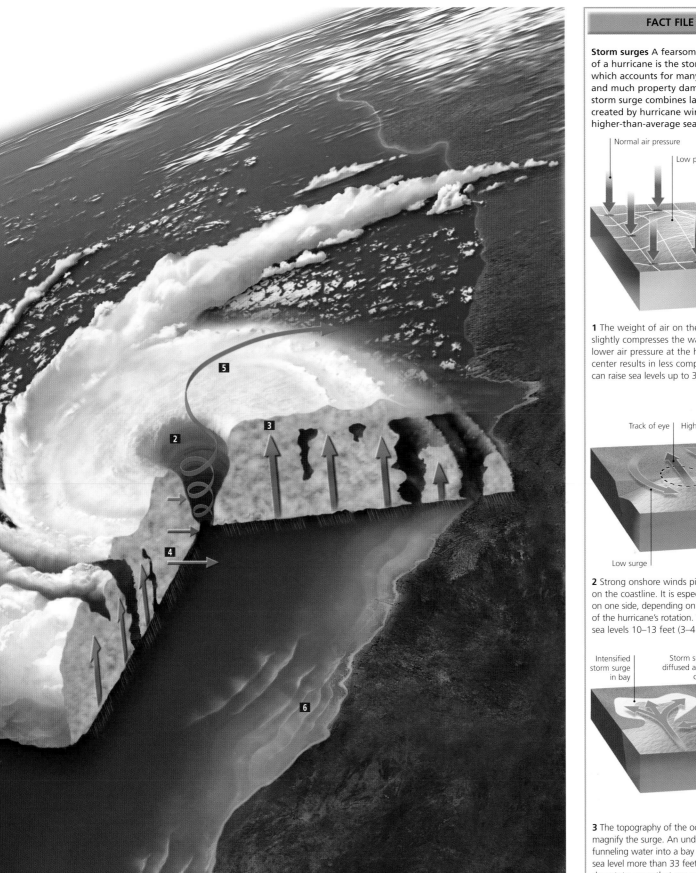

Storm surges A fearsome property of a hurricane is the storm surge, which accounts for many deaths and much property damage. A storm surge combines large waves created by hurricane winds with higher-than-average sea levels.

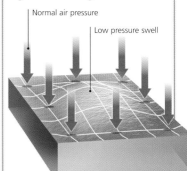

Normal air pressure

Low pressure swell

1 The weight of air on the ocean slightly compresses the water. Much lower air pressure at the hurricane center results in less compression. This can raise sea levels up to 3.3 feet (1 m).

Track of eye | High surge

Low surge

2 Strong onshore winds pile up water on the coastline. It is especially high on one side, depending on the direction of the hurricane's rotation. This can raise sea levels 10–13 feet (3–4 m).

Intensified storm surge in bay

Storm surge diffused along coast

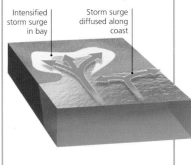

3 The topography of the ocean floor can magnify the surge. An underwater valley funneling water into a bay can raise the sea level more than 33 feet (10 m), and devastate areas that are normally safe.

Hurricane Climatology

Tropical cyclones form over ocean areas that are warm enough to provide the huge quantities of water vapor required to power the storm. Other factors need to be in place as well. The location must be sufficiently distant from the Equator for the rotating winds to form, and the surrounding atmosphere must have weak winds.

Hurricane conditions Hurricanes are powerful storms but they are picky about where they form. In order to organize a group of thunderstorms into a hurricane, five conditions need to be in place.

1 Ocean Temperature Hurricanes need a lot of moisture, so they form where the air is warm enough to hold sufficient water vapor. Typically, this occurs where ocean surface temperatures are greater than 79°F (26°C). Most tropical oceans are warm enough, but the ocean off Chile and Peru is too cold.

2 Convergence/divergence A hurricane forms from a pre-existing group of thunderstorms that is maintained by the presence of surface convergence and upper level divergence. In the Atlantic, this commonly occurs with a line of thunderstorms, known as an "easterly wave," which moves westward out of Africa.

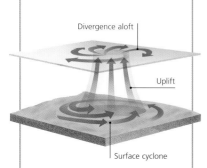

Divergence aloft

Uplift

Surface cyclone

3 Off the Equator Hurricanes cannot form on the Equator because the Coriolis effect is zero there (see page 39). The Coriolis effect determines the rotation around the cyclone. Tropical cyclones rotate counter-clockwise and clockwise in the Northern and Southern Hemispheres, respectively.

4 Deep convection Hurricanes are an organized system of deep convection, so an environment favoring tall convective clouds is needed. That environment occurs where temperature decreases rapidly with increasing elevation through the troposphere, such as from heating of the air in the lower atmosphere, cooling of the air aloft, or both.

5 Low wind shear In conditions of high wind shear, the upper and lower parts of a thunderstorm complex will move differently and the group will never organize into a coherent cyclone. Too much wind shear suppresses hurricanes in the central Pacific and southern Atlantic oceans.

Tropical cyclone tracks (below) This map shows tropical cyclone tracks from 1985 through 2005. Warmer colors indicate stronger storms. Individual tracks are irregular, but most head west then polewards.

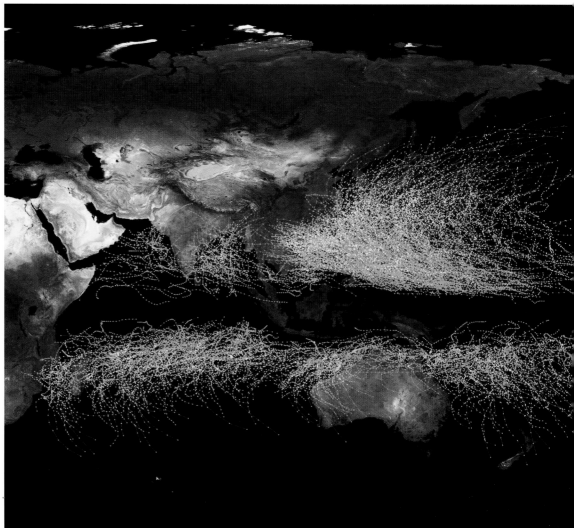

In the firing line (right) Fishermen in Yehliu, Taiwan bail out their boats after the passage of a typhoon. Being surrounded by very warm water and weak summertime wind shear makes Taiwan a frequent target of typhoons.

THE SAFFIR-SIMPSON SCALE

Category	Pressure, mb	Wind speed, mph (km/h)	Storm surge, ft. (m)	Damage
1	980	74–95 (119–153)	4–5 (1.2–1.5)	Minimal
2	965–979	96–110 (154–177)	6–8 (1.8–2.4)	Moderate
3	945–964	111–130 (178–209)	9–12 (2.7–3.7)	Extensive
4	920–944	131–155 (210–249)	13–18 (4.0–5.5)	Extreme
5	Less than 920	More than 155 (249)	More than 18 (5.5)	Catastrophic

Saffir-Simpson Scale The intensity of a hurricane, as determined by its maximum sustained wind speed, is given using a five-point scale. Sustained wind speed also defines earlier stages: disturbance, <23 mph (<37 km/h); depression, 23–39 mph (37–63 km/h); and tropical storm, 40–74 mph (64–119 km/h).

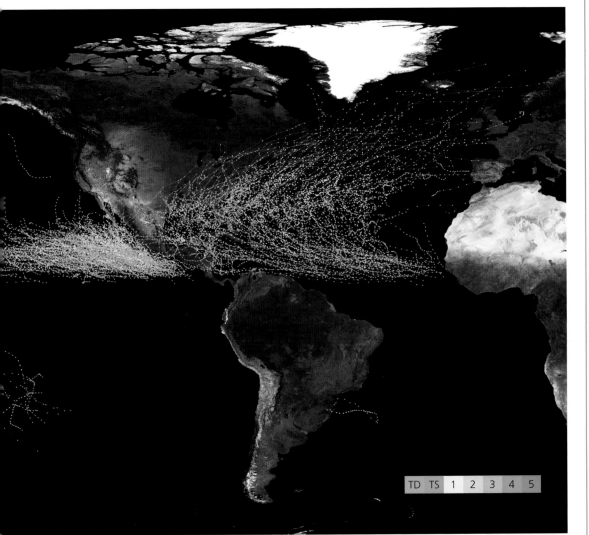

TD TS 1 2 3 4 5

Hot water (right) This map shows sea surface temperatures on August 29, 2005, when Hurricane Katrina reached its peak intensity over the Gulf of Mexico. Temperatures above the critical threshold of 79°F (26°C) for tropical cyclone formation are colored yellow and orange.

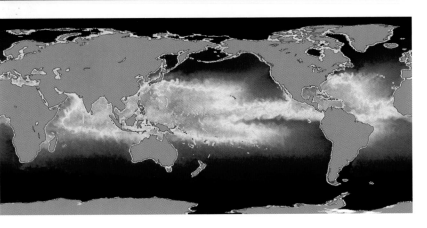

Sea surface temperature

°F 23 32 41 50 59 68 77 86 95

°C −5 0 5 10 15 20 25 30 35

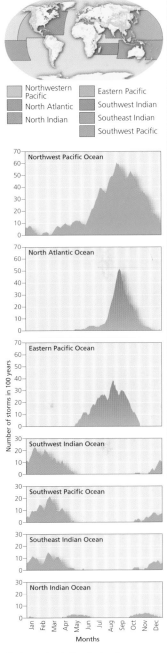

FACT FILE

Cyclone seasons Tropical cyclone frequency is greatest in late summer, when the ocean has been heated over several months. Other factors (like wind shear) control how long the hurricane season lasts in each region.

- Northwestern Pacific
- North Atlantic
- North Indian
- Eastern Pacific
- Southwest Indian
- Southeast Indian
- Southwest Pacific

Northwest Pacific Ocean

North Atlantic Ocean

Eastern Pacific Ocean

Number of storms in 100 years

Southwest Indian Ocean

Southwest Pacific Ocean

Southeast Indian Ocean

North Indian Ocean

Months

Jan Feb Mar Apr May Jun Jul Aug Sep Oct Nov Dec

Cyclone charts Tropical cyclones occur most frequently in the northwestern Pacific, where sea temperature is sufficiently hot for a long time. North Indian Ocean temperatures are also hot, but Asian monsoon wind shear suppresses mid-summer cyclones.

NOTABLE HURRICANES

Hurricanes are given names so there can be no confusion when talking about a particular storm. A consequence of this practice has been the clear identification of notorious storms. Just saying "Katrina" in the United States or "Nargis" in Burma evokes a cascade of images and emotions. Notorious storms uncover weaknesses in our preparedness but ultimately lead to better safeguards.

NOTORIOUS STORMS

Tropical cyclones span a range of sizes and intensities. Some are famous for the damage they caused, while others are remembered for their unusual behavior or meteorological extremes.

The deadliest cyclone on record slammed into what is now Bangladesh in 1970. The combination of primitive prediction and warning systems and a dense population living in a low-lying region proved particularly deadly. The "Bhola" cyclone was estimated to have killed more than 300,000 people, nearly all by drowning in the storm surge. Relief efforts were hampered by political and logistical factors and this added to the death toll. Dissatisfaction with the central government's reaction contributed to the secession of Bangladesh from Pakistan a year later.

Today, satellite images are far superior and ubiquitous, and computer models that predict storm tracks are infinitely better, but hurricanes can still cause large loss of life when warnings are not disseminated or relief efforts are mishandled. In 2008, Cyclone Nargis struck the Irrawaddy Delta of Burma, causing more than 140,000 deaths—a toll made worse by delays in accepting foreign relief. Hurricane Mitch (1998) caused the greatest loss of life by a hurricane in the Western Hemisphere during modern times. Mitch was a very powerful storm that dropped record rainfall on Honduras over a six-day period. The storm surge was unremarkable, but the incredible flooding, coupled with landslides, made Mitch devastating.

The costliest hurricane on record was Katrina (2005), with damage exceeding US$125 billion—about five times as costly as the second most damaging hurricane in U.S. history, Hurricane Andrew, in 1992. Extreme winds were the primary factor in fast-moving Andrew's trashing of southern Florida. Katrina eclipsed Andrew with massive destruction along the Louisiana, Mississippi, and Alabama coasts. Katrina's storm surge breached levees in New Orleans. Because some of the city lay below sea level, it filled like a bowl, resulting in astronomical damage.

Other hurricanes are notable for their meteorological factors. Wilma (2005) holds the record for the lowest estimated sea level pressure (882 hPa) in the Atlantic, though the deepest pressure occurred in the Western Pacific when Typhoon Tip reached 870 hPa. Tip (1979) packed sustained winds greater than 190 miles per hour (305 km/h), as have some other storms, notably Hurricane Camille (1969). Camille made landfall with ferocious winds a little east from where Katrina struck. Tip was very large, with hurricane-force winds extending for over 1,300 miles (2,100 km). Hurricane John (1994) is the longest-lasting tropical cyclone ever observed. It crossed and made loops around the North Pacific for a full month. Tropical Cyclone Catarina (2004) is notable as the only accepted hurricane to have occurred in the South Atlantic.

Katrina storm track (below)
After crossing the southern tip of Florida, Hurricane Katrina headed over unusually hot water in the central Gulf of Mexico. As it, did it rapidly intensified into the highest category hurricane, with an extraordinarily enormous eye.

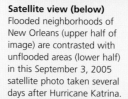

| TD | TS | 1 | 2 | 3 | 4 | 5 |

Satellite view (below)
Flooded neighborhoods of New Orleans (upper half of image) are contrasted with unflooded areas (lower half) in this September 3, 2005 satellite photo taken several days after Hurricane Katrina.

City submerged (above) The storm surge from Hurricane Katrina moved up shipping canals and found weak spots in levees protecting New Orleans, much of which is below sea level.

Nargis storm track Tropical Cyclone Nargis followed an unusual northeastward track in the Indian Ocean, striking southern Burma, a region ill-prepared for the storm surge, high winds, and flooding rains.

After Nargis (above) Those lucky enough to be evacuated ahead of Cyclone Nargis await boats to return them to their villages. Many villages were devastated by the cyclone in the low-lying Irrawaddy Delta.

Hurricane Mitch (left and below) This powerful hurricane stalled just north of the Honduran coast (below). Consequently, bands of intense rain slowly crossed Honduras over several days. The resulting extreme flooding (left) demolished nearly all the country's infrastructure.

FACT FILE

Insider's view Radar gives a detailed picture of the structures hidden within the cloud mass of a hurricane. Meteorologists use ground-based, aircraft-based, and satellite-based radars to study and predict hurricane development and movement.

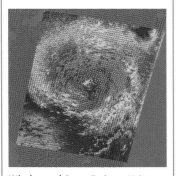

Wind speed Super Typhoon Haitang was observed by the QuickSCAT satellite on July 15, 2005. The wind speed on the ocean surface is estimated by the roughness of the sea. Color indicates speed, and barbs show direction.

Rainfall The TRMM satellite measured rainfall rates for Hurricane Felix. The rates are depicted as colors overlaid onto the corresponding satellite cloud image. Felix was a Category 5 storm that hit Central America on September 4, 2007.

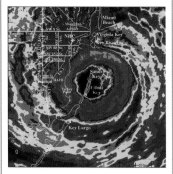

Features Hurricane Andrew made landfall south of Miami, Florida, U.S.A., on August 24, 1992. This ground-based radar image shows the eye wall as a solid red ring of extremely intense rain. Other concentric bands are also visible.

Hurricane Watch

The ability to forecast and issue warnings of hurricanes has improved greatly in the past 50 years. The number of deaths from hurricanes has generally declined, but exceptions occur mainly due to human activities and poor planning. While fatalities have decreased, damage costs have soared, primarily due to expanded development in areas impacted by hurricanes.

Looking ahead (below) Modern tools have improved forecasting. Video loops from satellite images can predict where a storm is heading. A composite image shows Hurricane Andrew over the Gulf of Mexico on three successive days in August 1992.

Improved forecast (below) Satellite data and computer modeling are the primary tools used to track and forecast severe storms. Computer models solve mathematical equations that represent weather processes.

Public warnings (below) Storm forecasts are disseminated over the Internet and broadcast media to alert the public to possible danger. The black line shows the most probable track of the storm at 12-hourly intervals. The white area allows for the uncertainty of its track.

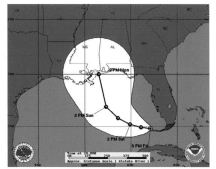

No barrier (right) When Hurricane Ike barreled across this Texas barrier island in 2008, hundreds of homes were completely destroyed. This home survived because the owners built it using construction methods designed to withstand a moderate hurricane.

FACT FILE

What went wrong Hurricane Katrina, which hit New Orleans in August 2005, was a calamity many expected. Katrina was a very powerful storm, but the damage it inflicted was exacerbated by poor preparation and an inadequate emergency response.

Crowding People were stranded for days at the Superdome and Convention Center. Facilities were overwhelmed and lawlessness emerged.

Poor infrastructure Bridges that failed during Hurricane Camille in 1969 and were rebuilt to a similar standard, failed again during Hurricane Katrina.

Bad planning Much of New Orleans lies in a low basin. Buses intended to evacuate people were rendered useless when parked in areas below sea level.

Weak spots Erosion and canal-building reduced the natural protection of the Mississippi delta. Katrina's storm surge flooded large tracts of New Orleans.

Evacuation (below) Proximity to the forecast track of a hurricane triggers either a voluntary or mandatory evacuation. Roads near the Gulf of Mexico and Atlantic coasts of the U.S.A. are often signposted to indicate evacuation routes, which generally head inland and to higher ground.

Safe house (below) If evacuation is not possible, residents must be prepared to live without basic utilities, such as electricity and running water. Essential items include battery-powered radios, flashlights, food, and water.

HURRICANE EVACUATION ROUTE

FLOODS

Several types of severe weather can unleash the power of flooding waters. Flash floods normally result from unusually intense rainfall from thunderstorms. The type and condition of the land surface, along with topography, greatly influence the severity of these floods. Large-scale floods occur primarily as a result of persistent rainfall or from rain falling upon snow and melting it. Flooding can also be a benign seasonal event that is essential for revitalizing soil fertility.

TYPES OF FLOODS

Large-scale floods can be months in the making. Persistent rain can saturate the soil, directing all further rainfall into rivers that may breach their banks and overflow into the surrounding countryside. Over the summer of 1993, a persistent circulation built thunderstorms day after day over the north-central U.S. Severe flooding developed and became the second most costly weather disaster in U.S. history.

Persistent extratropical cyclone tracks can also deliver unusual amounts of rainfall. During the fall of 1966, rainfall saturated soils in Tuscany, Italy. On November 4, after two days of particularly intense rainfall, the worst flood in the city's history inundated Florence. Irreplaceable documents and artworks were damaged.

Warm rain falling on snow adds snowmelt to rainwater flowing into streams and rivers. A 1936 flood that destroyed thousands of buildings in Pittsburgh, Pennsylvania resulted from this combination.

The deadliest floods in human history affected China's Yellow, Yangtze, and Huai rivers during much of 1931. They were the product of heavy snows, followed by warm spring rains, and finally the passage of several typhoons. The death toll may have been as high as four million people.

Accumulated ice can impede river flow. Ice dam flooding doused the Canada–U.S. border region during the spring of 1997, recurring in 2009; the former was worst in southern Manitoba, Canada, the latter the worst in adjacent North Dakota's history. Flooding in extreme cold poses peculiar problems: the floodwaters freeze in place, rescue boats cannot operate, and wading in such floodwaters risks hypothermia.

Some floods are at least partly caused by human activity. The town of Johnstown, Pennsylvania was devastated by a notorious flood on May 13, 1889, when heavy rains breached an earthen dam upstream. In early August 1975, an extratropical trough merged with spent Typhoon Nina over central China to produce record rainfall. The rains exceeded the design capacity of several dams, most notably the Banqiao Dam.

A cascade of several dozen dam failures ensued, with the loss of thousands of lives.

Dry creek beds, rocky canyons, saturated soils, frozen soils, and burned areas cannot absorb moisture and flood more easily. Flash floods are common in rocky deserts that experience summer thunderstorms.

It is impossible to outrun a flash flood, even in a car. A number of motorists drowned trying to escape the 1976 Big Thompson Canyon flood in Colorado, U.S.A. Now, road signs throughout the Rocky Mountains implore people to park and climb up to safety.

However, sometimes human memory is too short. Intense rainfall pounded northern Venezuela in late 1999. Massive landslides spilled out of coastal mountains and buried many coastal communities. A similar, smaller magnitude event had struck the same area in 1951, but the danger had been forgotten during the development and population growth of the intervening years.

Flash flood (left) The village of Boscastle, Cornwall, England lies at the base of a narrow valley. On August 15, 2004 heavy rainfall was funneled into the valley, causing a flash flood. Debris, including 14 cars, blocked the river and caused the waters to deepen.

Ice flood (right) An ice jam on the Yukon River backed up water along part of the Canada–U.S. border. The old town of Eagle, Alaska was destroyed when the flood peaked on May 5, 2009.

Torrent in Turin (left) The Po River rages through Turin, Italy on October 16, 2000. A front had stalled over the Piedmont region and three days of rain saturated the soil. The rain finally culminated in thundershowers over most of the region on this day. This frightening deluge was a one-in-200-year event.

Landslide (above) Intense rainfall in December 1999 saturated the steep hills behind Caraballeda, Venezuela. The resulting landslides flowed out of the narrow valleys like large alluvial fans and buried much of the town.

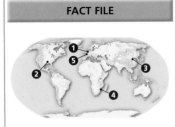

1. The Netherlands Much of The Netherlands is below sea level and for centuries the Dutch have struggled against the power of the sea. On February 1, 1953 a powerful storm surge, combined with a high tide, overwhelmed flood defenses and 1,835 people drowned. Parts of Belgium and England were also affected.

The Netherlands, 1953

2. Mississippi In 1927, after several months of exceptionally heavy rains, the Mississippi River and many of its tributaries flooded. Over 27,000 square miles (70,000 km²) of land were inundated. It remains the most damaging river flood in U.S. history.

Mississippi, U.S.A., 1927

3. Yellow River The Yellow River is often referred to as "China's Sorrow." Major floods along it have been recorded 1,500 times over the past 3,500 years, often with enormous loss of life. A flood in 1887 may have killed up to 2.5 million people.

Yellow River, China, 1887

4. Mozambique On February 22, 2000, Tropical Cyclone Eline reached Mozambique. The country, which was already suffering from disastrous flooding after weeks of rain, was hit by a further deluge from the cyclone. Nearly one million people were left destitute.

Mozambique, 2000

5. Paris After three months of heavy rain and snow, the Seine flooded the French capital in January 1910. The floodwaters peaked at over 20 feet (6 m) above the normal level and lasted a week. About 20,000 buildings were flooded in the city with further damage experienced up and downstream.

Paris, France, 1910

Floods continued

While flash floods can be devastating, they are usually short-lived, confined to a small area, and disperse quickly. In contrast, large scale floods can take weeks to develop and inundate thousands of square miles. Such flooding often begins along a river, which may break its banks and overflow. At other times, tides and storm surges bring water from the sea inland.

Wet feet (below) In November 1966, the Arno River overran its banks and flooded Florence, Italy. Historic documents and artworks were damaged by water containing mud and oil.

Elbe River, 2002 The image (top right) shows the Elbe River, upstream of Magdeburg, Germany, as it is normally viewed by satellite. The second image (bottom right) was captured on August 20, 2002, when the worst floods to strike Central Europe in over 100 years were at their peak.

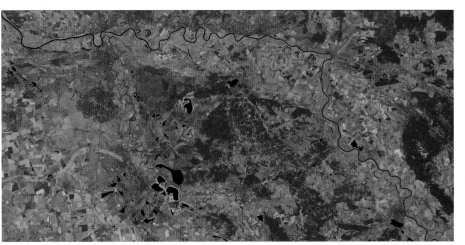

Thames Barrier (below) A series of 10 floodgates were constructed across the River Thames between 1974 and 1984 to protect London from flooding caused by exceptionally high tides associated with storm surges in the North Sea. The barrier is raised an average of four times a year.

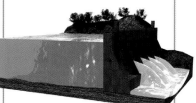
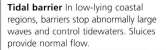

EXTREME FRONTAL SYSTEMS

Frontal systems, also known as extratropical frontal cyclones, are the dominant weather generators in the midlatitudes. No other weather system can produce such a diversity of extreme conditions: strong winds, flash floods, hail, and tornadic phenomena. Frontal systems may spawn vigorous convection, forming thunderstorms along a front, or as an intense squall line ahead of the front. Contrasting polar and tropical air masses provide the energy to bring heavy rains, develop bands of disruptive freezing rain, and produce blizzards that blanket large areas. This may drive frigid air toward the Equator, well outside the air's normal boundaries.

White hurricane (right) Heavy snowfall and strong winds hit New York during the superstorm of March 1993. Two feet (60 cm) of snow fell on parts of New York City. Nearby, winds gusted to 89 miles per hour (143 km/h) at Fire Island and 71 miles per hour (114 km/h) at LaGuardia Airport.

the U.S.A. and tropical air over the Gulf of Mexico. Record-breaking snowfalls extended from Alabama to Maine. A quarter of all flights within the U.S.A. were grounded as airports from Atlanta to Boston were closed. Severe squall lines ahead of its cold front cast 11 tornadoes onto Florida. Wind gusts over 100 miles an hour (160 km/h) were recorded in several places, and winds equivalent to a Category 3 hurricane spread over much of the U.S. east coast. Record cold for springtime followed in the storm's wake.

The Great Storm of 1987 was also notable for its winds. It struck on the night of October 15–16,

Wild winds (left) The Great Storm hit southern England in October 1987. The passenger ferry *Hengist* broke its moorings in the furious winds and ran aground near Folkestone.

bearing the strongest winds to hit England and parts of northern France in centuries. It toppled trees that had stood for hundreds of years, including six of the namesake trees in the town of Sevenoaks and historic trees in Kew and other royal gardens across southern England. Gusts exceeding 90 miles per hour (145 km/h) blew roofs off buildings in London. It was the second-costliest storm in British history. Advance warning of this fast-developing storm was issued for shipping in the English Channel but the storm took a slightly more northerly track than was predicted. Scientific inquiry after the storm led to improvements in weather forecasting.

On April 10, 1968, the remnants of Cyclone Giselle invigorated an extratropical trough over New Zealand's South Island, bringing severe winds. The strongest gusts ever experienced in the Wellington area were recorded at 167 miles per hour (269 km/h). The storm is remembered as the "Wahine Disaster" after the inter-island ferry that ran aground and sank in Wellington Harbour. During the peak of the storm, wind-generated waves in the harbor reached heights of 33–39 feet (10–12 m).

In late January 1953, strong winds from an intense frontal cyclone combined with high tides to flood portions of Belgium, the Netherlands, and eastern England. Dikes were breached and seawater covered nine percent of Dutch farmland. Several thousand people, mainly in low-lying parts of the Netherlands, drowned in the storm surge, which was 6.5–18.4 feet (2–5.6 m) high. The inability to warn people of the impending disaster late at night contributed to the tragedy.

AGENTS OF CHANGE

Masses of warm and cold air are moved around the globe by winds, with one air mass displacing another. The boundaries between warm and cold air masses are called fronts. The atmosphere is most active in these zones and it is here that weather changes occur. The interaction of warm and cold air masses may produce systems of low pressure that give rise to unsettled weather, including storms. Sometimes a storm develops into a severe frontal system or cyclone. Unlike a hurricane, the frontal cyclone has a cold core. Its main energy source is the large temperature differences across the storm. A hurricane has a warm core, its energy comes entirely from rainfall, and there is no temperature gradient across the system. Frontal systems are most common in the midlatitudes, where temperature differences between air masses are most extreme.

One of the most well-known extratropical frontal cyclones was the so-called Storm of the Century, which occurred in March 12–14, 1993. It formed at the boundary between very cold Arctic air over

Destructive forces (left) An estimated 15 million trees were toppled in southern England during the Great Storm of October 1987. High winds followed a wet period that saturated the soil.

Extreme elements (below) This illustration identifies some of the typical weather manifestations experienced during an extreme frontal system over Western Europe. These are produced by a cold front (blue), a warm front (red), and an occluded front (purple)—where a tongue of warm air rides over cold air.

Squall line Ahead of the cold front, an intense squall line forms over Spain. It brings heavy rains, strong straight winds, hail, and a tornado.

Cold An intense cold front stretching from northeastern France to Portugal brings heavy rain and some snow.

Blizzard Blizzard conditions prevail over Great Britain. Strong winds ahead and behind the occluded front blow snow horizontally.

Ice storm Freezing rain falls in a band ahead of and roughly parallel to the warm front.

Blizzards and Ice Storms

Blizzards are caused by a combination of heavy snow, low temperatures, and strong winds. They can dramatically reduce visibility. A typical blizzard occurs behind an intense surface low-pressure area. When a layer of warm air is sandwiched between two layers of sub-freezing air, freezing rain can develop and an ice storm may form. These storms can cause severe damage.

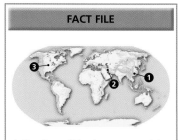

NEW YORK CITY

The Great Blizzard of 1888 dumped over 40 inches (102 cm) of snow that hurricane-force winds then piled into 40-foot (12 m) drifts. It changed New York's cityscape: communication and power lines were put underground and the subway system initiated.

Ice glaze (above) Rain falls on an object of sub-freezing temperature and freezes as a layer of ice. An accumulation of glaze can cause structural damage, bringing down tree branches and power lines.

Ice storm (below) Frozen water droplets first melt as they pass through a layer of warmer air. They then fall into sub-freezing air—as occurs along a warm front—to form freezing rain and produce an ice storm.

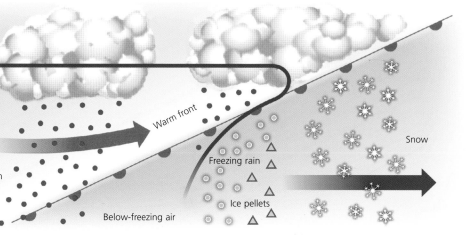

32°F/0°C

Above-freezing air

Warm front

Rain

Freezing rain

Ice pellets

Below-freezing air

Snow

Disorder (above)
Blizzards can disrupt daily life, snarling traffic, closing airports, and stopping trains. Strong winds can pile snow into drifts that far exceed the amount of actual snowfall.

Big chill (right) A heavy blanket of snow was left over Afghanistan after a blizzard crossed the region on January 31, 2006. While such snowfall is disruptive, it is essential in providing sufficient water for the long, dry summers.

Living with blizzards People who live with freezing rain and blizzards have various strategies for coping with the conditions. Suitable building design, securing access through the snow, and maintaining power supplies are all primary concerns.

Digging deep In Antarctica, snowfall is surprisingly small but fierce winds pile up the snow into large drifts that require frequent shoveling.

Building design In cold climate areas, roofs are steep to help shed snow, and reinforced to withstand the weight of a thick layer of snowfall.

Traffic flow Even during a blizzard, keeping the roads open is a high priority. Municipalities in snowy areas allocate significant resources for plowing roads.

Electricity Repairing breaks in power lines caused by falling branches is a high priority after an ice storm.

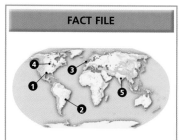
1. Texas and Florida In late December 1983, a severe cold-air outbreak killed most citrus trees in Texas and Florida. Many were never replanted. In Lakeland, Florida, the event lasted four nights,

two of which were 28°F (16°C) lower than normal, yet the monthly average was 1.5°F (0.8°C) above normal.

Texas and Florida, U.S.A.

2. Brazil The world's largest exporter of coffee is Brazil. Severe freezes strike its coffee-growing regions roughly once every decade. Coffee seeds on

the plant are frost sensitive. Frosts in 1975 and 1994 caused coffee prices to spike as supplies were affected.

Brazil

3. U.K. A series of cold-air outbreaks made 1962–63 the coldest winter in the U.K. since 1795. The outbreaks followed in the wake of several blizzards. Temperatures fell to 3°F

(–16°C) at London's Gatwick Airport. Major rivers froze over, including the Thames, in London, and the Avon.

U.K.

4. North Dakota In January 1997, the Northern Plains states of the U.S.A. had one of the windiest winters on record. Temperatures fell to –40°F (–40°C)

but the windchill temperature was –80°F (–62°C). Up to 90 inches (2,286 mm) of snow fell on North Dakota.

North Dakota, U.S.A.

5. Bangladesh and India Cold-air outbreaks that strike heavily populated parts of the world can cause great loss of life. In January 2003, one such event caused 500 fatalities in

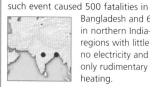

Bangladesh and 600 in northern India— regions with little or no electricity and only rudimentary heating.

Bangladesh and India

Cold-air Outbreaks

A potent frontal cyclone may drag a cold air mass over a region that rarely experiences cold weather. This anomaly can have serious consequences when the local population and infrastructure are not adapted to the cold. A temperature of 14°F (–10°C) during a central Siberian winter may be a unusually warm, but in central Florida this would be record cold.

Mountain barrier (below) Normally, the Himalayas block cold Arctic air from blowing across India. However, during December 2007 and January 2008, northern India shivered under a cold-air outbreak, with the poor and homeless worst affected.

Frozen fruit (left) Many types of citrus fruit are damaged by below-freezing temperatures. Ironically, a standard protection method is to spray water on the trees. The freezing process releases heat and the coating of ice provides some insulation, preventing colder temperatures from reaching the fruit.

Snow date (right) Snow fell in Syria— including a light dusting on these date palms in a Damascan park—during an unusual cold spell in January 2008. The first snow to fall in a dozen years also blanketed other parts of the Middle East.

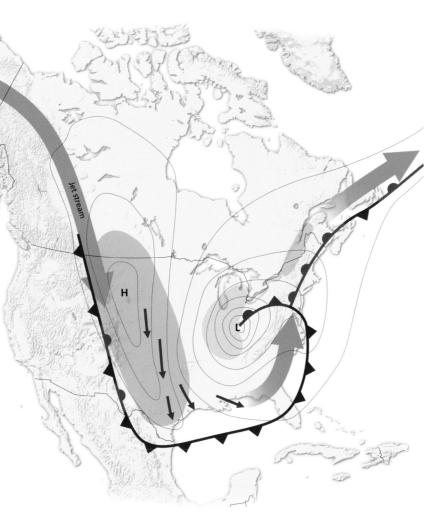

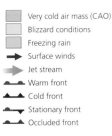

	Very cold air mass (CAO)
	Blizzard conditions
	Freezing rain
→	Surface winds
→	Jet stream
⌒	Warm front
▲	Cold front
▲⌒	Stationary front
⌒	Occluded front

Icy associates (right) This frontal cyclone over North America generates three types of severe icy weather. An intense warm front creates a band of freezing rain. Strong winds with heavy snow combine to produce a blinding blizzard. Strong winds well behind the low drag polar air down toward Mexico, creating a cold-air outbreak.

FACT FILE

Windchill The combination of strong wind and low temperature can produce much greater loss of body heat than the air temperature alone would suggest. The figure often used to express this is the windchill temperature.

Exposure to cold Loss of heat from the body can lead to hypothermia. The elderly are at particular risk from extremes of heat and cold.

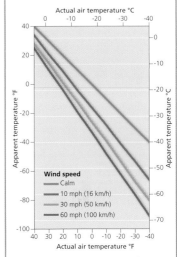

Windchill graph This shows the perceived temperature produced by the combination of actual temperature and wind speed.

Frostbite In extreme cold, windchill can cut the time needed for frostbite to begin from about half an hour to less than five minutes.

DROUGHTS AND HEAT WAVES

Drought is the oddest form of extreme weather. It does not shout its presence with fierce winds, flashing lightning, or lashing rains. Quite the opposite; drought is the absence of something, namely precipitation. Drought develops slowly and silently. Its withering effects may also bring about a heat wave and maybe a duststorm. Of course, heat waves and duststorms can also occur in the absence of drought. Heat waves happen when a hot air mass is displaced. Duststorms need only have exposed, fine soil to be picked up by vigorous winds to form.

THE NATURE OF DROUGHT

Drought is not merely low rainfall. Rain is unevenly distributed throughout the world and some areas will always receive less than others. Drought, therefore, is a relative term, based on the expected rainfall for an area at a given time. In the U.K., a drought is defined as a period of at least 15 consecutive days during which no measurable daily precipitation has fallen. In contrast, two years must pass without rain before a drought is declared in Libya.

In midlatitudes, droughts are generally caused by a large-scale pattern that directs a jet stream and its associated weather well away from its normal path. That pattern is called a "blocking ridge." In the tropics, the large-scale pattern may divert converging winds away from a region, as happens over Indonesia and northern Australia during a strong El Niño event. In its place there is surface divergence that encourages sinking air, which, in turn, suppresses clouds. Sinking air will heat up as it is compressed by the weight of the air above. So droughts are often associated with warmer temperatures.

Drought is one of several factors that contribute to the severity of a heat wave. In addition to heating by compression, the suppression of clouds allows more sunlight to strike the ground. Ground heated by the sun can release it either by heating the air above or indirectly through the evaporation of water. During a drought, there is less water, hence more of the solar heating must be removed by directly heating the air. Finally, winds during such conditions are often (but not always) light; light winds can't mix the heating so easily through a deep layer of the atmosphere.

There is no universal definition of a heat wave. People have adjusted their lifestyle and expectations to fit the normal range of temperatures expected in the place where they live. A summer maximum temperature of 100°F (38°C) in London would be headline news, whereas the same temperature in Marrakesh, Morocco would hardly be mentioned. Local definitions incorporate knowledge of the local norms and may use thresholds that measure the amount of heat stress a person feels, such as the "apparent temperature."

Duststorms may be associated with drought, as vegetation holding the soil withers away.

Drought map (below) This drought hazard map is derived from the length of time rainfall was less than half of its long-term median value for three or more consecutive months between the years 1980 to 2000. Areas with less than 0.04 inch (1 mm) average rainfall per day are excluded.

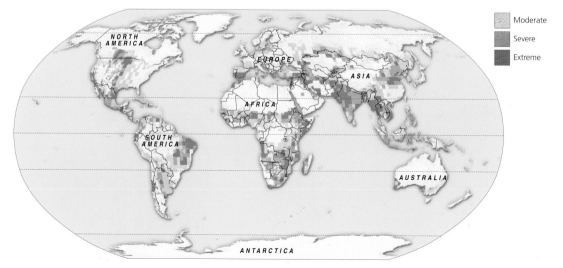

Moderate

Severe

Extreme

NORTH AMERICA

EUROPE

ASIA

AFRICA

SOUTH AMERICA

AUSTRALIA

ANTARCTICA

Thirsty land A woman carries water back to a camp in the Ogadin region, Ethiopia, April 2000. Drought is all too frequent in this part of the world.

Dust and bones A farmer in South Australia, surveys the bones of horses and cattle that perished in a prolonged drought that gripped much of Australia from 2003. Scant vegetation is seen for miles in this photo from June 7, 2005.

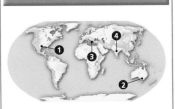

Heat Waves

Heat waves are periods of higher-than-average temperature. They occur when an air mass is transported from its normal location for several days or weeks. Other factors can make the heat more severe, including a lack of clouds and dry ground. People can adapt to constant heat, but an unexpected heat wave can make them sick and can even kill.

Scorched Earth This satellite view from August 4, 2003 records infrared radiation emitted from the ground or the tops of clouds. France and the Iberian Peninsula sizzle with Saharan heat (colored yellow). The cooler adjacent Atlantic waters are red and orange.

Deadly summer The 2003 European heat wave claimed an estimated 35,000 lives. The elderly living in areas ill-prepared for such heat were particularly at risk. Here French workers install cots in a refrigerated warehouse to store corpses.

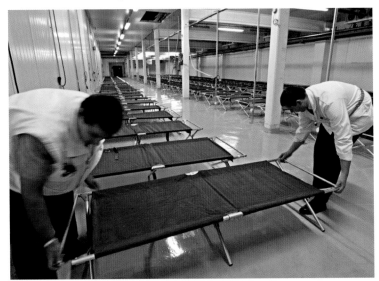

Cooling off Children in Perpignan, France play in a pool to keep cool during the 2003 European heat wave. Submersion in cool water is an effective way to reverse the early effects of heat stress.

Wildfire Fires explode across the landscape near Melbourne, Australia, on "Black Saturday"—February 7, 2009. In temperatures over 115°F (46°C), numerous fires raged through drought-ravaged forests. About 2,000 homes were destroyed and 173 people perished.

Hot spot This map shows how surface temperatures differed from the midsummer average during the 2009 southeastern Australian heat wave. Temperatures in the north were lower than average due to unusually heavy rain during the wet season.

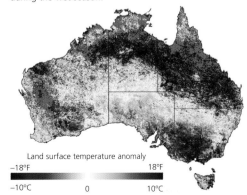

Land surface temperature anomaly

−18°F   18°F

−10°C   0   10°C

FACT FILE

Coping with heat There are several ways to keep cool on a very hot day. Drinking lots of water, soaking in a cool pool, and sheltering in air conditioned buildings all help to beat the heat.

Water replacement Dehydration is a danger in hot weather, when the body keeps cool by sweating. It is important to replenish fluids the body loses.

Temp.	Relative humidity (%)					
°F (°C)	90%	80%	70%	60%	50%	40%
80 (27)	85	84	82	81	80	79
85 (29)	101	96	92	90	86	84
90 (32)	121	113	105	99	94	90
95 (35)		133	122	113	105	98
100 (38)			142	129	118	109
105 (40)				148	133	121
110 (43)						135

Comfort zone   **Risk**

80°F–90°F (27°C–32°C)	Fatigue possible with prolonged exposure and physical acitivity
90°F–105°F (32°C–40°C)	Sunstroke, heat cramps, and heat exhaustion possible
105°F–130°F (40°C–54°C)	Sunstroke, heat cramps, and heat exhaustion likely; heat stroke possible
130°F (54°C) or greater	Heat stroke highly likely with continued exposure

Heat index chart Actual temperature and humidity are combined to produce a temperature related to physical comfort. For example, 100°F (38°C) with 60 percent humidity feels like 129°F (54°C) in dry air.

Air conditioning Power-hungry air conditioning is an increasingly popular way to keep cool around the world.

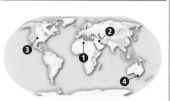

Dust Storms

Strong winds are capable of lifting topsoil and scattering it over wide areas, but occasionally certain conditions combine to produce huge walls of moving dust that transport thousands of tons of soil, sand, and debris. Dust storms are most common in deserts but can occur in other places including landscapes where glaciation has ground rocks into fine, dry sand.

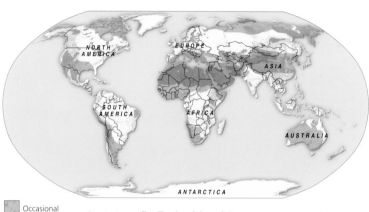

Occasional

Frequent

Dust storm distribution (above) Dust storms can occur in many places around the globe but the major source tends to be the subtropical desert regions. Once carried upward, the plumes of sand and dust can travel for great distances.

Business as usual (below) In regions where dust storms are common, people have adapted their dress, dwellings, and activities. In some parts of Saudi Arabia, dust storms occur on more than 200 days of the year, including this storm in Riyadh on March 10, 2009.

Sandstorm (above) A wall of dust half a mile (0.8 km) high approached Al Asad, Iraq, on April 26, 2005. This haboob marked the leading edge of a gust front from a thunderstorm downdraft, forming a turbulent curtain of raised dust and sand.

Dust bowl (below) This April 1936 dust storm in Cimarron County, Oklahoma, U.S.A., was during the Dust Bowl period in North America. Prolonged drought caused disastrous conditions: choking dust filled the air and vegetation was buried.

High and far Some dust storms are so large and thick that they are observable from space. When lifted by convection and strong winds, the dust can rise two miles (3 km) into the sky, and travel for several thousand more.

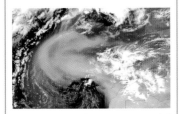

Sahara A plume of desert sand extended 1,000 miles (1,610 km) into the Atlantic Ocean on February 26, 2000.

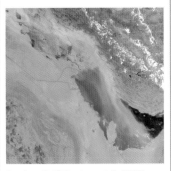

Persian Gulf On August 8, 2005, a veil of dust created by a shamal covered the Persian Gulf and Iraq.

China Strong winds behind a frontal cyclone whipped up thick clouds of dust over central China in April 2001.

Iceland Cold areas can also have dust storms. Fine, glacial soil in Iceland was carried by winds on October 5, 2005.

RECORD-BREAKING WEATHER

Particular combinations of circumstances produce extreme weather conditions. For example, Vostok Station is exposed to temperatures lower than those found anywhere else on Earth because of its latitude—78.46°S—and high elevation. Other extreme phenomena develop by chance. The biggest hailstone that formed inside a storm cloud could have grown in any of a number of places, while the strongest sustained winds could blow in any part of the world that is exposed to tropical cyclones.

EXTREME WEATHER

Tall tales and true have been told about the weather for as long as there have been people around to tell them. Sorting the tall from the true is not easy—few accounts have been confirmed by reliable meteorological measurements.

Systematic observations started only in 1814, when the Radcliffe Observatory in Oxford, England, began recording changes in weather. In the United States, daily records started in 1885, in an observatory founded by Abbott Lawrence Rotch in Milton, Massachusetts. This observatory—the Blue Hill—continues to keep meteorological records and is the longest continuously operating weather-observing station at the same location in the U.S.

Extremes at a weather station are officially cited as records only if the weather station that recorded them has a long-term set of weather measurements. The extremes recorded during the first year of readings should generally not be called records. Just how long weather stations should measure data before declaring records is debated, but the consensus is that at least 10 years' measurements are required before an extreme reading is declared to be a record.

The accurate records maintained by today's weather stations across the world are being extended by new technologies, including satellites, aircraft, automated stations, and radar. By measuring more places, new records are likely to follow. Together, these extremes reveal the enormous power of the forces that contribute to our weather.

Extreme dry The lifeless Valley of the Moon in Chile's Atacama Desert is part of the world's driest desert. The constant wind erodes the rock to form a Moon-like landscape.

Extreme cold This Russian weather station is located on the Severnaya Zemlya archipelago off the north coast of Siberia. Winters are long, dark, and extremely cold.

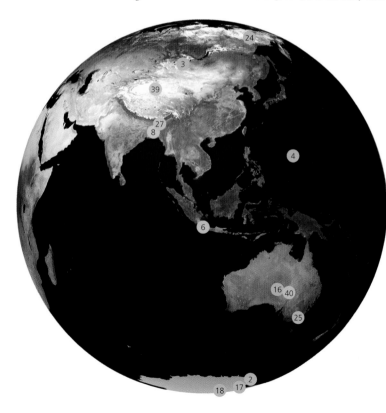

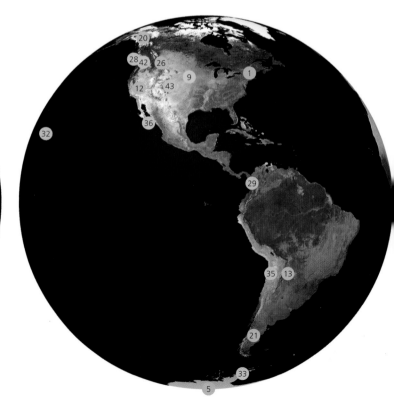

AIR

1. Highest surface wind speed, world
231 miles per hour (372 km/h)
Mt. Washington Observatory, New Hampshire, U.S.A.

2. Highest average wind speed, world
50.1 miles per hour (80.6 km/h)
Cape Denison, Antarctica

3. Highest air pressure, world
1085.7 hPa
December 18, 2001
Tosontsengel, Mongolia

4. Lowest sea-level air pressure, world
870 hPa
October 12, 1979
Eye of Typhoon Tip, near Guam, Pacific Ocean

5. Lowest average humidity, world
0.03 percent
South Pole, Antarctica

STORMY WEATHER

6. Greatest number of thunderstorm days, world
Average 322 days per year
Bogor, West Java, Indonesia

7. Most lightning, world
409 strikes per square mile per annum (158/km² / yr)
Near Kifuka, Democratic Republic of the Congo

8. Heaviest hailstone, world
2.25 pounds (1.02kg)
April 14, 1986
Gopalganj, Bangladesh

9. Largest hailstone, world
18.75 inches (47.62 cm) in circumference
June 22, 2003
Aurora, Nebraska, U.S.A.

TEMPERATURE

10. Highest temperature, world
136°F (57.8°C)
September 13, 1922
Al Azizyah, Libya, Africa

11. Highest annual average, world
94°F (34.4°C)
Between 1960 and 1966
Dallol, Ethiopia, Africa

12. Highest temperature, North America
134°F (56.7°C)
July 10, 1913
Death Valley, California, U.S.A.

13. Highest temperature, South America
120°F (48.9°C)
December 11, 1905
Rivadavia, Salta, Argentina

14. Highest temperature, Europe
118.4°F (48°C)
July 10, 1977
Athens, Greece

15. Highest temperature, Asia
129°F (53.9°C)
June 22, 1942
Tirat Tsvi, Israel

16. Highest temperature, Oceania
123.3°F (50.7°C)
January 2, 1960
Oodnadatta, South Australia, Australia

17. Highest temperature Antarctica
59°F (15°C)
January 5, 1974
Vanda Station, Scott Coast

18. Lowest temperature, world
−128.6°F (−89.2°C)
July 21, 1983
Vostok Station, Antarctica

19. Lowest annual average, world
−72°F (−58°C)
Pole of Inaccessibility, Antarctica

20. Lowest temperature, North America
−81.4°F (−63°C)
February 3, 1947
Snag, Yukon, Canada

21. Lowest temperature, South America
−27°F (−32.8°C)
June 1, 1907
Sarmiento, Chubut, Argentina

22. Lowest temperature, Europe
−72.6°F (−58.1°C)
December 31, 1978
Ust-Shchugor, Russia

23. Lowest temperature Africa
−11°F (−23.9°C)
February 11, 1935
Ifrane, Morocco

24. Lowest temperature, Asia
−90°F (−67.8°C)
February 7, 1892/February 6, 1933
Verkhoyansk, Russia/Oymyakon, Russia

25. Lowest temperature, Oceania
−9.4°F (−23°C)
June 29, 1994
Charlotte Pass, NSW, Australia

26. Greatest one day temperature change, world
From 44°F (6.7°C) to −56°F (−49°C)
January 23–24, 1916
Browning, Montana, U.S.A.

PRECIPITATION

27. Highest annual average rainfall, world
467.3 inches (11.87 m)
Mawsynram, Meghalaya, India

28. Highest annual average rainfall, North America
262 inches (6.65 m)
Henderson Lake, British Colombia, Canada

29. Highest annual average rainfall, South America
354 inches (8.99 m)
Quibdo, Colombia

30. Highest annual average rainfall, Europe
183 inches (4.65 m)
Crkvice, Bosnia-Herzegovina

31. Highest annual average rainfall Africa
405 inches (10.29 m)
Debundscha, Cameroon

32. Highest annual average rainfall, Oceania
460 inches (11.64 m)
Mt. Waialeale, Kauai, Hawaii

33. Highest annual average rainfall equivalent, Antarctica
>31.5 inches (>800 mm)
Along the coast of East and West Antarctica, and the Antarctic Peninsula

34. Highest 24 hour precipitation, world
71.85 inches (1,825 mm)
January 7–8, 1966
Foc-Foc, La Réunion, Indian Ocean

35. Lowest annual average rainfall, world
Parts of the Atacama Desert in Chile, South America, have received no rain in over 400 years.

36. Lowest annual average rainfall, North America
1.2 inches (30.5 mm)
Batagues, Baja California Sur, Mexico

37. Lowest annual average rainfall, Europe
6.4 inches (162.6 mm)
Astrakhan, Russia

38. Lowest annual average rainfall, Africa
<0.1 inches (<2.5 mm)
Wadi Halfa, Sudan

39. Lowest annual average rainfall, Asia
1 inch (25 mm)
Ruoqiang, China

40. Lowest annual average rainfall, Oceania
4.1 inches (102.9 mm)
Mulka Bore, South Australia, Australia

41. Lowest annual average rainfall equivalent Antarctica
0.08 inch (2 mm)
Amundsen–Scott South Pole Station

42. Greatest one-year snowfall, world
1,140 inches (28.95 m)
1998–99
Mount Baker, Washington, U.S.A.

43. Greatest one-day snowfall, world
76 inches (1.93 m)
April 14–15, 1921
Silver Lake, Colorado, U.S.A.

Earth's Coldest Places

Antarctica is Earth's capital of cold. It is noted for its long, savage, and perpetually dark winters, when freezing temperatures and hurricane-force winds make it dangerous to venture outside. However, other locations above and below the Arctic circle and at high elevations, come close to matching Antarctica's extremely cold temperatures and windchill.

Antarctica (below) The Antarctic Plateau has an average elevation close to 10,000 feet (3,000 m), which compounds the cold experienced there. Coastal Antarctica has a considerably warmer climate than the interior.

1. Vostok Station On July 21, 1983, the coldest temperature ever recorded on Earth of −128.6°F (−89.2°C) was observed at the Russian Vostok Station, near the South Magnetic Pole. About 13 staff brave the winter at this remote outpost opened in 1957.

Vostok, Antarctica

2. Yakutsk This city of 210,000 people in the Russian Far East is the coldest city on Earth. The average high temperature in January is −40°F (−40°C). Children are excused attendance at school when the temperature is below −67°F (−55°C). During the short summer, the days are frequently hot and humid.

Yakutsk, Sakha Republic, Russia

3. Ifrane, Morocco Situated 5,367 feet (1,636 m) above sea level in the Middle Atlas Mountains, Ifrane is the coldest town in Africa. It is surrounded by forests and offers winter skiing in the nearby hills. On February 11, 1935, the temperature was −11°F (−23.9°C).

Ifrane, Morocco

4. Snag, Yukon In February 1947, the temperature at Snag Airport fell to −81.4°F (−63°C). This small village and military airfield in Canada is the coldest inhabited place in North America. Only parts of Antarctica, Siberia, and very high elevations are colder.

Snag, Yukon, Canada

5. Mount Everest Judging by weather balloon data measured nearby, winter conditions on the summit of Earth's tallest peak are extreme. Temperatures likely drop 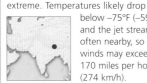 below −75°F (−59°C) and the jet stream is often nearby, so winds may exceed 170 miles per hour (274 km/h).

Mount Everest, Nepal and China

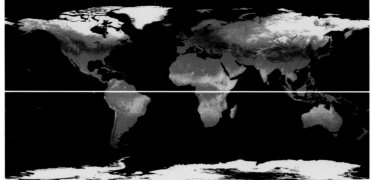

Winter world This satellite image shows daytime surface temperatures in January (Northern Hemisphere) and August (Southern Hemisphere). Outside of Antarctica, the southern landmasses do not extend beyond the midlatitudes. Cold polar air must cross large expanses of ocean which moderates winter temperatures.

°F −13	50	113
°C −25	10	45

Yakutsk bus queue (above) A group of women wait for a bus protected from the cold by reindeer coats and fur hats. It is January, midday, and the temperature is about –49°F (–45°C).

Greenland (below) The Greenland Ice Sheet may be the coldest place in the Western Hemisphere. A record low of –87°F (–66°C) was recorded at the North Ice research station in 1954.

FACT FILE

Polar clothing Traditionally, polar clothing is made from the fur of animals such as reindeer and polar bear. Modern polar explorers and scientists wear clothes made mostly of synthetic materials.

Waterproof and windproof synthetic parka

Reindeer-skin jacket

Fur trousers

Windproof outer layer over thick trousers and thermal long johns

Fur boots

Insulated boots with rubber compound shell

Eye protection The small slits in Inuit snow goggles and the tinted lenses of modern goggles protect the wearer from snow blindness.

Moisture from sweat

Water

Modern materials Modern synthetic fabrics allow moisture from sweating to escape but provide an impermeable barrier to water, keeping the wearer dry.

Earth's Hottest Places

Equatorial regions are hot, and continental climates far from the ocean experience scorching summers, but the highest temperatures occur in subtropical deserts. Solar heating cannot evaporate water from dry land, so all the heat is radiated back into the air. The subtropics is also where sinking occurs in the Hadley cells, further heating the air by compression.

Surviving the heat (right) The Tuareg are nomads who thrive in the Sahara. Their loose clothing, traditionally blue, shades the body and allows dry air to pass through.

Summer world This satellite image shows daytime surface temperatures in August (Northern Hemisphere) and January (Southern Hemisphere).

| °F −13 | 50 | 113 |
| °C −25 | 10 | 45 |

Death Valley (below) Death Valley, in California, U.S.A., is the hottest place in North America. It has an average summer temperature of 98°F (36.7°C). It also contains the lowest point in the Americas, 282 feet (86 m) below sea level.

FACT FILE

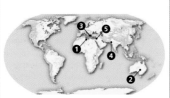

1. Al Aziziyah The highest air temperature ever recorded—admittedly questioned by some—was 136°F (57.8°C) at Al Aziziyah, Libya on September 13, 1922. Surprisingly, this

town is not located deep in the Sahara; in fact, it is only about 20 miles (32 km) from the Mediterranean coast.

Al Aziziyah, Libya

2. Oodnadatta Located in the heart of the desert, 628 miles (1,011 km) north of Adelaide, this is the hottest place in Australia. On January 2, 1960, the temperature rose to 123.3°F

(50.7°C), the highest ever recorded there. January maximum temperatures average 100°F (37.7°C).

Oodnadatta, South Australia, Australia

3. Athens The record for Europe's highest recorded temperature is debated: it may have been set in Spain in the 2003 heat wave or in Sicily in

1999. On July 10, 1977, it reached 118.4°F (48°C) in Athens. The urban location contributed to this temperature.

Athens, Greece

4. Pad Idan Average maximum temperatures are above 100°F (37.8°C) from April to October, but the temperature has been known to reach

120°F (48.9°C). Though hot in summer, winter temperatures can drop close to freezing at night.

Pad Idan, Pakistan

5. Tirat Tsvi This town in Israel's North District is about 50 miles (80 km) from Jerusalem, near the Jordan border, and 722 feet (220 m) below sea level. On June 6, 1942, the temperature rose

to 129°F (53.9°C), the highest ever recorded in Asia. Despite the hot climate, the area is cultivated.

Tirat Tsvi, Israel

Coping with heat Animals living in hot deserts seek shelter during the middle of the day to prevent overheating and conserve fluids. Some satisfy all their water needs from the food they eat.

Desert travelers Hot deserts often have areas of dry, very loose sand. To cross such areas efficiently, some snakes use a sideways action.

Heat transfer Blood vessels near the surface of a desert-dwelling jackrabbit's large ears release excess body heat efficiently.

Heat keepers A camel's body can withstand large differences in temperature. During the day, it acts as a heat sink; during the cool desert night, the body heat is dissipated.

Danakil The Danakil Depression, Ethiopia, has the world's highest average temperature, of 94°F (34.4°C). In places it is 380 feet (116 m) below sea level. Air sinking down to it warms by compression as it descends to this volcanically active area. The landscape pictured is colored by sulfur and salt from hot springs.

Earth's Wettest Places

Rainfall is not distributed evenly around the world. Moist sea air, rising as it crosses mountains, delivers heavy rain to the windward side of coastal ranges. In southern Asia, the summer monsoon brings torrential rain, while heavy rainstorms batter islands near the Equator. In the future, rising global temperatures may alter the rainfall in such places.

		Place	Rainfall	Intensity per hour
1.	1 minute	Barot, Guadeloupe	1.5 inches (3.8 cm)	89.76 inches (228 cm)
2.	15 minutes	Plumb Point, Jamaica	7.8 in. (19.8 cm)	31.18 in. (79.2 cm)
3.	1 hour	Shangdi, China	15.78 in. (40.1 cm)	15.78 in. (40.1 cm)
4.	3 hours	Smethport, U.S.A.	28.5 in. (72.4 cm)	9.50 in. (24.13 cm)
5.	6 hours	Muduocaidang, China	33.07 in. (84 cm)	5.51 in. (14 cm)
6.	1 day	Foc-Foc, Réunion	71.85 in. (182.5 cm)	2.98 in. (7.60 cm)
7.	1 week	Commerson, Réunion	212.6 in. (540 cm)	1.17 in. (2.98 cm)
8.	1 month	Cherrapunjee, India	366.14 in. (930 cm)	0.49 in. (1.25 cm)
9.	1 year	Cherrapunjee, India	1041.78 in. (2646.1 cm)	0.12 in. (0.3 cm)

Record rainfall (above) Over short periods, the heaviest rain falls from intense thunderstorms. Records from one to several days are associated with tropical cyclones. The wettest places, by annual average, receive either heavy monsoon rain over a few months or steady, year-round rain resulting from moist airflow up a topographic slope.

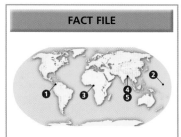

1. Lloró This small town in Colombia, located approximately 520 feet (159 m) above sea level, is the wettest place in South America and certainly one of the wettest in the world. Its estimated average annual rainfall is 523.6 inches (13.3 m).

Lloró, Columbia

2. Pohnpei Island This Micronesian island receives more than 300 inches (7.62 m) of rain a year. Pohnpei is the tip of an extinct volcano, lying in an oceanic warm pool, where the water temperature is almost 85°F (29°C).

Pohnpei Island, Micronesia

3. Debundscha Point This is on the coast, at the foot of the southwest side of Mount Cameroon, in Cameroon, West Africa. It receives an annual average rainfall of 405 inches (10.29 m), mostly between May and October. The upper slopes of Mount Cameroon are above the clouds and dry.

Debundscha Point, Cameroon

4. Cherrapunjee Located at 4,500 feet (1,370 m) above sea level, on the windward side of India's Khasi Hills, Cherrapunjee receives monsoon rains that approach across the Bay of Bengal. It has an average annual rainfall of 450 inches (11.43 m), of which 366 inches (9.3 m) falls between May and September.

Cherrapunjee, India

5. Mawsynram This village in the Khasi Hills, Meghalaya State, India, reportedly receives 468 inches (11.87 m) of rain a year. However, this figure is unofficial. As there is no meteorological office, the village cannot claim to be wetter than its, neighbor, Cherrapunjee.

Mawsynram, India

Mount Waialeale, Kauai, Hawaii (above) With an average rainfall of more than 460 inches (11,680 mm), this mountain is one of the wettest places on Earth.

Bergen, Norway (right) With an average annual rainfall of 88.6 inches (2,250 mm) Bergen is far from the world's wettest place, but its relentless year-round rainfall can make it feel that way. It once rained on the city for 85 consecutive days.

Mount Baker, U.S.A. (left) This mountain in the Cascade Range, Washington State, holds the world's record for the most snowfall during a year: 1,140 inches (28.96 m).

FACT FILE

Charting the deluge There are several contenders for the accolade of world's wettest place. Cherrapunjee and Mawsynram are subject to intense monsoon rains. Quibdo and Mount Waialeale receive year-round rain by virtue of their altitude and location.

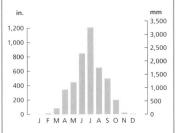

Cherrapunjee, India This town receives 98 percent of its extraordinary rainfall between March and October.

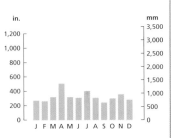

Mount Waialeale, Hawaii This conical Hawaiian peak is exposed on all sides to moist ocean winds. It receives rain almost every day of the year.

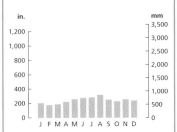

Quibdo, Colombia This town is located where prevailing warm moisture-laden winds meet the Andes. The nearby town of Lloró is even wetter but it does not have an official weather station.

WEATHER WATCH

Wettest no more? There are indications that Cherrapunjee—the wettest town in the world—is becoming drier. Since the turn of the century, rainfall has been consistently below average and the monsoon has been arriving later. Climate change and intensive deforestation may be to blame.

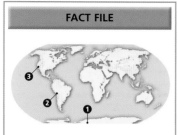

1. South Pole This is a long way from the ocean and high in elevation. The very cold air contains almost no water vapor. Permanent high pressure means

surface air flows outward, suppressing clouds and blocking milder, moister air from penetrating the interior.

South Pole, Antarctica

2. Arica In the north of the Atacama Desert, the city of Arica experiences prevailing winds that blow across South America from the northeast. Before reaching Arica, air crosses the Andes Mountains, where it condenses, loses

moisture, sinks and produces extreme aridity. Arica's average annual rainfall is just 0.03 inches (0.76 mm).

Arica, Chile

3. Mojave Desert This arid region lies to the south and east of the Sierra Nevada mountain range and is in its rain shadow. As air from the ocean crosses

the mountains, it dumps its moisture as rain. As a result, the desert receives five inches (127 mm) of rain a year.

Mojave Desert, U.S.A.

OASES

Even in the driest of deserts, water flows below ground. In certain places, rock strata bring it to the surface as an oasis, People, animals, and plants concentrate around these isolated water holes. Oases were critical links on important trade routes for centuries.

Earth's Driest Places

The driest climates are found in large subtropical deserts, continental interiors, and polar areas. In deserts, dry, subsiding air suppresses clouds. Prevailing winds that shed moisture on the windward side of mountains can make downwind continental interiors very dry. Polar air is very cold and contains little moisture even when saturated.

Antarctic dry valley Mountains divert large glaciers flowing off the polar plateau exposing dark ground in the McMurdo Dry Valleys. What little snow falls, does not accumulate but melts and evaporates.

Atacama Desert (above) Parts of the Atacama Desert in Chile have not seen rain in recorded history. However, more than one million people live in the region today, mostly along the Pacific coast.

Fleeting beauty (below) When rain does come to a desert, a profusion of colorful wildflowers may appear. These plants only germinate after rain and complete their life cycle within a few months.

FACT FILE

Conserving water Plants and animals that live in very dry areas have developed ways of making the most of every drop and of surviving dry times. Some species store water. Others remain mostly dormant, reviving after rain.

Thorny devil A network of channels in the skin of this Australian lizard directs water toward its mouth.

Brine shrimp The eggs of desert brine shrimp can lay dormant for several years before conditions are suitable for hatching.

Cactus Cacti have a shallow but extensive root system to collect as much water as possible from dew or infrequent rains.

SAHARA ROCK ART

Rainfall patterns have varied in the course of human history. The Sahara is now among the world's driest places, but rock art records a time when rain and life was much more abundant.

WATCHING

WATCHING

TRACKING WEATHER 218

STUDYING WEATHER

Meteorology comes from the Greek word *meteoros,* meaning "lofty," and is the modern term given to the study of weather. The affairs of humanity have always been intimately linked with weather: drought can devastate agriculture, storms sink ships at sea, floods inundate dwellings, and hurricanes damage even the largest cities. Subsistence societies and developed countries can have their food production, communications, and power infrastructure dislocated by the vagaries of the atmosphere. Not surprisingly, the study of weather has been of major importance throughout history.

Aristotle Greek scholar Aristotle (384–322 B.C.) believed all matter was composed of five elements: fire, earth, air, water, and ether. He attempted to explain weather through these elements.

GODS TO GIGABYTES
Humanity's effort to understand and predict the behavior of our weather has been one of the more remarkable odysseys of history. In many early civilizations, the state of the weather was believed to be a direct reflection of the mood of the gods, who could punish human misdeeds with storms, drought, and floods. Elaborate systems of prayer and ritual were constructed in order to appease these gods, sometimes involving human sacrifice.

The first weathermen were usually high priests, witchdoctors, or medicine men, whose duties involved not only foretelling the weather, but also ensuring that the gods were placated, so as to guarantee favorable conditions into the future. The belief in the divine nature of weather continued across the millennia. In medieval Europe, witches—thought to be disciples of the devil—were sometimes burned following a hailstorm.

Nevertheless, even in ancient societies inquisitive minds were at work to try to identify natural causes that drove the weather. Such great intellects as Hippocrates, Aristotle, and Pliny the Elder all produced treatises on the subject, offering ideas that were later shown to be incorrect but which at least demonstrated a desire to look beyond the supernatural.

Another source of progress in the understanding of weather came from the two main groups most intimately affected by it—mariners and farmers. Both noted correlations between the weather and cloud patterns, the nature of the wind, and even changes in plant and animal behavior.

Progress was stifled in the Middle Ages by the growth of astrometeorology, but during the Renaissance exciting discoveries were made that would catapult meteorology into the world of science, where it flourished.

Instruments for measuring the state of the atmosphere were invented in the 1600s. This was followed over the next 100 years by important discoveries about the atmosphere's chemical composition and temperature structure.

Knowledge expanded during the 19th century with the discovery that large belts of high and low atmospheric pressure constantly circulated the globe. The invention of the electric telegraph allowed this information to be transmitted faster than the pressure systems themselves could move.

The 20th century saw progress accelerate at an astounding rate, with national meteorological services forming around the world, and the beginnings of weather forecasting using mathematics.

The first meteorological satellite was launched in 1960. This was accompanied by a steady increase in weather observing stations and radar installations right around the world.

By the beginning of the 21st century, meteorology had evolved into a complex interdisciplinary science. Weather forecasts were being produced up to seven days ahead by supercomputers linked to an array of international automatic weather stations, meteorological satellites, and floating oceanic buoys.

The long march from gods to gigabytes has been one of the epic journeys of humanity, and provides a fascinating insight into weather and also the almost infinite inventiveness of the human mind.

Divine winds (above right) Tropical cyclones, or typhoons, have a special place in Japanese history. Kublai Khan's invading fleets were twice smashed by typhoons in the 13th century, prompting the Japanese to call them *kamikaze,* or "divine winds."

Seasons in stone (right) The Aztec calendar stone is a large sculpture that records how the Aztecs measured time. It also depicts the seasons that were part of Aztec culture and ritual. The stone was dedicated to the sun god, Tonatiuh, whose face is thought to be depicted at its center.

350 B.C. One of the great publications on meteorology was produced—Aristotle's *Meteorologica.* It explained the weather through natural processes, and gave birth to the term meteorology.

50 B.C. Roman scholar Pliny the Elder assembled a compendium of ancient knowledge, including a treatise on weather. Called *Historia Naturalis,* it was said to be heavily influenced by the works of Aristotle.

1452–1647 During this period, Leonardo da Vinci, Galileo Galilei, and Evangelista Torricelli invented three of the key instruments of meteorology: the hygrometer, thermometer, and barometer.

1816 German physicist Heinrich Brandes discovered that belts of high and low pressure were constantly circling Earth around the midlatitudes. This laid the foundation for today's synoptic chart, or weather map.

1835 A fundamental force acting on all weather systems is produced by the rotation of Earth itself. This was first recognized by Frenchman Gaspard-Gustave de Coriolis and is called the Coriolis effect.

Star science Astrolabes were early computers used to calculate sunrise and sunset times, and to estimate the time of day. This was done by setting the instrument to coincide with the current positions of the Sun and stars. Astrolabes represent an early example of applying science to the skies.

Radio detection The invention of radar enabled scientists to study many weather phenomena, in particular approaching rain areas, thunderstorms, and hail. Radar can also be employed aboard ships, for scientific missions at sea, and fitted to satellites, to measure the state of the atmosphere below.

1843 American academic Samuel Morse invented the electric telegraph. The ability to transmit information quickly across vast distances meant that weather warnings could be issued for the first time.

1873 The International Meteorological Organization (IMO) was established. In 1950 it became the World Meteorological Organization (WMO) and now has more than 180 member nations.

1922 *Weather Prediction by Numerical Process* was published by L. F. Richardson. The techniques outlined could not effectively be utilized until the advent of electronic computers some 25 years later.

1960 On April 1, the first meteorological satellite was placed into orbit above Earth. TIROS 1 (Television Infrared Observation Satellite) soon began beaming back primitive images of cloud patterns.

1990s The Internet transformed meteorology. The latest satellite and radar images, along with up-to-date observations from automatic weather stations around the world, became readily available.

Ancient Beginnings

Many of the great early civilizations had their own distinctive pantheon of weather gods. They were revered as all-powerful deities who ruled the skies and controlled the elements, including the Sun, wind, and rain. It was believed that these weather gods could, if provoked or displeased, produce inclement weather as a punishment for human transgressions.

Ancient beliefs Many weather gods of the ancient world were depicted as magical animals, or animal–human hybrids, having special powers. Prayers, rituals, and sometimes sacrifices—even human sacrifices—were offered to placate them.

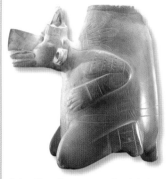

Aztec Ehecatl was a prominent Aztec deity who was accorded the status of god of the wind. He was one of over 100 gods in the Aztec pantheon.

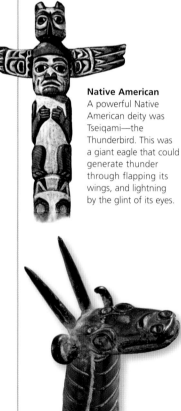

Native American A powerful Native American deity was Tseiqami—the Thunderbird. This was a giant eagle that could generate thunder through flapping its wings, and lightning by the glint of its eyes.

Babylon Marduk was the chief god of the ancient Babylonians. Like Zeus in Greek mythology, he was the god of the sky, and particularly thunderstorms.

Egypt (right) Osiris was one of the more important of the Egyptian gods, and evidence of his presence in Egyptian society goes back as far as 2500 B.C. He had several roles, including that of lord of the dead, but was also believed to be responsible for the periodic flooding of the Nile River.

Noah's ark (far right) This is a popular story of divine intervention in the weather, relating to a flood in the Middle East. God warned Noah of an impending disastrous flood that he was about to unleash, allowing Noah to build the ark and save humanity and the animal kingdom.

Heavenly Greek (above) Zeus was the Greek god of the heavens who determined the nature of various weather elements, including rain, clouds, thunder, and lightning. He was sometimes depicted as a bearded giant hurling thunderbolts down to Earth from his vantage point on Mount Olympus.

Mayan magic (right) El Castillo is a large Mayan temple in the ancient city of Chichen Itza, in present-day Mexico. It was dedicated to the feathered serpent god Quetzalcoatl, who was god of the wind. At the spring and fall equinoxes, the shadow cast on the stone staircase evokes a serpent.

FACT FILE

Birth of logic Attempts to use natural forces to explain weather stretch well back into antiquity. This is somewhat surprising, as many of the early societies still relied on religious or supernatural explanations for most phenomena.

Hippocrates The Greek physician Hippocrates (c. 460–375 B.C.) was one of the first to pioneer logic, rather than religion, to explain natural phenomena.

Pliny Pliny the Elder (A.D. 23–79) wrote *Historia Naturalis*, one of the last attempts to look at weather scientifically before the fall of the Roman Empire.

Galileo One of history's foremost astronomers, Galileo Galilei (1564–1642) also invented the thermometer, for measuring air temperature.

WEATHER LORE

Many early civilizations turned from religion to nature in their quest to understand weather phenomena. Often the two systems ran in parallel: while continuing to pay homage to their weather gods, people also observed links between weather, the state of the sky, and the behavior of plants and animals. This desire to make connections between weather and the surrounding natural world generated a weather folklore. For many centuries, this lore attempted to describe and even forecast weather, and examples persist to this day. Some of the earliest developers of weather folklore were the indigenous peoples of the world.

Trees Pine cones are highly sensitive to changes in atmospheric moisture levels: they close when humidity is high and open when humidity is low. High humidity levels are often associated with the onset of rain.

Spiderwebs Spiders are said to slow their web production when rain is approaching and to increase it once rain has passed. Such behavior may be a response to sensed changes in atmospheric conditions.

NATURAL SIGNS

The Australian aboriginal people carefully observed the arrival of various migratory birds, and the flowering of certain plant species, to predict the onset and cessation of the wet season. The Samoans and Tongans were expert fishermen and predicted changes to weather through the appearance or absence of various species of fish and seabirds, the temperature of the seawater, and the nature of waves in the area at the time.

A piece of Pacific Island folklore is common to several island nations, including Fiji and the Cook Islands. It suggests that the early flowering of the islands' mango trees indicates a more active tropical cyclone season. There is some truth to this lore. The mango trees flower early when air temperatures are warmer than normal, and this occurs in the region when sea surface temperatures are raised. Tropical cyclones feed off warm ocean waters and when the seas are warmer than average, and the mango trees flower early, tropical cyclone activity tends to surge. This connection may have been recognized for many centuries across various Pacific Island nations. European weather lore evolved later, much of it originating from farmers and seafarers. As with the indigenous weather knowledge,

Sacred rock According to Australian aboriginal legend, Uluru, the giant stone monolith in central Australia, was made during the time of creation by two spirit boys, who built it from mud after a heavy rainstorm. Uluru has particular spiritual significance to the people of the area.

European weather folklore was based on the observed connections between weather, the appearance of the sky, and the behavior of plants and animals. Many examples of these beliefs and sayings are still familiar today.

In English folklore, the timing of the arrival of the migratory swallow is an important event. Its early arrival is said to be the harbinger of a dry summer.

Honeybees returning to their hives is supposedly a sign that rain and storms are approaching. This seemingly prophetic behavior is more likely to be an instinctive response bees have developed to changes in atmospheric pressure.

Weather lore developed as a curious mix of common sense and sheer superstition, and was often expressed in proverbs. Some sayings, for example those linking the Moon with cloud cover, are based on sound observations, but most weather lore has only limited application in forecasting weather.

Cattle In Europe, cows lying down has long been regarded a sign of approaching rain. The theory is that cows don't like to rest on wet grass, so they create a dry patch by lying down before the rain begins.

Insects Grasshoppers or crickets chirping loudly during the day supposedly indicates that fine weather will follow. As with much weather folklore, there is little scientific evidence to support this notion.

Red sky (left) "Red sky in the morning, sailors take warning; red sky at night, sailors delight." In midlatitudes, weather systems move from west to east. A red sky at night indicates that the western sky is clear and weather will be fine.

Halo (above) The ethereal ring sometimes seen around the Sun or Moon is caused by the bending of light through a thin veil of ice-crystal cloud. This formation sometimes runs ahead of a cold front, and cold fronts can certainly generate rain.

Groundhog Day (above) An element of North American weather folklore concerns the behavior of a small rodent. Legend has it that if a groundhog emerging from its burrow on February 2 casts a shadow on the ground, then the following six weeks will be cold.

Through the Ages

The Middle Ages was a low point for meteorology as astrometeorology—weather prediction based on the positions of the stars and planets—thrived. The Renaissance once again saw the flourishing of exploration, science, and the arts. It was a time of revolutionary ideas and discoveries, and the invention of several important meteorological instruments. In the late 17th and 18th centuries, the sciences of physics and meteorology advanced along a broad front.

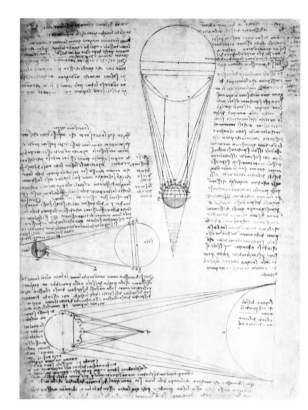

Renaissance spirit (right) Italian inventor and key Renaissance figure Leonardo da Vinci (1452–1519) was one of the brilliant minds of history. He is credited with the invention of the hygrometer, an instrument for measuring atmospheric humidity. This notebook page shows da Vinci's notations about Earth and the Moon, and their sizes and relationships to the Sun.

Age of discovery (below) Exploring the world became common in the Renaissance. By 1600, the Americas had been added to European maps, and explorers like Columbus and Magellan had recorded weather across the oceans. This magnificently illustrated plate is from *The Celestial Atlas*, published in 1660 by the Dutch mapmaker Johannes Janssonius.

Telescope Newton also dabbled in astronomy and constructed an early reflector telescope that used mirrors, rather than lenses, to magnify images.

Observatory In 1557, the Turkish Sultan Suleyman founded the Galata Observatory, dedicated to the study of astronomy. This was a period of scholarship in the Ottoman Empire, with Suleyman an enthusiastic patron of education, the arts, and sciences.

WEATHER WATCH

Standard temperature scales were proposed by German physicist Gabriel Fahrenheit (1686–1736) and Swedish astronomer Anders Celsius (1701–1744). In the Fahrenheit scale, the freezing point of water is 32°F and the boiling point is 212°F, while the Celsius equivalents are 0°C and 100°C.

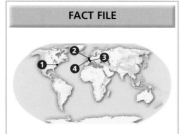
Toward the Modern Era

Progress in meteorology during the 19th and early 20th centuries was faster and more extensive than at any other time in history. An assortment of meteorological instruments was invented and refined, several national weather services established, and then linked by a system of international cooperation. These events formed the beginning of the modern era.

Balloon launch (right) In this 1965 photograph, a weather balloon is launched in the Arctic. Monitoring in the Arctic has provided valuable contributions to our understanding of weather extremes, assisting with the reconstruction of past climates.

Rainfall (above) This bulky instrument—from 1911—is an early pluviometer, a device that not only measures rainfall totals, but also records when the rain actually fell. This is important in providing estimates for rainfall intensity over a given area.

Weather dome (above) A scientist takes readings from an observation dome in Antarctica in 1957. Scientific bases were established in Antarctica by several nations during the 20th century, partly to monitor local weather conditions.

Scott's barometer (right) Explorer Robert Falcon Scott (1868–1912) headed two expeditions to Antarctica. In the second, Scott and his team died from cold and starvation after reaching the South Pole. This is the barometer he used on his final journey.

Weather map (right) This colorful illustration is in fact a synoptic chart, or weather map, showing the weather patterns across the United States on September 1, 1872. It was prepared by the U.S. Army Signal Service, a forerunner of the National Weather Bureau, when weather matters were regulated by the Department of War.

Modern Meteorology

The science of meteorology has changed enormously in the past 50 years. Information is collected frequently, sometimes in a continuous stream, from automatic weather stations, orbiting satellites, radar, unmanned aircraft, and ocean buoys. Data is processed by supercomputers, checked for accuracy, and used by meteorologists to produce maps, charts, and long-range forecasts.

Forecasting Research into atmospheric motion and fluid dynamics by Professor Vilhelm Bjerknes advanced the fields of meteorology and weather prediction.

ENIAC In 1950, scientists succeeded in producing a machine, the Electronic Numerical Integrator and Computer, that generated weather forecasts.

TIROS1 Launched in 1960, the first meteorological satellite, the Television and Infrared Observation Satellite, transmitted infrared and visible images.

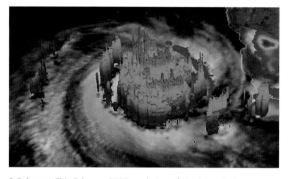

3-D image This February 2007 rendering of Hurricane Favio as it passed Madagascar, depicts cloud heights in a tropical storm system. In this case, the highest clouds, shown in red, are part of the thunderstorm cluster that surrounds the hurricane's eye.

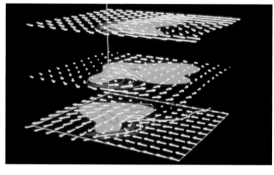

Layered image Doppler radar data can produce a computer simulation of a severe thunderstorm that also indicates whether tornado development is likely. This information can be used to issue weather warnings in advance of the actual storm.

Data collection Devices that monitor meteorological conditions have become more sophisticated over the past several decades, vastly improving the accuracy of weather forecasting.

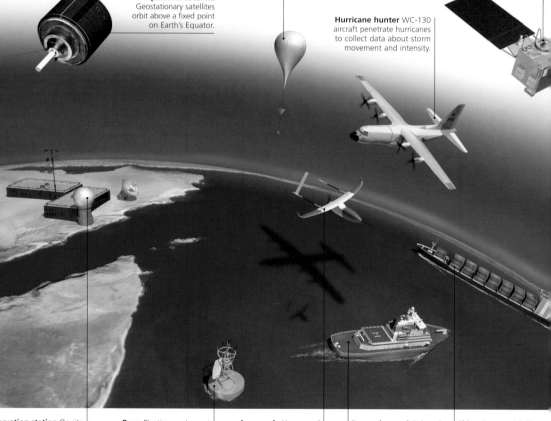

Equatorial satellite Geostationary satellites orbit above a fixed point on Earth's Equator.

Weather balloon Equipment attached to balloons monitors Earth's atmosphere.

At the poles Polar-orbiting satellites generate detailed images of cloud patterns.

Hurricane hunter WC-130 aircraft penetrate hurricanes to collect data about storm movement and intensity.

Observation station On-site instruments continuously record weather conditions.

Buoy Floating equipment platforms measure wind, pressure, and temperature.

Aerosonde Unmanned aircraft provide information about storm structure.

Research vessel Onboard radar units analyze and track storm development.

Ships Commercial ships routinely monitor weather and transmit data.

MetOp-A Europe's first polar-orbiting satellite, MetOp-A, became operational in 2007. Specialized equipment monitors atmospheric ozone levels and measures temperature, humidity, and wind speed and direction at various altitudes.

WEATHER WATCH

The Tropical Rainfall Measuring Mission (TRMM), a joint venture between NASA and the Japan Aerospace Exploration Agency (JAXA), studies hurricanes and their effect on regional climate. Data gathered by TRMM helps assess if climate change is increasing storm intensity and frequency.

TRACKING WEATHER

Being able to measure meteorological elements, such as rainfall, temperature, humidity, winds, and cloud-cover, was the first step in converting meteorology to a science. As the Scottish physicist Lord Kelvin was to remark in 1883: "When you can measure what you are speaking about, and express it in numbers, you know something about it." Many instruments that we use to measure weather today have been based on the same design for centuries. More recently, meteorology has advanced at an astonishing pace, thanks to the development of electronic instruments.

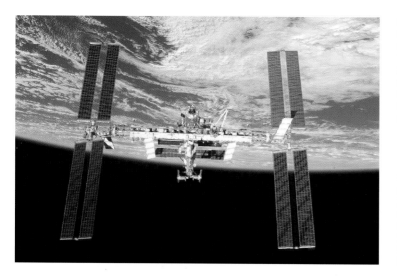

International cooperation (left) Space shuttle *Atlantis,* pictured in 2007, moves away from the International Space Station. The station is a research facility staffed by scientists from many countries. Information about aspects of Earth's atmosphere is collected, providing insights into weather and climate. The station is expected to remain operational until at least 2015.

Satellite sight (right) This 2002 view over Alaska's Aleutian Islands was taken by the Landsat 7 environmental satellite. Winds blowing across the islands from the upper right-hand corner of the image, break up into chains of long, spiraling eddies called von Kármán vortices, which are made visible by the extensive low cloud in the area.

MEASURING DEVICES

Various instruments for measuring meteorological variables have been under continuous development for centuries. Probably the oldest is the rain gauge, or pluviometer, which was used widely in the ancient world around 2,000 years ago.

Three major instruments were developed in the 16th and 17th centuries: the hygrometer, to gauge relative humidity; the thermometer, for temperature; and the barometer, for air pressure. Modern versions of these instruments have the ability not just to measure, but also to record, and take the suffix *graph*, Greek for "drawn" or "written." We now have pluviographs, thermographs, barographs, and many others.

The use of wind vanes was first noted in ancient Greece and Rome. However, accurate, scientific measurement of the wind was not achieved until the mid–18th century, using devices called anemometers.

Today, concurrent measurements of atmospheric variables—humidity, temperature, atmospheric pressure, and wind speed and direction—are taken manually every three hours from hundreds of locations around the world. Automatic weather stations take records even more frequently, typically every half hour.

Meteorological satellites monitor cloud cover, and the location of precipitation is determined through radar imagery. These 20th-century inventions provided a quantum leap forward in tracking weather.

This vast quantity of information is sent to national and international meteorological centers. Weather forecasts are then constructed using computer weather simulations.

Click of a button (below) National weather services now have a vast amount of observational data available: the latest satellite and radar images, and information from human observers and automatic weather stations, are readily accessible.

THROUGH THE AGES

300 B.C. The earliest known rain gauge dates back to the 4th century B.C. in India. These instruments were also used in Palestine about 2,000 years ago, and in China and Korea around the same time.

50 B.C. The Tower of the Winds, built in Athens in the 1st century B.C., had a bronze wind vane on top of the structure. Wind vanes were also common in ancient Rome and across medieval Europe.

A.D. 1500 The first hygrometer appeared in about 1500. The work of prolific Italian inventor Leonardo da Vinci, this tool allowed the moisture content, or humidity, of the air to be measured for the first time.

1600s Italian philosopher Galileo Galilei invented the first thermometer in the early 17th century. Measuring the temperature of the air was of vital importance in monitoring the state of the atmosphere.

1620 Evangelista Torricelli, a pupil of Galileo's, designed and built the first barometer, which allowed atmospheric pressure to be measured. This invention paved the way for weather forecasting.

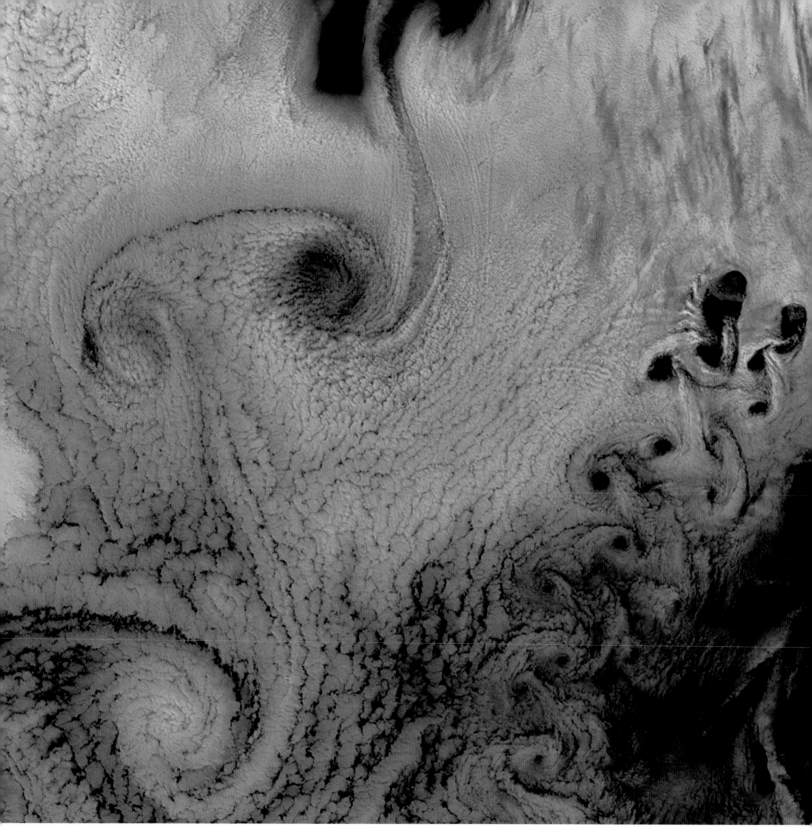

1654 The first meteorological observing network in Europe was established by Grand Duke Ferdinand II of Tuscany. This was the forerunner of the network that underpins today's World Meteorological Organization.

1853 One of the earliest sunshine recorders was the Campbell–Stokes instrument. It was invented by Scotsman John Francis Campbell and modified by mathematician Sir George Gabriel Stokes.

1929 Frenchman Robert Bureau developed the first operational radiosonde, which could transmit weather data back from a floating balloon. The suffix *sonde* is French for "probe."

1934 During World War II the British developed radar (radio direction and ranging) that could track aircraft, and later precipitation. Today, radar units are used to provide detailed analyses of local weather.

1960 TIROS 1, the first meteorological satellite, was launched. This vehicle beamed back images of cloud patterns that covered Earth, allowing observers to locate weather systems over remote areas.

Temperature and Pressure

Knowing the ambient air temperature is important to meteorologists for preparing tomorrow's forecast: monitoring bursts of hot and cold air is vital to the prediction of maximum and minimum temperatures. Tracking atmospheric pressure is also significant. The areas of high and low pressure that circle Earth are the "fingerprints" of weather.

Cold facts (below) Many early Antarctic expeditions kept records of ice pack and sea surface temperature. This photograph, taken during Scott's 1910–12 expedition, shows temperature measurements being taken through a hole cut in the ice.

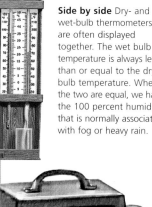
Thermal imaging This infrared image is from a geostationary satellite high above Australia. The coldest temperatures, nearly white, are thunderstorm tops. Lower clouds are orange and green; and the warmest temperature, the ground, appears as black.

THERMOGRAPHY

Temperature can be measured using infrared imaging. In this picture, a child drinks cold water. Colors vary from white (warmest), through red, green, and blue, to black (coldest).

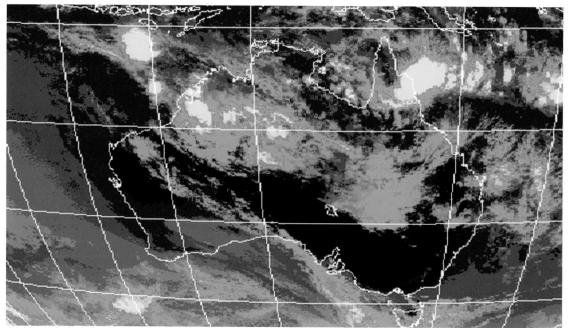

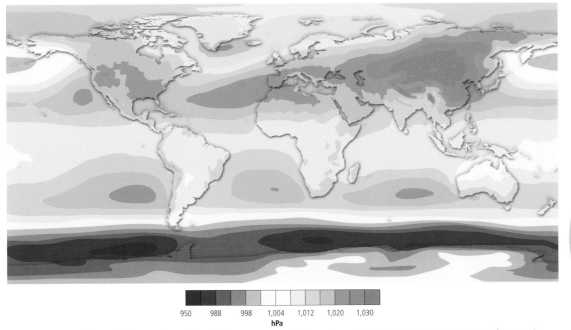

| 950 | 988 | 998 | 1,004 | 1,012 | 1,020 | 1,030 |

hPa

Pressure patterns (above) This map shows atmospheric pressure at sea level across the globe, averaged over the months of June, July, and August. The highest pressures are found over Siberia, where figures in excess of 1,030 hPa are common.

Extreme temperatures (below) Temperatures at the summit of Mount Everest are extreme all year round. In January, the temperature averages –33°F (–36°C) and in July, the warmest month, the average temperature is –2°F (–19°C).

FACT FILE

Barometers Atmospheric pressure is measured using a barometer, of which there are several types. A mercury-in-glass instrument is the oldest, but is fragile and bulky. More common is the clock-faced aneroid barometer.

Aneroid barometer This barometer displays the atmospheric pressure by means of a moveable needle operating across a dial. Instruments of this type are quite sturdy and accurate.

Barograph This instrument measures and records atmospheric pressure. Graph paper covers a revolving drum, and a pen continuously registers the pressure during each 24-hour period.

FRACTION OF 1 ATM	AVERAGE ALTITUDE	
	(ft.)	(m)
1	0	0
1/2	18,000	5,486
1/3	27,480	8,376
1/10	52,926	16,132
1/100	101,381	30,901
1/1,000	159,013	48,467
1/10,000	227,899	69,464
1/100,000	283,076	86,282

Variations Atmospheric pressure decreases with altitude. This NASA table shows air pressures (given as a fraction of one atmosphere, or 1 atm) at various heights above sea level.

Sun, Precipitation, and Wind

The detailed measurement of sunshine, rainfall, and wind requires specialized meteorological instruments, many of which were invented during the 19th and 20th centuries. Increasingly, environmental satellites are used for these purposes, having the advantage of being able to monitor large areas of Earth's surface, including remote areas where no land–based stations exist.

Sea fog (right) A satellite image shows large areas of sea fog covering the North Sea and the Skagerrak, a strait between Norway and Denmark. Sea fog forms when air blows over colder ocean water, causing the contained moisture to condense.

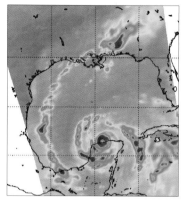

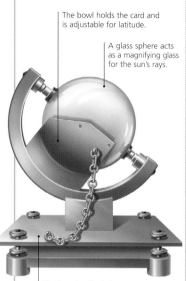

The bowl holds the card and is adjustable for latitude.

A glass sphere acts as a magnifying glass for the sun's rays.

The base is adjusted with leveling screws.

Sunshine recorder The Campbell–Stokes instrument measures the amount of sunlight received over a day by creating a scorch mark along a card.

Rainfall is collected and funneled into a collecting chamber.

A pen attached to the float records a line on a chart.

The collecting chamber fills, activating a float.

A paper chart is attached to a clockwork drum.

Pluviograph Rainfall totals and intensities can be retrieved using a pluviograph, which measures the amount of rain and the time it fell.

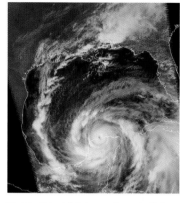

Visible This visible image from the U.S. Aqua satellite shows Hurricane Isidore storming across the Gulf of Mexico during September 2002. The cloud bands spiraling into the center of the system are visible, as is the clear eye area in the middle of the mass.

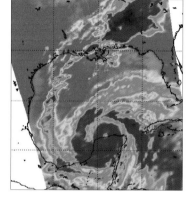

Infrared This image of Hurricane Isidore uses infrared wavelengths, which detect heat rather than visible light. The coldest temperatures occur in cloud tops (blue). The warmest areas are cloud-free parts of the land and ocean (red).

Microwave This is Hurricane Isidore as viewed at microwave wavelengths, which detect zones of humidity, clouds, and rain. The image defines areas of precipitation (blue), clouds and high-humidity air (green and yellow), and low-level moist air (orange).

Wind power (above) The revolutionary Bahrain World Trade Center has three huge wind turbines, two of which are shown here, that spin between the twin towers. These generate nearly 10 percent of the energy requirements of the entire complex.

FACT FILE

Measuring wind Because of its high variability, wind is a difficult weather element to measure. Wind vanes and wind socks are among the most simple instruments used; anemometers are far more complex but yield more detailed information.

Wind sock This simple device indicates from which direction the wind blows, as well as its approximate strength. If it is strong, the sock will stand out horizontally from the mast.

Anemometer A cup anemometer measures wind velocity by recording the spinning speed of a horizontally mounted rotor. Cups attached to the rotor catch the wind and produce spin.

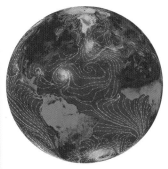

Satellite The QuikSCAT environmental satellite carries radar, and provides data on wind speed and direction around the world. The fastest speed is shown in orange and the slowest in blue.

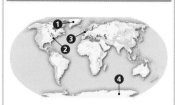

Meteorological Stations

The meteorological observing network that supports the daily weather forecasting effort around the world has become a massive, well-organized machine that gathers information from an astonishing number of sources. The platform underpinning this work is World Weather Watch (WWW), a core program of the World Meteorological Organization (WMO). It was initiated in 1963, following a proposal by U.S. President John F. Kennedy.

Mountain station (above) Mont Ventoux, a prominent mountain in France, is notoriously windy on its 6,273-foot (1,912 m) summit. A meteorological station was established here in 1882 but is no longer operational.

Desert station (above) A remote weather station is situated at Giles, in the central Australian desert, approximately 465 miles (750 km) from the nearest town. Temperatures regularly exceed 104°F (40°C).

1. Met Office The British national weather service, known as the Met Office, operates an extensive public weather service, delivered both through conventional media and a comprehensive website. The Met Office also offers tailor-made forecasts to industry on a commercial basis.

Exeter, England

2. CMA The China Meteorological Administration (CMA) is the national weather service of China. Forecasting is of vital importance because of China's dependence on agriculture. The CMA is represented in all provinces and has a staff of around 53,000 people.

Beijing, China

3. BOM Established in 1908, Australia's Bureau of Meteorology (BOM) has its headquarters in Melbourne. The BOM produces a range of forecasts and issues weather warnings for the Australian continent and adjacent waters. It has a network of field offices across the continent and in Antarctica.

Melbourne, Australia

4. SAWS The South African Weather Service (SAWS), headquartered in Pretoria, became a public organization in 2001 and is the official national weather service for the Republic of South Africa. It operates 23 regional weather offices and over 100 automatic stations.

Pretoria, South Africa

Windy station (left) Mount Washington in New Hampshire, U.S.A., has an altitude of 6,288 feet (1,916 m). The world's highest officially recorded surface wind speed—231 miles per hour (372 km/h)—was observed here in April 1934. While stronger winds have occurred elsewhere, they have damaged recording equipment in the process and are not documented.

Forecasting Weather

Lewis Fry Richardson's 1922 vision of predicting the weather using mathematical simulations was years ahead of its time, and could not be utilized until the advent of computers some 25 years later. Now, numerical weather prediction is one of the most powerful tools at the meteorologist's disposal, and forms the basis of short-term weather forecasting.

Weather warnings (left) Short-term forecasting—including the issuing of weather warnings—is of vital importance in saving lives. Tornado warnings, for example, allow people to evacuate or take shelter. This image graphically illustrates the destruction produced by a tornado, which occurred on May 26, 2008 in Iowa, U.S.A.

Computer modeling (below left) The course of Hurricane Floyd along the east coast of the U.S.A. in September 1999 was accurately predicted by computer modeling. Warnings could be issued to the public well ahead of time. This computer-generated, three-dimensional image shows the structure of the hurricane.

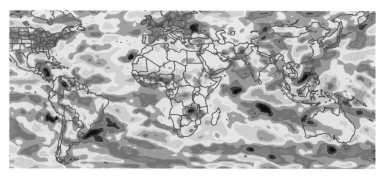

Rainfall This is a three-month rainfall forecast for the period June, July, and August 2002 that was generated by computer. Blue indicates above-average rainfall and red indicates below-average. Such predictions have considerable application in agriculture, particularly in planning crops for the upcoming season.

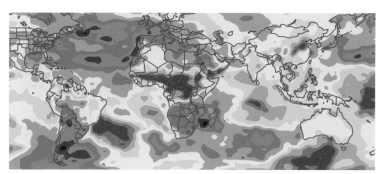

Temperature This is a computer-generated three-month temperature forecast for the period June, July, and August 2002. Warmer than average temperatures are shown in red, and cooler than average in blue. Energy companies are particularly interested in predictions of this type as it assists them in forecasting energy demand.

3-D simulation (below) The great Blizzard of 1993 produced record snowfalls and rain along the east coast of North America. Computer simulations of the storm provided excellent guidance for meteorologists who were tracking the system. This three-dimensional computer simulation shows surface features and the upper level structure of the storm.

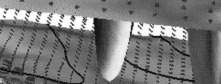

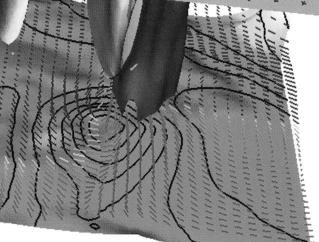

FACT FILE

Blizzards Because of the dislocation that blizzards can produce, early warnings are essential. Increasingly, computer simulations can successfully identify where blizzards will develop.

Sea breezes Particularly during the summer months, sea breezes have a dramatic effect on temperatures. These can drop by as much as 18°F (10°C) along coastal areas.

PROCESSING POWER

More powerful computers enable processing of the huge number of mathematical calculations needed for detailed weather modeling. Mesoscale weather effects, such as sea breezes and thunderstorm clusters, can now be forecast.

Mapping Weather

Synoptic and prognostic charts—more commonly known as weather maps—are the main tools of meteorology. The synoptic chart is a snapshot of an area's weather at any given time; the prognostic chart predicts future weather. They condense a huge amount of information into an internationally recognized format and have become a familiar part of our everyday life.

Prediction (right) A prognostic chart is a forecast weather map. It is a computer-generated prediction of where pressure systems and rainfall patterns are expected to be at some time in the future. This type of chart is the basis for forecasting the weather up to a week ahead.

Hand-made (above) Until the late 1970s, most synoptic charts were drawn by hand. This 1955 photograph shows a chart being manually prepared for a U.K. television presentation.

High-speed (below) The Internet has made synoptic charts portable. Here, the latest weather map is available to a storm chaser in his vehicle, allowing him to follow thunderstorm activity.

Winds

Weather front

Isobars

0.04 0.08 0.2 0.4 0.6 0.8 1.2 1.6 2 2.4 3 in

1 2 5 10 15 20 30 40 50 60 75 mm
Rainfall

Fronts Cold fronts are identified on weather maps as solid lines with "shark fins" attached. When a cold front passes across an area, there is usually a marked drop in temperature, often accompanied by rain and increased wind.

Isobars The chart shows areas of low pressure to the west of Iceland, Norway, Britain, and Italy. The isobars are close together in these areas, indicating that windy weather is likely. A high-pressure area lies to the southwest of Spain, with light winds expected for this area.

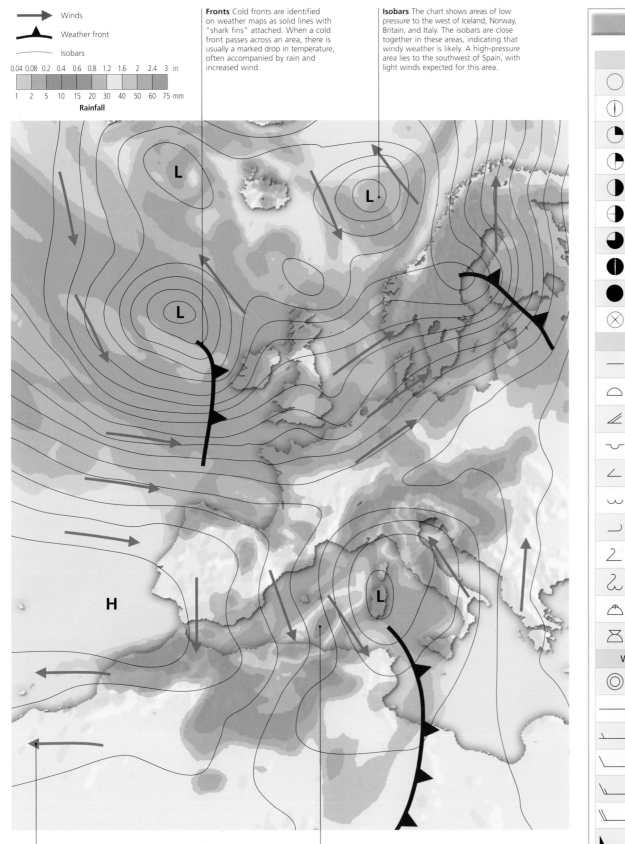

Wind Red arrows indicate the direction of the wind at various locations around Europe and Africa. They move counterclockwise around low-pressure cells, and clockwise around high-pressure cells. The winds are blowing from the west in the north of Spain and from the south over Italy.

Rainfall This prognostic chart also predicts areas of rainfall, with colors representing expected rainfall levels, as shown on the scale above. The heaviest falls (in red) are forecast to be over northern Italy and the western part of the Mediterranean Sea.

FACT FILE	
SKY COVERAGE	
	no clouds
	one-tenth covered
	two- to three-tenths covered
	four-tenths covered
	half covered
	six-tenths covered
	seven- to eight-tenths covered
	nine-tenths covered
	completely overcast
	sky obscured
CLOUDS	
	stratus
	cumulus
	nimbostratus
	stratocumulus
	altostratus
	altocumulus
	cirrus
	cirrostratus
	cirrocumulus
	cumulonimbus calvus
	cumulonimbus with anvil
WIND SPEED mph (km/h)	
	calm
	1–2 (1–3)
	3–8 (4–13)
	9–14 (14–23)
	15–20 (24–33)
	21–25 (34–40)
	55–60 (89–97)
	119–123 (192–198)

Information transmission Cameras mounted on satellites photograph weather systems over large areas, even on the dark side of Earth; radar reveals where precipitation occurs; Doppler radar gauges the movement of air within weather systems.

Radio beams Radar transmitters fire radio waves into the atmosphere. Some waves are reflected back off raindrops, are picked up by the radar's antenna, and converted into a computer image.

Around the Poles Polar-orbiting satellites rotate around Earth, at an altitude of approximately 530 miles (850 km), completing one orbit in about 100 minutes. Weather images from these vehicles are highly detailed.

Around the Equator Geostationary satellites orbit the Equator at an altitude of 22,300 miles (36,000 km). As it orbits at the same rotational speed as Earth, a geostationary satellite remains over the same part of Earth.

Radar and Satellites

Satellite imagery has revolutionized meteorology since the first one was launched in 1960. Basically an observational vehicle, a satellite carries both visible and infrared cameras, as well as radar, for recording meteorological elements. Cameras detect zones of clouds and cloud heights. Other instruments can predict where cloud is likely to develop before it becomes visible.

Doppler on wheels (right) Storm-chaser vehicles equipped with mobile Doppler radar can track severe thunderstorm activity. Doppler radar images are used to identify possible tornado development within severe thunderstorms.

International effort Several nations have meteorological satellites in orbit. Some orbit the Poles, while others are geostationary. Data from these sources is freely exchanged internationally for the common good of weather forecasting.

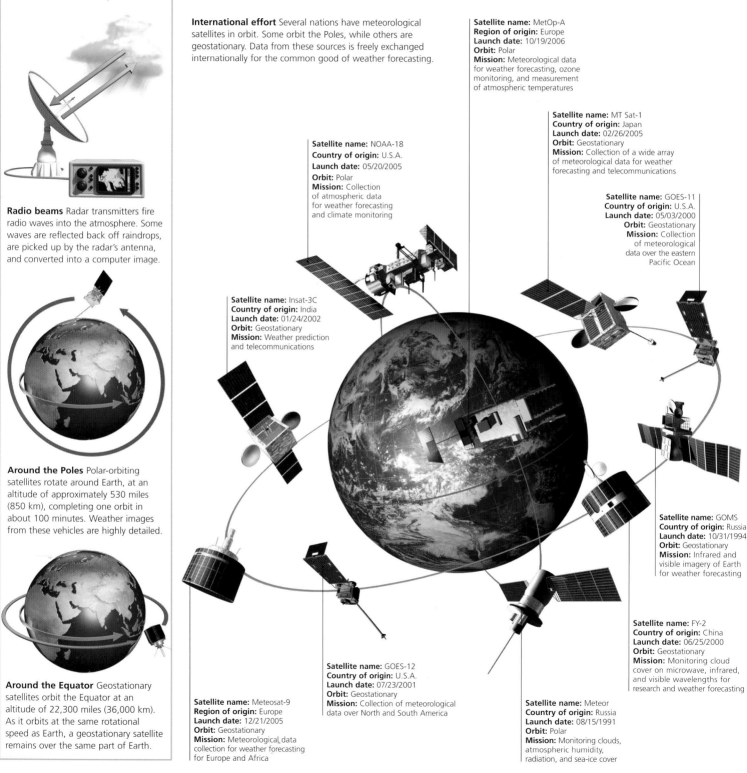

Satellite name: MetOp-A
Region of origin: Europe
Launch date: 10/19/2006
Orbit: Polar
Mission: Meteorological data for weather forecasting, ozone monitoring, and measurement of atmospheric temperatures

Satellite name: MT Sat-1
Country of origin: Japan
Launch date: 02/26/2005
Orbit: Geostationary
Mission: Collection of a wide array of meteorological data for weather forecasting and telecommunications

Satellite name: NOAA-18
Country of origin: U.S.A.
Launch date: 05/20/2005
Orbit: Polar
Mission: Collection of atmospheric data for weather forecasting and climate monitoring

Satellite name: GOES-11
Country of origin: U.S.A.
Launch date: 05/03/2000
Orbit: Geostationary
Mission: Collection of meteorological data over the eastern Pacific Ocean

Satellite name: Insat-3C
Country of origin: India
Launch date: 01/24/2002
Orbit: Geostationary
Mission: Weather prediction and telecommunications

Satellite name: GOMS
Country of origin: Russia
Launch date: 10/31/1994
Orbit: Geostationary
Mission: Infrared and visible imagery of Earth for weather forecasting

Satellite name: FY-2
Country of origin: China
Launch date: 06/25/2000
Orbit: Geostationary
Mission: Monitoring cloud cover on microwave, infrared, and visible wavelengths for research and weather forecasting

Satellite name: GOES-12
Country of origin: U.S.A.
Launch date: 07/23/2001
Orbit: Geostationary
Mission: Collection of meteorological data over North and South America

Satellite name: Meteosat-9
Region of origin: Europe
Launch date: 12/21/2005
Orbit: Geostationary
Mission: Meteorological data collection for weather forecasting for Europe and Africa

Satellite name: Meteor
Country of origin: Russia
Launch date: 08/15/1991
Orbit: Polar
Mission: Monitoring clouds, atmospheric humidity, radiation, and sea-ice cover

Radar images Radar enables the location and intensity of precipitation to be estimated, and severe weather phenomena to be detected. Satellite images reveal the movement of hurricanes, fronts, and storm cells, even in remote areas.

2-D thunderstorm This radar image shows a line of severe thunderstorm activity. Zones with the heaviest rainfall are shown in red and orange, with the darkest spots indicating hail.

3-D thunderstorm Modern radar can produce three-dimensional views of thunderstorm activity. This type of information can identify which storms are likely to become severe.

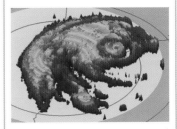

3-D tropical cyclone This radar view of Cyclone Bobby, located off the Western Australian coastline during February 1995, shows much of the internal structure of the storm.

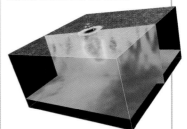

Sunspot The structure of a sunspot is revealed by the Michelson Doppler Imager, which measures magnetic fields and gas flow around the Sun.

CLIMATE

CLIMATE

WHAT IS CLIMATE?

Climate is the average pattern of weather. It is determined by measuring temperature, precipitation, atmospheric pressure, and wind over a long period of time. The climate of a region also results from the interaction of many other variables, the most important of which are latitude, height above sea level, topography, and proximity to oceans or large water bodies. Although climate is an average condition, short- or long-term changes can and do occur. Temperature, precipitation, atmospheric pressure, and wind speed increase or decrease through either the natural variability in Earth's climate or recent human activity.

CLIMATE VARIABLES

Perhaps the most reliable predictor of temperature is latitude. Regions that lie close to the Equator are generally the warmest, because at those latitudes solar radiation, our source of energy, is most intense and there are more daylight hours. Regions close to the Poles, however, are the coldest, because solar radiation is less intense there and for much of the year there are few daylight hours.

The effect of latitude on temperature is modified by altitude, because the higher we ascend in the troposphere, the colder it becomes. At the cooler temperatures of very high altitudes, precipitation occurs as snow, causing glaciers to form in the tropical Andes and on some mountains in eastern Africa—places

that would otherwise be warm all year round.

Large bodies of water also modify temperature. Because water warms and cools more slowly than land, regions near the coast tend to experience more moderate daily and annual temperature variations than those inland, even though they lie at the same latitude.

Persistent snow and ice covers force temperatures to be cooler than would otherwise occur. Precipitation tends to be greatest in warm regions with large sources of moisture, such as the tropics. Mountain barriers redistribute precipitation, since they force air to rise. Clouds form as moisture condenses in the rising air and the resulting precipitation falls on the windward mountain slopes. Much

drier conditions exist on the lee—the side of the mountains sheltered from the rain.

Atmospheric pressure also influences climate. Most regions that experience high pressure have low precipitation, while low-pressure regions tend to have higher precipitation.

CLASSIFYING CLIMATE

Attempts to classify climate date back to the 4th century B.C., when the ancient Greek scholar Aristotle used the geometry of the Sun's position in relation to Earth to divide Earth into three climate zones. Much later, in the early 20th century, Köppen based his system on links between vegetation and climate. Then, as climate knowledge increased, Thornthwaite used the relationship between precipitation and evaporation, while Strahler's system used air masses. Each method is useful, but no single classification can include every aspect of a region's climate. Köppen's classification, for example, does not cover the climates of the Pacific islands. The climate zone maps in this book are based on Köppen's classification.

Rain shadow (left) Low annual precipitation in the lee of the Rocky Mountains supports grasses rather than trees. This provides extensive pastures for cattle ranching in the rolling foothills.

Gulf Stream (right) Thanks to this warm current (red and brown in this satellite image), which flows northeast across the Atlantic Ocean, the climates of east-coast North America and western Europe remain mild, despite their northerly latitudes.

NORTH AMERICA

SOUTH AMERICA

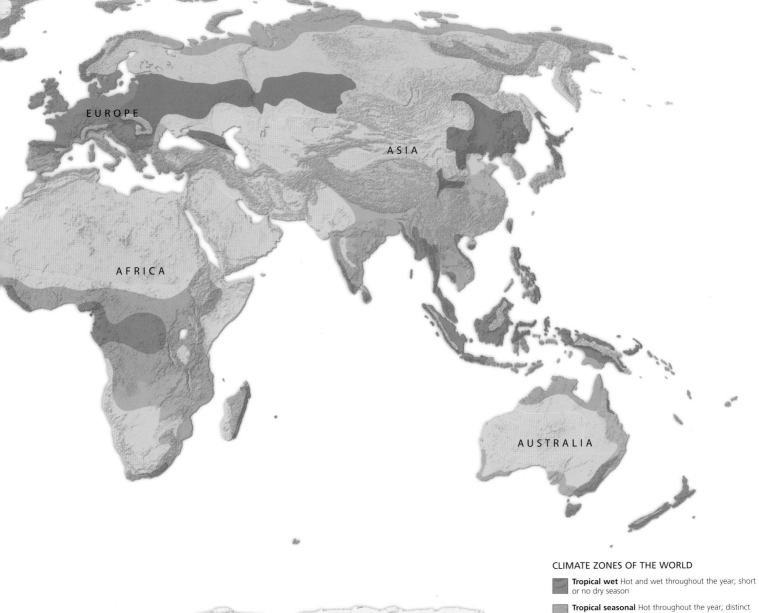

EUROPE

ASIA

AFRICA

AUSTRALIA

ANTARCTICA

Pacific Ocean (right) Changes in surface temperatures of the tropical Pacific Ocean influence weather. In La Niña events, unusually cold currents, shown by high chlorophyll concentrations (light blue in this image), bring dry conditions to northwest South America and floods to eastern Australia. Warmer than normal currents (El Niño) reverse these weather conditions.

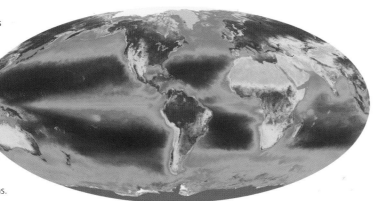

CLIMATE ZONES OF THE WORLD

Tropical wet Hot and wet throughout the year; short or no dry season

Tropical seasonal Hot throughout the year; distinct wet and dry seasons

Arid Little or no precipitation year round; hot days and cold nights

Semi-arid Low precipitation; smaller diurnal temperature variation than arid climates

Mediterranean Hot, dry summers; mild, moist winters, occasionally below freezing

Subtropical Warm and moist; hot summers and cooler, drier winters

Temperate Four distinct seasons; precipitation year round; warm summers and cold winters

Continental Cool and moist; warm summers and severe winters

Boreal Cool summers and very severe winters with snow; evergreen forest vegetation

Subpolar Very cold throughout the year; no true summer; tundra vegetation

Polar Extremely cold and dry throughout the year; permanent ice cover

Mountain Colder than low-level locations found at the same latitude

Tropical Climate Zones

Straddling the Equator between latitudes 20°N and 20°S, tropical climates experience high temperatures as a result of intense solar radiation. In the tropical wet climate, the hot, moist air causes daily thunderstorms and high rainfall all year. Tropical seasonal regions have a wet season in summer, but experience drier conditions in the winter.

Tropical crop (below) Rice grows best in warm conditions, with plenty of water. The high rainfall and constant warm temperatures of Indonesia's tropical wet climate are ideal for successful rice cultivation, often on terraced hillsides.

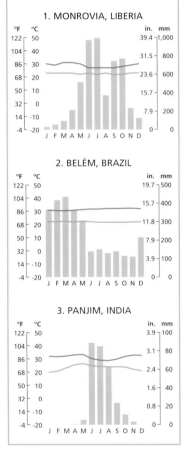

Monsoon floods (right) In India's tropical seasonal climate, warm, moist monsoon winds move onshore in summer, bringing with them extremely heavy downpours and sudden floods that can cause havoc for city commuters.

Foggy forests (below) Luxuriant rain forests thrive in the hot, moist conditions of tropical wet climates. In Malaysia's high-altitude rain forests, humidity levels can be so high that fog often forms over the canopy and persists for some time.

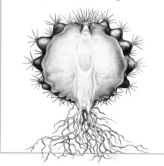
Arid Climate Zone

Arid climates, found between 15° and 35° north and south of the Equator, may result from high air pressure near the tropics, mountain barriers to moist ocean air, locations far inland, or proximity to cold ocean currents. With mainly clear skies and many sunshine hours, they have high daytime temperatures, low rainfall, high evaporation, and strong, drying winds.

Barren landscape (above) In the Thar Desert of Rajasthan, India, high daily temperatures, strong convectional winds, and infrequent rain create hot, windswept desert landscapes that are almost completely bare of vegetation.

Wind-shaped dunes (below) Sand in the Karakum Desert, Turkmenistan, central Asia, is shaped into an extensive system of dunes by the strong winds that blow daily in this hot region.

Cold desert (above) High pressure at the Tropic of Capricorn and a cold, northward-flowing ocean current combine to give Valle de la Luna, in Chile's Atacama Desert, an extremely dry, cold climate.

FACT FILE

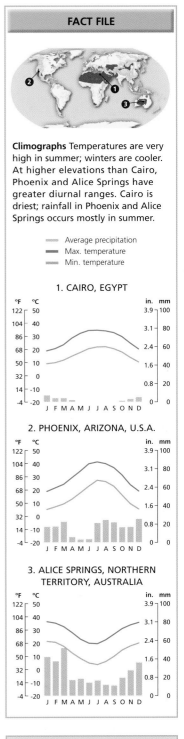

Climographs Temperatures are very high in summer; winters are cooler. At higher elevations than Cairo, Phoenix and Alice Springs have greater diurnal ranges. Cairo is driest; rainfall in Phoenix and Alice Springs occurs mostly in summer.

- Average precipitation
- Max. temperature
- Min. temperature

1. CAIRO, EGYPT

2. PHOENIX, ARIZONA, U.S.A.

3. ALICE SPRINGS, NORTHERN TERRITORY, AUSTRALIA

WEATHER WATCH

Expanding deserts With global warming, hot deserts are expected to become even hotter. Annual precipitation, already low, is likely to decrease further. The ensuing drier conditions may in turn lead to expansion of the dune fields of deserts like Africa's Kalahari, so extending hot desert boundaries.

Semi-arid Climate Zone

A transitional band between arid and tropical seasonal climates, the semi-arid zone is influenced by their climatic processes and has similar aspects. Rainfall is higher than in arid regions, but it is seasonal and can support savanna-like grasses and isolated small trees. Temperatures can be as hot as in arid regions, although semi-arid areas experience a wider annual range.

Seasonal rains (below) Semi-arid grasslands support large grazing mammals, such as these buffaloes in Botswana. In the rainy season, animals roam the plains to feed on the grasses; in the dry season, they migrate to permanent watering holes.

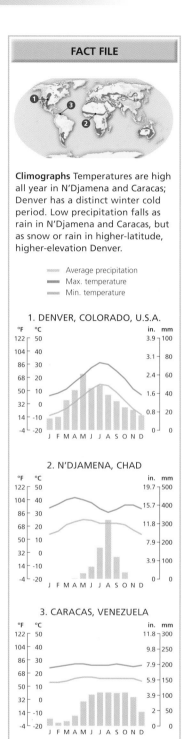

Climographs Temperatures are high all year in N'Djamena and Caracas; Denver has a distinct winter cold period. Low precipitation falls as rain in N'Djamena and Caracas, but as snow or rain in higher-latitude, higher-elevation Denver.

Average precipitation
Max. temperature
Min. temperature

1. DENVER, COLORADO, U.S.A.

°F	°C		in.	mm
122	50		3.9	100
104	40		3.1	80
86	30		2.4	60
68	20		1.6	40
50	10		0.8	20
32	0		0	0
14	-10			
-4	-20			

J F M A M J J A S O N D

2. N'DJAMENA, CHAD

°F	°C		in.	mm
122	50		19.7	500
104	40		15.7	400
86	30		11.8	300
68	20		7.9	200
50	10		3.9	100
32	0		0	0
14	-10			
-4	-20			

J F M A M J J A S O N D

3. CARACAS, VENEZUELA

°F	°C		in.	mm
122	50		11.8	300
104	40		9.8	250
86	30		7.9	200
68	20		5.9	150
50	10		3.9	100
32	0		2	50
14	-10		0	0
-4	-20			

J F M A M J J A S O N D

More droughts and floods With their highly variable precipitation, semi-arid regions are prone to droughts and floods. These extremes of weather are expected to become more frequent as temperatures rise with global warming. Expansion of the deserts, or desertification, is also a threat to these regions.

Pampas (right) In the Argentinian pampas, much of the natural grassland has been replaced by agriculture. The vast plains and the semi-arid climate are well suited to the cultivation of wheat, a crop that originated in Asia's semi-arid regions.

FACT FILE

Wildlife Like animals in arid climates, those living in the semi-arid zone are adapted to the hot, dry conditions. They can survive for long periods without drinking, acquiring much of their water from their food.

Gerenuk The long neck and pointed snout of this antelope allow it to nibble the moist, small leaves of thorn bushes in semi-arid East Africa, but avoid the thorns.

Emu Australia's large flightless bird lives in semi-arid open woodland and eats the seeds, fruit, and growing shoots of plants, as well as insects.

Giant anteater Found in South American semi-arid grasslands, this animal obtains water mainly from its diet of ants, with some fruit and larvae.

Storms (center left) Summer rainfall comes in the form of frequent thunderstorms that tend to develop in the late afternoon or early evening, when the air is most unstable. Here, ominous storm clouds hang over the U.S. prairies as the sun sets.

Floating dust (left) In semi-arid regions, seasonal rainfall can be variable. During extended dry conditions, exposed soil is lifted and blown over long distances, creating dust storms such as this one in Beijing, China, in March 2008.

FACT FILE

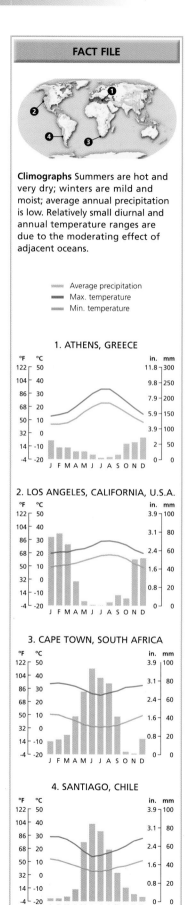

Climographs Summers are hot and very dry; winters are mild and moist; average annual precipitation is low. Relatively small diurnal and annual temperature ranges are due to the moderating effect of adjacent oceans.

- ▬ Average precipitation
- ▬ Max. temperature
- ▬ Min. temperature

1. ATHENS, GREECE

2. LOS ANGELES, CALIFORNIA, U.S.A.

3. CAPE TOWN, SOUTH AFRICA

4. SANTIAGO, CHILE

Mediterranean Climate Zone

Regions with Mediterranean climates are found on the west of continents, between 30° and 40° north and south of the Equator. In summer they experience high pressure, clear skies, moderate-to-high temperatures, and low rainfall from isolated afternoon thunderstorms. Most rainfall occurs in the mild winters, when cyclones bring moisture from adjacent oceans.

Western Cape, South Africa (above) This region enjoys a Mediterranean climate with many sunshine hours. Its heath- and shrubland, or "fynbos," supports a huge diversity of plant species that thrive in the winter rainfall and survive well during the dry summer.

Grape harvest (right) Many of the world's wines are produced in regions with a Mediterranean climate. The mild, moist winters and hot, dry summers are key elements in the cultivation of grapes. Here, French workers harvest mature grapes for Beaujolais wine.

Wildfire hazard (below) In late summer, regions with a Mediterranean climate are prone to wildfires like this one pictured raging in northeastern Spain. Flammable trees such as pine and eucalyptus provide fuel for the fires in the hot, dry condiitions.

Clouded leopard This medium-sized cat lives in the trees of Southeast Asia's humid subtropical forests, where it preys on birds, squirrels, monkeys, deer, and wild pigs.

Koala Native to the eucalypt woodlands of eastern Australia, the koala seldom drinks water, feeding only on eucalyptus leaves, to which its digestive system is specially adapted.

American alligator The wetlands of southeastern U.S.A. are home to this reptile. Once considered endangered, the alligator population has now recovered.

HURRICANES

Tropical Storm Edouard, shown off the east coast of Florida, U.S.A., in September 2002, is an example of the hurricanes, or tropical cyclones, that affect subtropical regions in summer, contributing to high precipitation. These storms develop mostly in late summer or early fall.

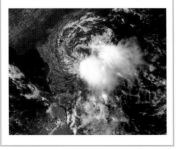

Subtropical Climate Zone

Lying roughly between latitudes 20° and 40° north and south of the Equator, regions with moist subtropical climates are found on the eastern margins of continents. Warm ocean currents passing off their coasts and maritime air masses from the east produce warm summers with high rainfall, but cyclones bring cooler and drier weather in winter.

Humid environment (right) At Lake Bistineau, Louisiana, U.S.A., the subtropical climate provides ideal conditions for the Spanish moss growing on these cypress trees. This herb absorbs the moisture it needs from dew, mist, fog, or rain.

Subtropical wetlands (left) The high annual precipitation and mild temperatures of the humid subtropical climate can produce a high water table. As a result, much water may remain at the surface, supporting dense vegetation and extensive swamps, such as these in Everglades National Park, Florida, U.S.A.

Seasonal variations (below left) Subtropical summers can be hot and humid, and the winters cold and dry. Spring and fall are more moderate, ensuring that the climate of cities like Shanghai, China, is mild.

Summer storms (below) The coastal location of Sydney, Australia, provides a constant supply of moist, unstable air. Intense heating of the land and warm ocean currents produce strong convection, often resulting in severe thunderstorms.

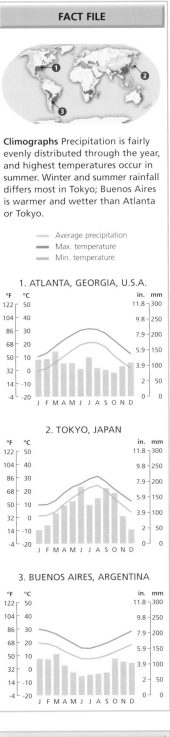

Climographs Precipitation is fairly evenly distributed through the year, and highest temperatures occur in summer. Winter and summer rainfall differs most in Tokyo; Buenos Aires is warmer and wetter than Atlanta or Tokyo.

— Average precipitation
— Max. temperature
— Min. temperature

1. ATLANTA, GEORGIA, U.S.A.

°F	°C		in.	mm
122	50		11.8	300
104	40		9.8	250
86	30		7.9	200
68	20		5.9	150
50	10		3.9	100
32	0		2	50
14	-10		0	0
-4	-20			

J F M A M J J A S O N D

2. TOKYO, JAPAN

°F	°C		in.	mm
122	50		11.8	300
104	40		9.8	250
86	30		7.9	200
68	20		5.9	150
50	10		3.9	100
32	0		2	50
14	-10		0	0
-4	-20			

J F M A M J J A S O N D

3. BUENOS AIRES, ARGENTINA

°F	°C		in.	mm
122	50		11.8	300
104	40		9.8	250
86	30		7.9	200
68	20		5.9	150
50	10		3.9	100
32	0		2	50
14	-10		0	0
-4	-20			

J F M A M J J A S O N D

Increased intensity If the global climate continues to warm, an increase in the peak wind speed of tropical cyclones and in summer rainfall is expected. If sea levels rise as projected, subtropical regions will become even more vulnerable to the flooding that often results from tropical storm surges.

FACT FILE

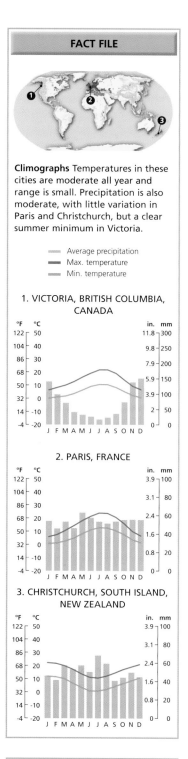

Climographs Temperatures in these cities are moderate all year and range is small. Precipitation is also moderate, with little variation in Paris and Christchurch, but a clear summer minimum in Victoria.

— Average precipitation
— Max. temperature
— Min. temperature

1. VICTORIA, BRITISH COLUMBIA, CANADA

°F °C   in. mm
122 50   11.8 300
104 40   9.8   250
86   30   7.9   200
68   20   5.9   150
50   10   3.9   100
32   0   2   50
14 -10   0   0
-4 -20
    J F M A M J J A S O N D

2. PARIS, FRANCE

°F °C   in. mm
122 50   3.9   100
104 40   3.1   80
86   30   2.4   60
68   20   1.6   40
50   10   0.8   20
32   0   0   0
14 -10
-4 -20
    J F M A M J J A S O N D

3. CHRISTCHURCH, SOUTH ISLAND, NEW ZEALAND

°F °C   in. mm
122 50   3.9   100
104 40   3.1   80
86   30   2.4   60
68   20   1.6   40
50   10   0.8   20
32   0   0   0
14 -10
-4 -20
    J F M A M J J A S O N D

WEATHER WATCH

Species displacement Regions with temperate climates are expected to become warmer and have increased precipitation, with negative impacts on the health of birds and fish. Species that prefer warmer temperatures will potentially displace present species, which are adapted to cooler temperatures.

Temperate Climate Zone

Temperate climates are found at latitudes between 40° and 60° in both hemispheres. Prevailing westerly winds flow off the oceans and bring cool, moist, maritime air to these regions. They have moderate year-round temperatures that are seldom below freezing, precipitation is evenly spread throughout the year, and they experience four distinct seasons.

Winter fog (below) The maritime air of temperate climates has a high water vapor content and is always close to saturation. In winter, cool morning and evening air causes the water vapor to condense into very small droplets of water, forming fog.

Spring renewal (right) Flowering cherry trees announce the arrival of spring in Vancouver, British Columbia, Canada. With longer daylight hours and warmer temperatures, trees that have been bare all winter produce festoons of pink blossom.

FACT FILE

Wildlife Animals in the temperate zone must adapt to the changing seasons and especially to the long, cold winter. Some hibernate right through the winter; others store food in summer and fall to sustain them in winter, when food is scarce.

European mole The vegetation in temperate climates is a rich source of food for moles, which eat earthworms, succulent plant parts, seeds, and fungi.

European robin Also known as robin redbreast, this bird is common in temperate Europe, both as a resident and as a migrant from colder northern climes in winter.

Platypus This egg-laying, nocturnal, duck-billed mammal frequents the small streams and waterways of eastern and southeastern Australia.

Fall color (left) With cooler fall weather, the chlorophyll that colors deciduous leaves green disappears, leaving the russet hues spectacularly displayed by these plane trees in France's Loire Valley before they drop their leaves in winter.

WARM OCEAN CURRENT

In the eastern Atlantic, temperate climates are influenced by the North Atlantic Drift, an extension of the Gulf Stream. Thanks to this warm ocean current, regions north of the Arctic Circle are warmer than others at the same latitudes. Norwegian fjords remain ice-free year round, but at the same latitude Alaska is frozen in winter.

ADAPTING TO CLIMATE

Plants and animals are directly and strongly influenced by climatic conditions. They survive only because they have adapted to the climate of the region in which they live. This adaptation is a long, complex process, which takes place quite slowly over millions of years as species gradually make the adjustments that help them to cope with small changes in climate. Those species that adapt successfully continue to exist, and pass on their traits to the next generation. This process is what underpins Darwin's theory of natural selection, which suggests that the organisms that survive are better suited to their environment.

SURVIVING EXTREMES

To thrive in a particular region, plants and animals must be able to cope with extreme climate conditions such as the heat and aridity of deserts, the frigid temperatures of polar regions, or severe rain, snow, or wind events. They must also be able to adjust to the seasonal changes in weather. The mechanisms used to survive often involve modifying appearance and behavior.

In extremely arid environments, plants such as cacti and baobab trees have become succulents, which are able to store water in their fleshy stems or leaves. The ocotillo plant of the Californian and Mexican deserts sheds its leaves during dry periods and becomes dormant. Since much water is lost from leaves through transpiration, this is an effective way to conserve water. Bulbs like the desert lily can lie dormant for years until rain brings them briefly back to life.

Desert animals sleep in shady locations during the heat of the day and emerge in the cooler dusk-to-dawn hours to hunt or forage for food. Some desert animals, such as the camel, have physiological adaptations that allow them to function for long periods without drinking. Desert frogs lie dormant in burrows during the long, dry, hot months, emerging only after heavy rain.

Plants in very cold, subpolar regions are usually small and grow in clumps close to the ground to conserve heat. Some are dark, even red, in color, which optimizes their absorption of solar energy, while others may be covered in hair, which helps to slow down heat loss.

Deciduous trees growing in climates with cold winters shed their leaves in fall to conserve energy. Evergreen conifers, however, have downward-sloping branches that allow snow to slide off before it causes any damage. Needle-like leaves withstand freezing temperatures—their dark color absorbs solar heat well and they have a waxy coating.

Animals in cold regions develop thick fur to keep them warm. Some increase what they eat in summer and autumn, storing the food as fat, and then hibernate during the winter months to survive the cold.

Flexibility (right) Palm trees have flexible trunks, allowing them to sway dramatically in hurricane-force winds without breaking. Their fronds can also bend backward, avoiding damage from strong winds.

Swollen trunks (below) These baobab trees are well adapted to the arid climate of Madagascar. Their huge trunks can store large quantities of water, enabling them to survive long periods without rain.

A long sleep The hazel or common dormouse, resident in Europe and northern Scandinavia, hibernates from October to April to escape the cold and conserve energy when the seeds, flowers, fruits, and insects on which it feeds are scarce.

The wind is less able to penetrate the ground-hugging plant, reducing heat loss.

Dark-colored leaves positioned parallel to the sun's rays increase the amount of energy a plant receives from sunlight.

Compact growth conserves heat generated by plants and gained from their surroundings.

Reflective leaves parallel to the sun's rays reduce heat absorption.

Loose growth allows air to pass through foliage and cool the plant.

Little contact with soil reduces the amount of heat absorbed from the ground.

Arctic plant (above) Cushion plants in Arctic and alpine climates grow in compact clumps close to the ground. This conserves warmth gained from the sun's rays and reduces the loss of heat to the wind.

Desert plant (right) In hot, dry climates plants need to avoid overheating in the high temperatures. An open growth form allows wind passing through the stems and foliage to dissipate heat.

Shared heat (below) Emperor penguins and their chicks huddle together for warmth in the Antarctic. A layer of air between their skin and plumage also insulates them against the extreme cold.

Escaping the cold (right) Canada geese migrate in late fall from their northern breeding grounds to the warmer climates of the southern United States and Mexico, where they spend the winter.

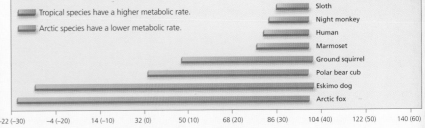

Tropical species have a higher metabolic rate.

Arctic species have a lower metabolic rate.

Sloth
Night monkey
Human
Marmoset
Ground squirrel
Polar bear cub
Eskimo dog
Arctic fox

Environmental temperature °F (°C)

−22 (−30) −4 (−20) 14 (−10) 32 (0) 50 (10) 68 (20) 86 (30) 104 (40) 122 (50) 140 (60)

Conserving energy (above) Arctic species can maintain their metabolic rate over a wider range of air temperatures than tropical species. Adaptations such as thicker fur, rounded body shape, and white coloring allow them to conserve energy in very low temperatures.

TORNADOES

Continental Climate Zone

As the name suggests, continental climates are found in the interior of continents, far from the moderating effects of the oceans. Summers are warm and moist, as a result of the warm air masses that develop when the sun is high. Winters are cold and dry, influenced by the cold air masses with embedded cold fronts that move south from higher, colder latitudes.

Storms (below) Manitoba, Canada, is prone to the severe summer thunderstorms common in continental climates. Surface heating of unstable, warm, moist air triggers strong convection, and storms develop by late afternoon.

Vegetation (right) Trees in regions with continental climates, like those in temperate climates, are mainly deciduous, dropping their leaves in winter and renewing them in spring. Forests are characterized by five main plant zones: the tallest trees (**1**) shelter the smaller trees and saplings (**2**). Beneath them are the shrubs (**3**) and herbs (**4**), while on the ground are mosses and lichen (**5**) among the decaying leaves and fallen branches.

Winter (above) In Jilin, northeastern China, bare trees are covered in ice when rain from a winter warm front falls on branches that are below freezing. Mist rises off the river when moist air comes in contact with its frozen surface.

Fall glory (below) The deciduous trees in Russia's Kaluga region produce a glorious display in fall. As days grow shorter and colder, leaves stop producing the chlorophyll that colors them green, and previously masked golden colors appear.

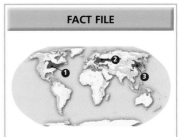
Climographs In winter, Lake Michigan has a moderating influence on Chicago, which is not as cold as Moscow or Shenyang. Precipitation occurs all year in the three cities, but peaks in summer. Shenyang has the driest winters.

— Average precipitation
— Max. temperature
— Min. temperature

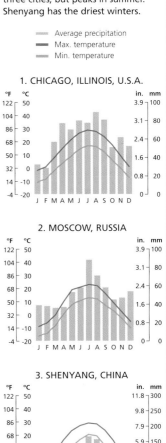

1. CHICAGO, ILLINOIS, U.S.A.

2. MOSCOW, RUSSIA

3. SHENYANG, CHINA

WEATHER WATCH

Changes to vegetation This climate zone is expected to become warmer and drier. Warmer temperatures will promote poleward migration of the treeline. Earlier, warmer springs and warmer days in summer will extend the growing season of plants that are affected more by temperature than by day length.

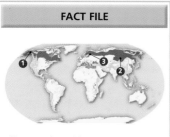

Climographs With a strong marine influence, Helsinki is not as cold as Fairbanks or Irkutsk and has year-round moderate precipitation, mostly as rain. Fairbanks and Irkutsk have low precipitation all year, with snow in winter.

- — Average precipitation
- — Max. temperature
- — Min. temperature

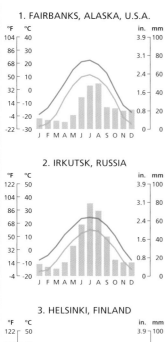

1. FAIRBANKS, ALASKA, U.S.A.

2. IRKUTSK, RUSSIA

3. HELSINKI, FINLAND

Boreal Climate Zone

Located between 50°N and 70°N, a band of boreal climates stretches from western Alaska across Canada to northern Europe and Siberia. Arctic air masses make the long, dark winters severely cold: up to six months are below freezing and the soil is frozen. Precipitation is low, but the short, cool summers remain moist because evaporation is also low.

Siberian taiga (above) In late winter and early spring, when the light is good but it is too cold for new growth to start, the evergreen trees of the vast boreal forests use their old leaves for photosynthesis.

Canadian wetlands (right) While winters are drier, cyclonic storms can develop as cold, dry air from Siberia crosses the Pacific, bringing precipitation to northern Canada as well as to regions farther south.

Severe winters (above) In Sweden's northern forests, spruce trees are well adapted to the long, snowy winters of the boreal climate. While snowfalls are heavy, the snow slides easily off their downward-sloping branches, preventing damage to the trees.

FACT FILE

Wildlife The boreal forest is home to large herbivores such as elk, wapiti, and moose, as well as smaller mammals such as lynxes, weasels, and beavers. It also shelters a large variety of birds, some permanent and others migratory.

Moose The largest of the deer family, the moose eats the leaves, twigs, and buds of hardwood and softwood trees and shrubs that grow in the boreal forest.

Hairy woodpecker This medium-sized woodpecker is found in the boreal forests and also across much of North America. It eats insects, fruit, and nuts.

American black bear The multi-layered boreal forest is an abundant source of food for this herbivore, which is widespread in North America.

BIRD NURSERY

The summer abundance of insects, seeds, and fruit make North America's boreal forest an ideal bird nursery. Some 300 species breed there before heading south to spend the winter in the southern U.S.A. The red-necked grebe breeds on small lakes in Canada and Alaska but winters along the southern east and west coasts of North America.

FACT FILE

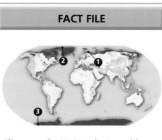

Climographs McMurdo Sound has below-freezing temperatures all year, but in Vardo temperatures are below freezing for only six months. Precipitation, which mostly occurs as snow, is highest in Vardo; Alert and McMurdo Sound are much drier.

--- Average precipitation
--- Max. temperature
--- Min. temperature

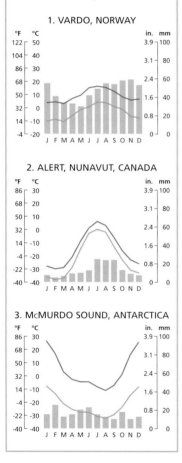

WEATHER WATCH

Dramatic changes to come Glaciers, sea ice, and permafrost regions are already warming and melting. Ice floes are melting faster and earlier, and pack ice is becoming thinner. Species such as the giant walrus that live on large, thick ice floes are being forced onto land, where it is more difficult for them to survive.

Subpolar and Polar Climate Zones

Inside the Arctic and Antarctic circles, climate is dominated by frigid air masses. Winter is very long, cold, and dark, while summer is very short, with near-freezing temperatures. In polar regions, vast expanses are permanently ice-covered, while on the subpolar fringes low temperatures and little precipitation can support only low-growing perennial plants.

Antarctica (below) Very strong winds and severely cold temperatures occur daily in this icebound continent, where average yearly precipitation of only 2 inches (50 mm) falls as snow. These Adelie penguins live in the milder coastal regions.

Floating ice (right) A layer of frozen seawater, called pack ice, expands in winter to cover extensive areas of the northern and southern oceans. When it melts in summer, as in this fjord in subpolar Greenland, ice floes dot the sea's surface.

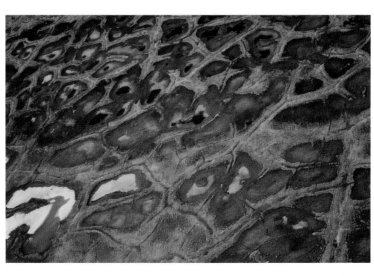

FACT FILE

Wildlife Animals that live in subpolar and polar climates must adapt to the long, severely cold winters or migrate to warmer regions in winter. All-year residents have thick fur and an insulating layer of fat under the skin.

Musk ox This very large herbivore of the Alaskan Arctic is protected from the cold by extremely thick fur covering its whole body, including the udder.

Arctic hare Deep fur and short ears help this native of the North American tundra to conserve heat. Groups dig into the snow instead of hibernating.

Polar bear A layer of fat and a thick coat of fur insulate this Arctic resident against the cold. Underneath, its black skin soaks up the sun's warmth.

Short summers (center left) The growing season lasts only 50 to 60 days but about 400 varieties of flowers bloom in spring and summer in the subpolar tundra. This honey-producing fireweed grows where there has been human activity or fires.

Polygons (left) These patterns occur in subpolar permafrost. When this thin soil layer freezes in winter, the ground contracts and cracks. During the summer thaw, the cracks fill with water and snow, which refreeze when winter returns.

Mountain Climate Zones

If they are tall enough, mountains create climates that would otherwise not exist at the latitudes at which they are located. They redirect winds and affect temperature, pressure, and precipitation. Mountain climates depend on latitude, altitude, and exposure to the wind, and range from the lowest-level montane climate to the highest-level alpine climate.

Alpine versus temperate (right) Mont Blanc, in the French Alps, lies in the temperate climate zone, but its peak and upper slopes have a cold, alpine climate. It is snow-capped even in summer, while the green valley below enjoys milder weather.

Guanaco This herbivore lives at all altitudes in South America's Andes Mountains. Its large heart and lungs allow it to survive where oxygen levels are low.

Brown bear The many habitats of the North American brown bear, one of the largest living carnivores, include alpine meadows and mountain forests.

Mountain goat A long beard and a thick, long coat protect this North American agile climber from the cold alpine climate in which it lives.

Puya Thriving on the cool, dry slopes of South America's Andes, this plant folds its leaves up around its stem at night to protect itself from the cold.

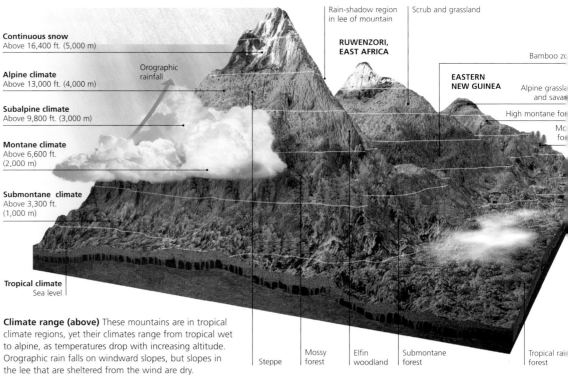

ANDES, CENTRAL PERU

Rain-shadow region in lee of mountain

Scrub and grassland

RUWENZORI, EAST AFRICA

Bamboo zo

EASTERN NEW GUINEA

Alpine grassl and savar

High montane for

Mo for

Continuous snow Above 16,400 ft. (5,000 m)

Alpine climate Above 13,000 ft. (4,000 m)

Orographic rainfall

Subalpine climate Above 9,800 ft. (3,000 m)

Montane climate Above 6,600 ft. (2,000 m)

Submontane climate Above 3,300 ft. (1,000 m)

Tropical climate Sea level

Steppe

Mossy forest

Elfin woodland

Submontane forest

Tropical rai forest

Climate range (above) These mountains are in tropical climate regions, yet their climates range from tropical wet to alpine, as temperatures drop with increasing altitude. Orographic rain falls on windward slopes, but slopes in the lee that are sheltered from the wind are dry.

Stark contrast (above) Morocco's fertile Dades River Valley contrasts with the dry southern High Atlas Mountains, which run between the pleasant Mediterranean coastal climate and the Sahara desert.

Nepal (right) High in the Himalayas, the summer monsoon winds bring cloudy conditions and heavy precipitation to villages in Nepal from May to September, but in winter it is dry.

FACT FILE

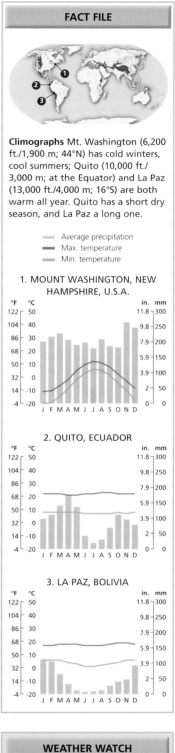

Climographs Mt. Washington (6,200 ft./1,900 m; 44°N) has cold winters, cool summers; Quito (10,000 ft./ 3,000 m; at the Equator) and La Paz (13,000 ft./4,000 m; 16°S) are both warm all year. Quito has a short dry season, and La Paz a long one.

— Average precipitation
— Max. temperature
— Min. temperature

1. MOUNT WASHINGTON, NEW HAMPSHIRE, U.S.A.

2. QUITO, ECUADOR

3. LA PAZ, BOLIVIA

WEATHER WATCH

Less drinking water From the European Alps to Africa's Mount Kenya, mountain glaciers have been shrinking. If current trends continue, they will disappear by the end of the 21st century, affecting the water supply of millions of people who depend on the water stored in them.

REGIONAL CLIMATE GUIDE

The twelve main climate zones described in the preceding pages may lie within or span political and geographical boundaries. This regional climate guide examines in more detail how these climate zones affect the weather around the world in 43 regions and within six major geographical units: North America and Mesoamerica, South America, Europe, Asia, Africa, and Oceania and Antarctica. Some regions belong to one climate zone, such as North Woods in North America, while others embrace a number of zones, such as southern Africa.

Cold desert (right) The Namib Desert was formed by the same large-scale processes as the Sahara, but its climate is much cooler, influenced by the cold Atlantic Ocean current that washes its shores.

Dry coast (below) The cold Humboldt Current makes Lima, on the Peruvian coast, very dry, unlike the west African coast at similar latitudes. El Niño events bring heavy precipitation every two to six years.

ZONAL DIFFERENCES

Closer scrutiny of the climate zones by region shows that the same climate zone may have different weather in different parts of the world. Factors that cause these differences include the characteristics of the surface, the proximity to warm or cold ocean currents, and the influence of periodic large-scale atmospheric processes such as the El Niño–Southern Oscillation (ENSO) phenomenon or monsoons.

Parts of the Canadian and Finnish Arctic lie in the subpolar climate zone. However, while the Canadian Arctic surface is largely covered by water, the Finnish Arctic is landlocked. As a result, seals abound in the Canadian Arctic, but in Finland there are reindeer.

Differences like these are not limited to the Arctic. In India, Cherrapunjee has recorded the world's highest rainfall but it is not located in the rainiest climate zone, the tropical wet zone. Cherrapunjee is wet because moist summer monsoon winds bring extremely heavy rainfall to northeastern India.

The climate along the coast of Peru is dry and mild, but at the same latitudes the climate of the Pacific islands is warm and moist. The difference between these two regions is the temperature of the ocean currents. Peru is influenced by the the cold Humboldt Current, while the temperature of the western Pacific Ocean is always warm. Ocean current temperatures also affect weather in Africa's Sahara and Namib deserts. Both of these arid regions were the result of descending, drying air in the semi-permanent subtropical high-pressure systems. The Namib Desert is cold, however, while the Sahara is hot. The difference in temperature is due to the cold Benguela Current that flows off the coast of southwestern Africa.

Climate zones are determined by long-term averages, but these values are likely to alter as climate change occurs. Average temperatures are expected to increase. This could result in a poleward shift of climate zones that will be more pronounced at high latitudes, where the temperature increase is expected to be greatest. Such shifts will affect existing vegetation and agricultural patterns in both negative and positive ways.

Two different Arctics While climate is defined by atmospheric elements, it is modified by Earth's surface. The surface in the polar Arctic may consist of land, water, or ice. In Lapland's Arctic winter **(above)**, the snowmobile is the best way to travel over vast expanses of snow. At a similar latitude at Queen Maud Gulf in Canada's Northwest Territories **(right)**, an icebreaker is needed to navigate the ice-covered water.

NORTH AMERICA AND MESOAMERICA

Together, North America and Mesoamerica have the greatest range of climates, covering all 12 zones. The area's wide latitudinal span and the long north–south extent of the Rocky Mountains create a variety of temperature and precipitation regimes. Continentality—land warming and cooling faster than water—causes temperature extremes in summer and winter, and contributes to the North American monsoon. On the coast, the Pacific and Atlantic oceans and the Caribbean Sea moderate temperatures and provide moisture for precipitation.

ROCKY MOUNTAINS

The Rocky Mountains play a major role in determining the climate of North America. This range of mountains soars higher than 13,800 feet (4,200 m) in places, extends from Alaska to Mexico, and varies in width from 70 to 300 miles (110 to 480 km). The high altitudes mean that climates that would normally be found at polar or subpolar latitudes can exist at subtropical latitudes. Very moist climates, due to orographic precipitation, form on the windward slopes of the mountains, while in the lee (the side sheltered from the wind) extensive dry zones exist.

In summer, a long, narrow band of high-speed winds, or a jet stream, occurs low in the troposphere in the lee of the Rockies. This moves out over the Great Plains and creates conditions that lead to the formation and development of extensive and extremely severe thunderstorms during spring and summer nights. Tornadoes are often spawned by the thunderstorms and cause widespread devastation. Heavy precipitation accompanying the thunderstorms makes a significant contribution to the average annual rainfall of western North America.

In winter, the lee of the Rocky Mountains favors the formation of low-pressure systems known as cyclones. High-pressure systems, or anticyclones, develop in the wake of the cyclones. The surges of cold air associated with the anticyclones penetrate far south, creating the so-called northers of Texas—cold fronts that bring sudden drops in temperature, and precipitation followed by blue skies—and sometimes influencing places as far south as the gulfs of Tehuantepec, Mexico, and Panama.

The presence of the Rockies has a marked effect on airflow over North America. They deflect the westerly winds to the north and force the development of a wave in the airflow. On average, location of the crest of this wave lies over the mountains; its trough lies to the southeast. The effect of this wave is to bring cold air from north to south over the continent, particularly in winter, while warm air is transported from the southeast to the North Atlantic and northern Europe. As a result, these regions are warmer than they would be without the Rocky Mountains.

This major mountain chain also contributes to the North American monsoon. In summer the Rocky Mountain plateau becomes very warm. The lower pressure that develops draws in moist air, which converges to produce intense rainfall over southern slopes.

Coastal versus inland Located at similar latitudes, Arizona (**right**) has a semi-arid climate, but the New Orleans region (**below**) is subtropical. Both are influenced by the same high-pressure system. Because New Orleans is coastal, the system causes warm, moist air to flow from the Gulf of Mexico. In inland Arizona, however, the high pressure keeps it dry for most of the year, except during the summer monsoon.

Climate and population Along with access to transportation, climate type is arguably the most influential factor in determining the location of major population centers. The temperate region of North America's west coast is home to large cities like Vancouver in British Columbia, Canada (**opposite**), while Alaska's inhospitable subpolar climate (**below**) leaves the tundra relatively unpopulated.

CLIMATE ZONES OF
NORTH AMERICA AND MESOAMERICA

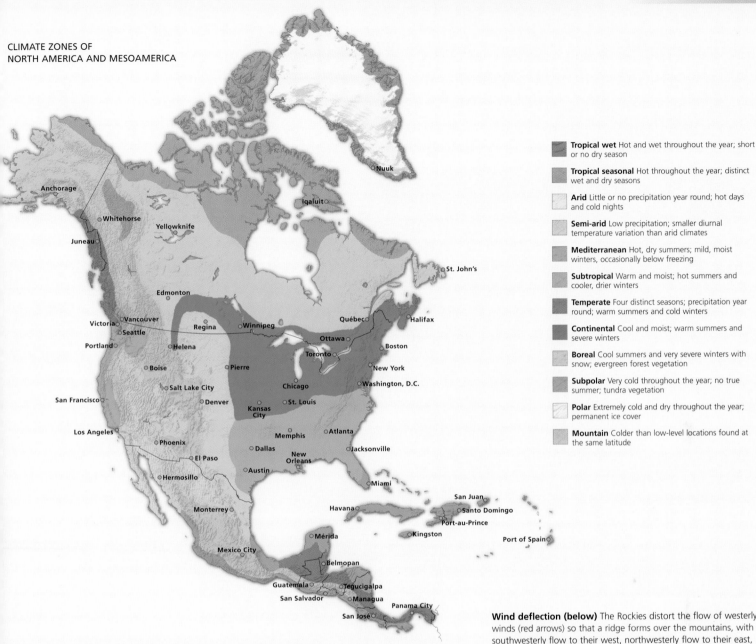

Tropical wet Hot and wet throughout the year; short or no dry season

Tropical seasonal Hot throughout the year; distinct wet and dry seasons

Arid Little or no precipitation year round; hot days and cold nights

Semi-arid Low precipitation; smaller diurnal temperature variation than arid climates

Mediterranean Hot, dry summers; mild, moist winters, occasionally below freezing

Subtropical Warm and moist; hot summers and cooler, drier winters

Temperate Four distinct seasons; precipitation year round; warm summers and cold winters

Continental Cool and moist; warm summers and severe winters

Boreal Cool summers and very severe winters with snow; evergreen forest vegetation

Subpolar Very cold throughout the year; no true summer; tundra vegetation

Polar Extremely cold and dry throughout the year; permanent ice cover

Mountain Colder than low-level locations found at the same latitude

Wind deflection (below) The Rockies distort the flow of westerly winds (red arrows) so that a ridge forms over the mountains, with southwesterly flow to their west, northwesterly flow to their east, and a trough over southeastern U.S.A.

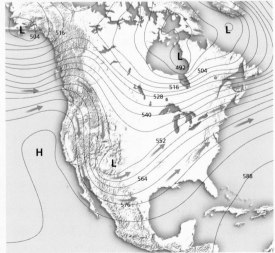

North American Arctic

North America's Arctic region lies generally north of 66.5°N, extending from northern Alaska across Canada's Northwest Territories and Nunavut to Greenland. Dominated by cold, dry Arctic air masses, it experiences long, sunless, and severely cold winters, with short, sunny, cool summers. Low precipitation falls mainly as snow in winter and as rain in summer.

Tundra (below) With frozen soil for much of the year, low precipitation, and cool weather, the Arctic tundra supports only mosses, grasses, and shrubs. These low-growing plants sustain the musk oxen that inhabit Alaska's coastal plains and valleys.

FACT FILE

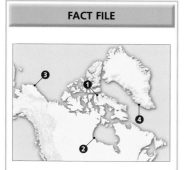

Climographs Temperatures are below freezing for most of the year, as in Arctic Bay. Summers range from cold in Barrow to mild in Churchill. Precipitation is usually small—greatest in Godthaab—and peaks in late summer.

— Average precipitation
— Max. temperature
— Min. temperature

1. ARCTIC BAY, NUNAVUT, CANADA

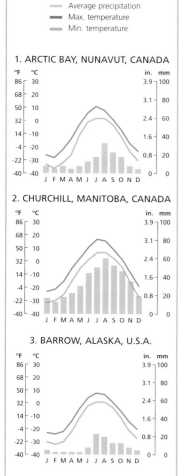

2. CHURCHILL, MANITOBA, CANADA

3. BARROW, ALASKA, U.S.A.

4. GODTHAAB (NUUK), GREENLAND

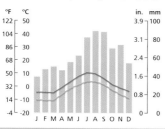

Severe winters (right) Although there is very little snowfall in the Arctic, blizzards can be a frequent occurrence. In winter, very strong winds pick up the little snow that exists on the surface, stirring it into intense blizzards and large drifts.

Icebergs (below) In winter, icebergs that originate from Greenland's glaciers form part of the pack ice in Arctic waters. When the pack ice melts in spring, they are often seen off the Newfoundland coast, steered south by winds and ocean currents.

Far north (above) Ellesmere Island and the northern coast of Greenland, together with Alert Island, are Earth's northernmost land areas. They are continually influenced by the frigid, dry air masses typical of the polar climate.

Permafrost and ice (right) In the severely cold conditions of the Arctic, permanently frozen ground, or permafrost, underlies most of the land. Around the Arctic Ocean ice cap is seasonal ice, forming in winter but melting in summer. This ice is either moved around the ice cap by the Beaufort Gyre, or carried south by the Transpolar Drift and the East Greenland and Labrador currents to the Atlantic Ocean. Global warming has dramatically reduced sea ice, opening the Northwest Passage between the Pacific and Atlantic oceans, with potential global economic and security implications.

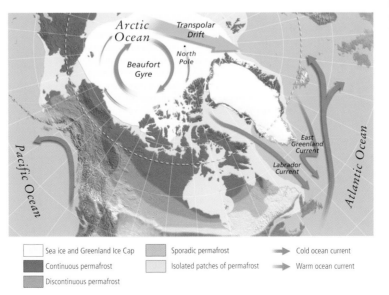

Arctic Ocean

Transpolar Drift

North Pole

Beaufort Gyre

Pacific Ocean

Atlantic Ocean

East Greenland Current

Labrador Current

☐ Sea ice and Greenland Ice Cap	☐ Sporadic permafrost
■ Continuous permafrost	☐ Isolated patches of permafrost
■ Discontinuous permafrost	

→ Cold ocean current
→ Warm ocean current

Pacific Northwest

In the north of this region, the coastal fringes of southeast Alaska, Yukon Territory, and British Columbia are temperate. In the south, Washington, Oregon, and part of Idaho have Mediterranean climates. Topography and the cool, moist prevailing westerlies produce a number of variations. Temperatures are warm inland, but moderated by the ocean on the coast.

Washington State, U.S.A. (below)
The state's topography strongly influences precipitation. High amounts occur along the coast and on the windward side of the coastal ranges. Further inland, the climate is quite dry, due to the rain-shadow effect.

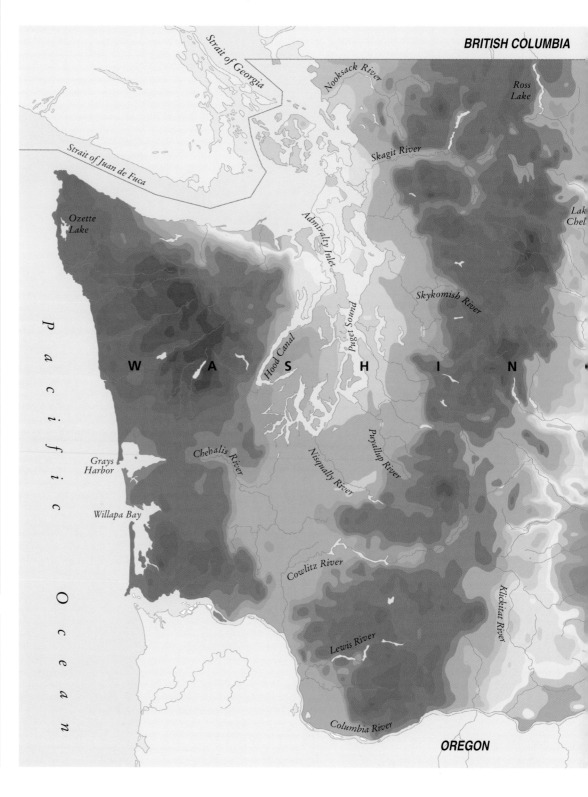

British Columbia (right) Sea fog, or advection fog, occurs frequently along the coast of British Columbia and on the west coasts of the Queen Charlotte Islands. It develops when moist Pacific air moves over the cold coastal water.

BRITISH COLUMBIA

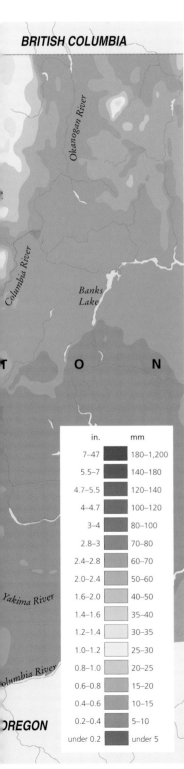

Okanogan River

Columbia River

Banks Lake

T O N

Yakima River

olumbia River

OREGON

in.	mm
7–47	180–1,200
5.5–7	140–180
4.7–5.5	120–140
4–4.7	100–120
3–4	80–100
2.8–3	70–80
2.4–2.8	60–70
2.0–2.4	50–60
1.6–2.0	40–50
1.4–1.6	35–40
1.2–1.4	30–35
1.0–1.2	25–30
0.8–1.0	20–25
0.6–0.8	15–20
0.4–0.6	10–15
0.2–0.4	5–10
under 0.2	under 5

Mountain climate (above) Skiing is popular in the Cascade Range in Washington and Oregon states, where abundant winter snowfall is the result of the orographic effect of the mountains on the moisture-laden westerly winds that blow off the Pacific.

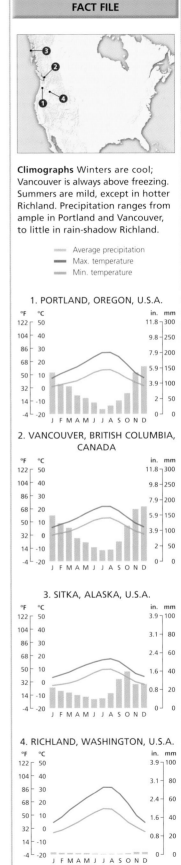

Climographs Winters are cool; Vancouver is always above freezing. Summers are mild, except in hotter Richland. Precipitation ranges from ample in Portland and Vancouver, to little in rain-shadow Richland.

- Average precipitation
- Max. temperature
- Min. temperature

1. PORTLAND, OREGON, U.S.A.

2. VANCOUVER, BRITISH COLUMBIA, CANADA

3. SITKA, ALASKA, U.S.A.

4. RICHLAND, WASHINGTON, U.S.A.

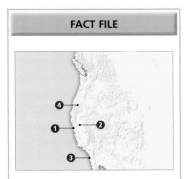

Climographs As is typical of Mediterranean climates, summers are hot and dry, winters cool and moist. Sacramento is hottest and driest in summer. San Francisco has the most winter rainfall, and Ashland is driest all year.

Average rainfall
Max. temperature
Min. temperature

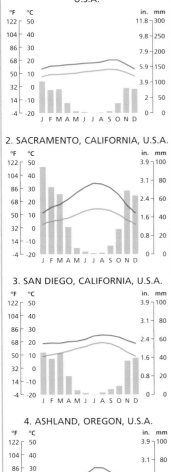

1. SAN FRANCISCO, CALIFORNIA, U.S.A.

°F °C in. mm
122 50 11.8 300
104 40 9.8 250
86 30 7.9 200
68 20 5.9 150
50 10 3.9 100
32 0 2 50
14 -10 0 0
-4 -20
J F M A M J J A S O N D

2. SACRAMENTO, CALIFORNIA, U.S.A.

°F °C in. mm
122 50 3.9 100
104 40 3.1 80
86 30 2.4 60
68 20 1.6 40
50 10 0.8 20
32 0 0 0
14 -10
-4 -20
J F M A M J J A S O N D

3. SAN DIEGO, CALIFORNIA, U.S.A.

°F °C in. mm
122 50 3.9 100
104 40 3.1 80
86 30 2.4 60
68 20 1.6 40
50 10 0.8 20
32 0 0 0
14 -10
-4 -20
J F M A M J J A S O N D

4. ASHLAND, OREGON, U.S.A.

°F °C in. mm
122 50 3.9 100
104 40 3.1 80
86 30 2.4 60
68 20 1.6 40
50 10 0.8 20
32 0 0 0
14 -10
-4 -20
J F M A M J J A S O N D

Mediterranean West Coast

This region extends from the Mexican border north through California to parts of southern Oregon, and inland from the coast to the Sierra Nevada foothills and the edge of the southwestern deserts. Most of the region has a Mediterranean climate, with wet winters and long, dry, warm summers. The north has lower average temperatures and higher rainfall than the south.

Santa Ana winds (below) Originating from high pressure (H) over the arid Great Basin, this gusty wind starts at the end of the hot, dry summer. It warms and dries as it descends toward the coast of southern California, and can cause or fan wildfires.

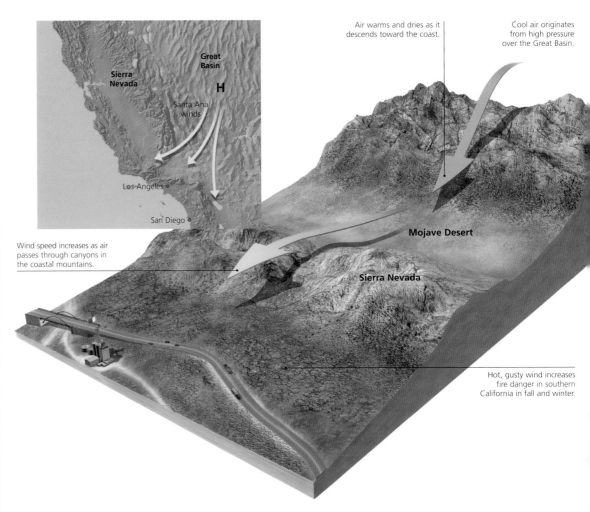

Air warms and dries as it descends toward the coast.

Cool air originates from high pressure over the Great Basin.

Great Basin

Sierra Nevada

H

Santa Ana winds

Los Angeles

San Diego

Mojave Desert

Sierra Nevada

Wind speed increases as air passes through canyons in the coastal mountains.

Hot, gusty wind increases fire danger in southern California in fall and winter.

Dry valley (left) Precipitation in the south of California's vast Central Valley is low, allowing the cultivation of grasses for cattle fodder. Irrigation, using water from the northern parts of the valley, makes agriculture possible in this dry climate.

Sea breezes (right) Along California's exposed shores, the prevailing westerlies make hang gliding and paragliding popular sports on clear summer days. Here, gliding enthusiasts take flight over the Pacific Ocean from rocky bluffs near San Diego.

Moist environment (right) Californian redwoods are the world's tallest trees. They depend on the moisture produced by the advection fog and the higher orographic rainfall of coastal regions, where they thrive on the western mountain slopes.

Climate variant (below) To the north, the mix of coastal, mountain, and valley influences in Mendocino County makes winters cooler and wetter, and summers less dry than in the south. This imparts a unique quality to the grapes grown there.

North Woods

The North Woods region extends across most of Canada and a large portion of Alaska. Its short, cool summers and long, extremely cold winters are typical of the boreal climate. Summer days are long and sunlit, but low sun angles keep temperatures down. Winter is cold and dry, dominated by the polar anticyclone—high atmospheric pressure over the North Pole.

Canada warbler This bird breeds in the southern reaches of the boreal forests, where it is estimated that 64 percent of the population nests.

Migration The Canada warbler flies north in spring (blue arrows) to its breeding grounds (blue), and then south in the fall (orange arrows) to winter in South America.

Newfoundland wetlands (right) Winter's snow melts in spring and, along with summer rains, feeds water into the boreal forest's lakes and wetlands. In the cool temperatures of summer, there is little evaporation, so the ground stays very wet.

Winter snow (above) In the boreal winter, fine, dry snow blankets the vast coniferous forests that extend over much of North Woods. Precipitation is relatively low, but below-freezing temperatures ensure the snow lasts most of the winter.

Stressed forests (left) Environmental stresses such as logging and drought can make coniferous forests susceptible to insect infestation. Vast tracts of dead trees will in turn fuel frequent summer wildfires, releasing carbon dioxide, reducing carbon storage, and exacerbating global warming.

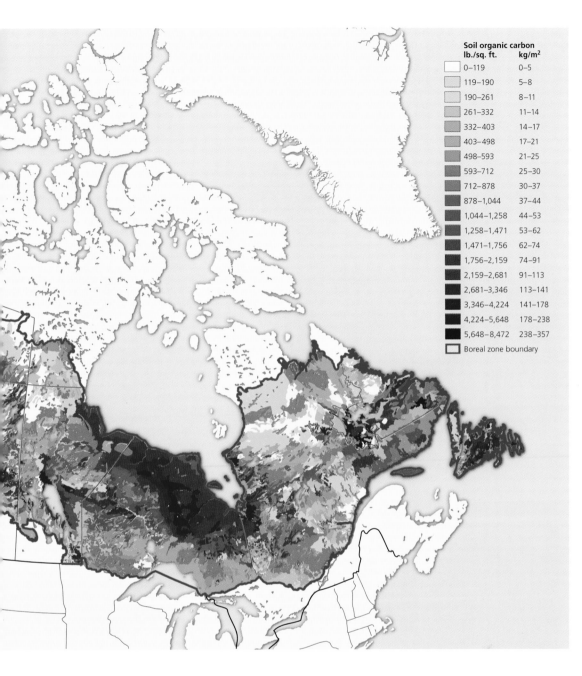

Soil organic carbon

lb./sq. ft.	kg/m²
0–119	0–5
119–190	5–8
190–261	8–11
261–332	11–14
332–403	14–17
403–498	17–21
498–593	21–25
593–712	25–30
712–878	30–37
878–1,044	37–44
1,044–1,258	44–53
1,258–1,471	53–62
1,471–1,756	62–74
1,756–2,159	74–91
2,159–2,681	91–113
2,681–3,346	113–141
3,346–4,224	141–178
4,224–5,648	178–238
5,648–8,472	238–357

▢ Boreal zone boundary

FACT FILE

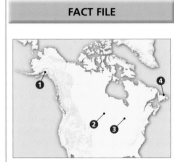

Climographs Winters are very cool or cold; least cold in Gander. Inland, summers are warm in Winnipeg and Sault Ste. Marie, and cooler in coastal Anchorage and Gander. Precipitation varies little, except in Anchorage. Gander is wettest.

— Average precipitation
— Max. temperature
— Min. temperature

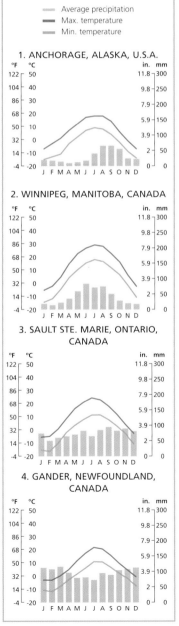

1. ANCHORAGE, ALASKA, U.S.A.

2. WINNIPEG, MANITOBA, CANADA

3. SAULT STE. MARIE, ONTARIO, CANADA

4. GANDER, NEWFOUNDLAND, CANADA

Carbon storage (above) Canada's boreal forest stores vast amounts of carbon, but is very sensitive to temperature fluctuations. With the predicted increased warming at high latitudes, much of this forest may disappear, releasing the stored carbon into the atmosphere.

Cool summers (right) Canoeing is a popular summer pastime on the many lakes and wetlands in the North Woods forests, where summers are pleasantly cool and encounters with moose and other wildlife are likely.

Rocky Mountains

Latitudinal span and elevation produce a variety of climates in this vast range. In the north, winters are longer and precipitation higher than in the south. Summers in the south are hotter and drier, with afternoon thunderstorms, while farther north they tend to be cool and moist. Large elevation differences allow alpine and desert climates to exist close to each other.

Precipitation (below) Orographic precipitation is an important source of moisture in the Rockies. Westerly airflow is forced to rise along the western slopes. As air rises it cools, water vapor condenses to form clouds, and rain or snow may fall.

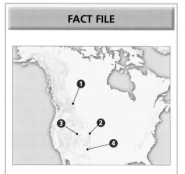

Climographs In the northern cities, winters are colder than in the south, with more precipitation. Banff has long, cold, snowy winters and short, mild summers. Summers are warmest in Santa Fe and Salt Lake City, but cooler in Laramie.

- Average precipitation
- ■ Max. temperature
- Min. temperature

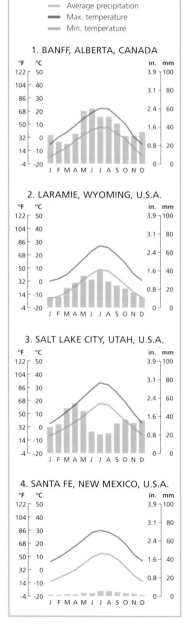

1. BANFF, ALBERTA, CANADA

2. LARAMIE, WYOMING, U.S.A.

3. SALT LAKE CITY, UTAH, U.S.A.

4. SANTA FE, NEW MEXICO, U.S.A.

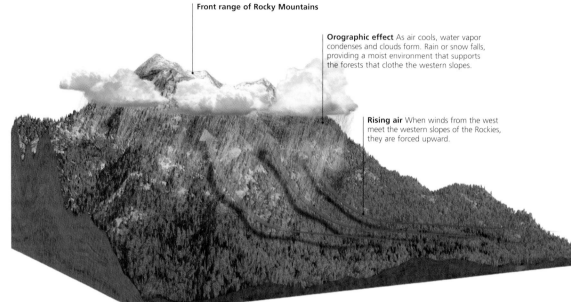

Front range of Rocky Mountains

Orographic effect As air cools, water vapor condenses and clouds form. Rain or snow falls, providing a moist environment that supports the forests that clothe the western slopes.

Rising air When winds from the west meet the western slopes of the Rockies, they are forced upward.

Permanent snow (above) Rising higher than 11,000 feet (3,350 m), the southern Rockies, Colorado, shown in this satellite image are snow-capped all year. Below the snowline, the trees are largely spruce and fir on the upper slopes (dark green), with shorter trees and bushes (light green) on the lower slopes and in the valleys.

Storms (left) Dramatic cloud-to-cloud lightning lights up peaks in Grand Teton National Park, Wyoming. These summer afternoon thunderstorms are frequent, sometimes extending into the night.

Glaciers (above) Lake Louise sits high in the Canadian Rockies, where cold winters and heavy snowfalls maintain the glaciers. In summer, glacial runoff creates waterfalls and preserves lake levels.

Alpine bloom (below) In the warm spring sunshine, Colorado's alpine meadows come alive with colorful flowers. Adapted to the cooler conditions, these last well into late spring or early summer.

FACT FILE

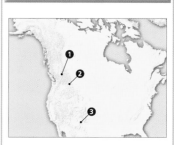

1. Jasper National Park High latitude (50°N), elevation, and rain shadow affect Jasper's climate. Winters are long and cold, summers short and cool. Precipitation is mostly orographic, especially in summer when the westerlies are strong; high altitudes receive more than valleys. Winter brings storms that form along the Arctic front, and cold, polar continental air.

Jasper National Park, Alberta, Canada

2. Montana The climate in the Montana Rockies is also influenced by elevation, latitude, and rain shadow. At around 46°N, winters are cold on the upper slopes, but warmer in the valleys. Summers are cool at higher altitudes and warmer in the lowlands. Precipitation occurs as snow in fall, winter, and spring, and increases with elevation. Summers are dry.

Rocky Mountains, Montana, U.S.A.

3. San Luis Valley This is a high desert area, with less than 8 inches (200 mm) of precipitation per year. Winters are dry, and daily temperatures can fluctuate between cold and warm. Precipitation occurs mainly in spring and summer. Summer days are hot, with cool nights. In late summer, this part of the Rockies often has afternoon thunderstorms linked to the monsoons.

San Luis Valley, Colorado, U.S.A.

WEATHER WATCH

Less snow As the climate changes, temperatures and precipitation in the Rockies are predicted to rise. Higher temperatures will cause more precipitation to fall as rain than snow, and there may be changes in its timing. Decreasing water storage in the snowpack will affect city water supplies.

Desert Southwest

This arid and semi-arid region extends over the lowlands of most of southwestern U.S.A. and northern Mexico. In summer, temperatures are very high by day and slightly cooler at night, when much of the daytime heat is lost to clear skies. In much of the region, daily late-summer storms—the summer monsoon—mark a break from the heat and aridity.

Spring color (above) Light winter rains still provide enough moisture for plants in Arizona's Sonoran Desert to bloom in spring. Able to store water, the tall saguaro cactus is likely to survive the long droughts predicted to accompany climate change.

Late summer rain (right) Spectacular lightning over the Grand Canyon, Arizona, is typical of the summer, or North American, monsoon. Climate change may affect the future timing and intensity of these thunderstorms.

DESERT CULTIVATION

Viewed from above, the green floor of Imperial Valley, California, contrasts starkly with the desert slopes that border it. Although the valley lies within the Sonoran Desert, it is an extremely productive agricultural region. This is because its warm, sunny climate is aided by water diverted from the Colorado River—seen in the lower part of this image—which is used to irrigate its fertile soils.

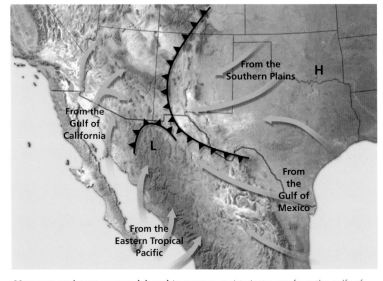

From the
Southern Plains   **H**

From the
Gulf of
California

L

From
the
Gulf of
Mexico

From the
Eastern Tropical
Pacific

Monsoon moisture sources (above) In summer, moist air streams from the gulfs of California and Mexico, the eastern Pacific Ocean, and the high pressure (H) in the Southern Plains converge on a thermal low-pressure region (L) that results from intense heating in the desert southwest. Frontal boundaries form, and air rises along them, producing the afternoon thunderstorms and rain that mark the North American monsoon.

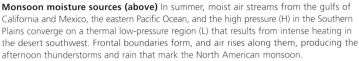

Joshua trees (above) Lying in the rain shadow of the Coast Ranges, California's Mojave Desert receives very low rainfall, mostly from November to April. The Joshua tree grows only in the Mojave's dry, sandy soils, blooming in spring after winter rains.

Cold winters (left) On a plateau nearly 6,200 feet (1,890 m) high, Monument Valley, Utah/Arizona, can be cold and snowy in winter, but temperatures may also stay above freezing. Summer days are hot, with very little rain and cooler nights.

FACT FILE

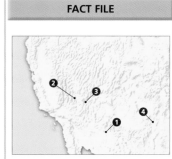

Climographs Albuquerque has very low rainfall, with a late-summer maximum. Phoenix has very little rain, evenly distributed throughout the year. Death Valley is very dry. Winters are mild, except in Albuquerque, where it is cooler.

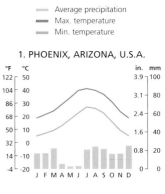

Average precipitation
Max. temperature
Min. temperature

1. PHOENIX, ARIZONA, U.S.A.

2. DEATH VALLEY, CALIFORNIA, U.S.A.

3. LAS VEGAS, NEVADA, U.S.A.

4. ALBUQUERQUE, NEW MEXICO, U.S.A.

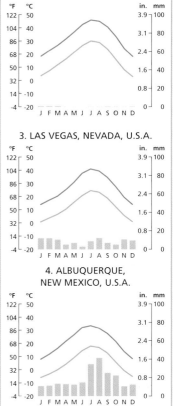

Great Plains

Lying east of the Rockies and west of the Mississippi River, this region in central U.S.A. has a mostly semi-arid climate, thanks to its inland location and the rain–shadow effect of the Rockies. Winters are cold and dry, with some snowstorms. Most precipitation occurs in spring and early summer, when atmospheric conditions often lead to supercell thunderstorms.

Supercell storms (right) A deep, rotating updraft produces the overshooting top and anvil shape typical of supercell clouds, with the main storm fed by a flanking line of smaller convective clouds. Tornadoes often form in the rear, downdraft region.

FACT FILE

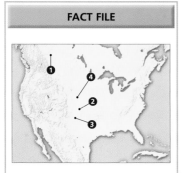

Climographs These cities are typical of semi-arid regions, with cold, dry winters and small amounts of precipitation in spring and summer. Edmonton, farther north, has a mild, short summer. The others have high summer temperatures.

― Average precipitation
― Max. temperature
― Min. temperature

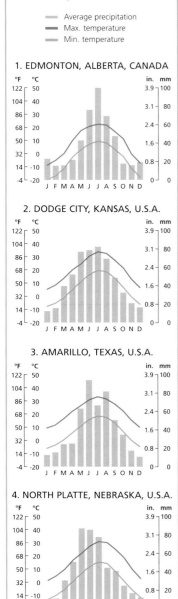

1. EDMONTON, ALBERTA, CANADA

°F	°C		in.	mm
122	50		3.9	100
104	40		3.1	80
86	30		2.4	60
68	20		1.6	40
50	10		0.8	20
32	0		0	0
14	-10			
-4	-20			

J F M A M J J A S O N D

2. DODGE CITY, KANSAS, U.S.A.

°F	°C		in.	mm
122	50		3.9	100
104	40		3.1	80
86	30		2.4	60
68	20		1.6	40
50	10		0.8	20
32	0		0	0
14	-10			
-4	-20			

J F M A M J J A S O N D

3. AMARILLO, TEXAS, U.S.A.

°F	°C		in.	mm
122	50		3.9	100
104	40		3.1	80
86	30		2.4	60
68	20		1.6	40
50	10		0.8	20
32	0		0	0
14	-10			
-4	-20			

J F M A M J J A S O N D

4. NORTH PLATTE, NEBRASKA, U.S.A.

°F	°C		in.	mm
122	50		3.9	100
104	40		3.1	80
86	30		2.4	60
68	20		1.6	40
50	10		0.8	20
32	0		0	0
14	-10			
-4	-20			

J F M A M J J A S O N D

Cattle-friendly climate (below) Limited precipitation supports the growth of grasses over the extensive, flat landscape of the Great Plains. These are optimum conditions for large-scale cattle ranching, as at this farm in Nebraska, U.S.A.

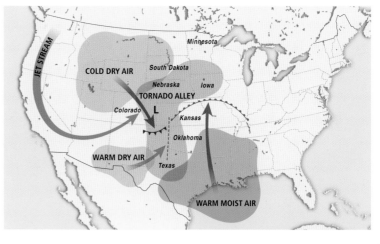

Tornado Alley (above) Over the Great Plains, cold, polar air masses and warm, dry, air sinking down the Rockies' eastern slopes meet warm, moist tropical air from the Gulf of Mexico. At their meeting place, or dryline (red broken line), conditions favor the development of tornadoes.

Grain harvest (below) Grasses, but not trees, thrive on the low levels of precipitation in the Great Plains. Much of the region, also known as the prairies, is given over to the cultivation of extensive fields of grain, like these in Saskatchewan, Canada.

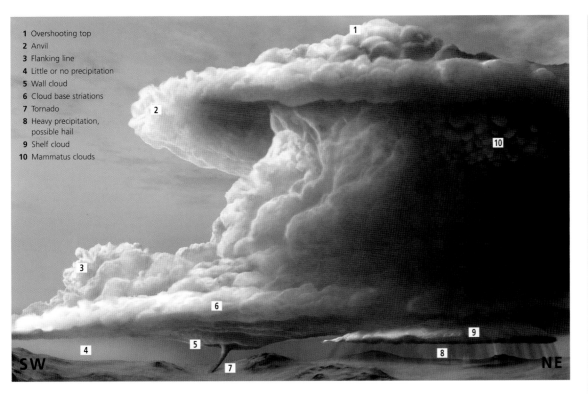

1 Overshooting top
2 Anvil
3 Flanking line
4 Little or no precipitation
5 Wall cloud
6 Cloud base striations
7 Tornado
8 Heavy precipitation, possible hail
9 Shelf cloud
10 Mammatus clouds

SW
NE

FACT FILE

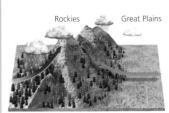

Rockies   Great Plains

Rockies' rain shadow In the lee of the Rockies, air warms as it descends, reducing cloud and rain development. This means the Great Plains to the east of the Rockies get little precipitation.

Chinook wind As this warm, dry wind descends the eastern slopes of the Rockies, the air compresses and warms up, very quickly causing temperatures to rise and snow to melt.

Dust Bowl The Great Plains are prone to drought. Lacking moisture, the dry topsoil is easily transported in huge windstorms, as seen here in the 1930s Dust Bowl drought, Texas, U.S.A.

Storm damage (left) Forming over the Great Plains, supercell thunderstorms can be very destructive. In this late-afternoon storm in Colorado, U.S.A., in spring 1999, winds approaching 60 mph (97 km/h) and hail destroyed much of the wheat crop.

WEATHER WATCH

Water shortages Although higher precipitation levels are expected in much of the Great Plains, increased evaporation caused by higher temperatures will make the region drier. This, coupled with depleted aquifer water levels, means that the Great Plains will experience serious water shortages in the future.

Continental Northeast

Stretching east from the Great Plains across to the east coast and from the Great Lakes to central U.S.A, this region has marked variations in climate. It experiences four distinct seasons, whose extremes are moderated by the influence of the Great Lakes in the north. Summers are hot and humid, with frequent thunderstorms. Winters are cold with snowstorms.

Northern fall (below) The reduced sunlight and cooler temperatures of fall prompt the maple leaves in Massachusetts, U.S.A., to turn brilliant red and orange. This change begins in early September and ends in late October or early November.

Spring (above) Flowering cherry trees mark the arrival of spring in Washington, D.C. Warmer temperatures attributed to climate change have been blamed for ever earlier blossoms each year.

Lake-effect snow (below) More snow falls downwind (to the east and south) of the Great Lakes than upwind because winds crossing the lakes pick up moisture and then deposit it as snow.

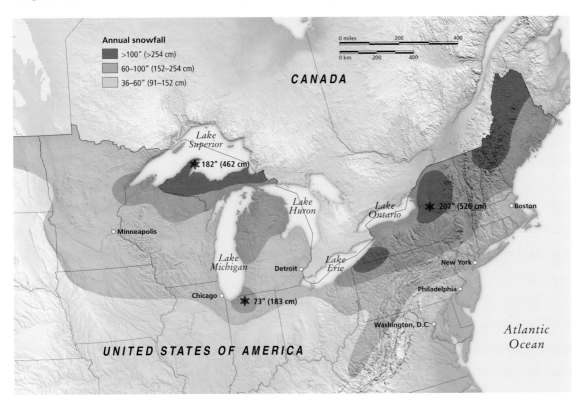

Annual snowfall
- >100" (>254 cm)
- 60–100" (152–254 cm)
- 36–60" (91–152 cm)

0 miles   200   400
0 km   200   400

CANADA

Lake Superior ❄ 182" (462 cm)

Lake Huron

Lake Ontario ❄ 207" (526 cm) ○ Boston

○ Minneapolis

Lake Michigan

Detroit ○ *Lake Erie*

New York ○

Chicago ○ ❄ 73" (183 cm)

Philadelphia ○

Washington, D.C. ○

UNITED STATES OF AMERICA

Atlantic Ocean

FACT FILE

Climographs Winters are cold, except in Richmond, where they are cool. Summers in all four cities are warm. Precipitation is mostly evenly distributed across the months, with a small summer maximum. Boston has the most annual precipitation.

- Average precipitation
- Max. temperature
- Min. temperature

1. CHICAGO, ILLINOIS, U.S.A.

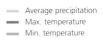

2. KANSAS CITY, MISSOURI, U.S.A.

3. BOSTON, MASSACHUSETTS, U.S.A.

4. RICHMOND, VIRGINIA, U.S.A.

Climographs New Orleans and Miami are warm all year, but New Orleans is warmer in summer. Charlotte and Asheville have cooler winters. Rainfall in Miami is highest in summer, but is evenly distributed in Charlotte and Asheville.

Average precipitation
Max. temperature
Min. temperature

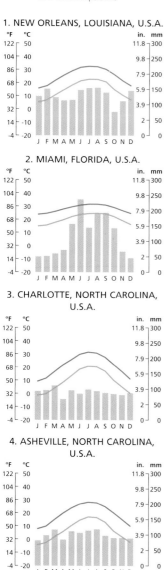

1. NEW ORLEANS, LOUISIANA, U.S.A.

2. MIAMI, FLORIDA, U.S.A.

3. CHARLOTTE, NORTH CAROLINA, U.S.A.

4. ASHEVILLE, NORTH CAROLINA, U.S.A.

Subtropical Southeast

Hot summers and mild, sunny winters characterize the subtropical climate of southeastern U.S.A. Warm, moist air from the Gulf of Mexico and the Atlantic Ocean brings year-round rainfall, but summer is wettest. In summer and fall, hurricanes sometimes come ashore, while all areas except Florida have infrequent winter storms with cold winds and freezing rain.

Atlanta, Georgia (below left) Summers in this large city are hot and humid, while its winters are cool and drier. In 2007–09, Atlanta experienced its worst water shortage to date, possibly caused by high demand from its expanding population.

High humidity (left) In South Carolina's humid climate, most of the rainfall occurs in summer. Because of the region's high water table and poor drainage, much of the rain that falls remains near the surface to form extensive swamps and marshes.

Balmy climate (right) Key West, southern Florida, at the edge of the tropics, is warm year round. Temperatures are moderated by the winds blowing off the surrounding water. Spring and summer are wet, while fall and winter are relatively dry.

Hurricane Katrina (below and inset) Hurricanes come ashore almost every year, guided by the warm waters of the Gulf of Mexico and the upper airflow. Hurricane Katrina moved north through the Gulf of Mexico and struck the city of New Orleans on August 29, 2005. High wind speeds and damaging rainfall combined to wreak havoc on the city and surrounding region. The inset (right) depicts the storm's wind speeds in color, with the highest speed around the center shown as purple and white barbs indicating areas of heavy rain.

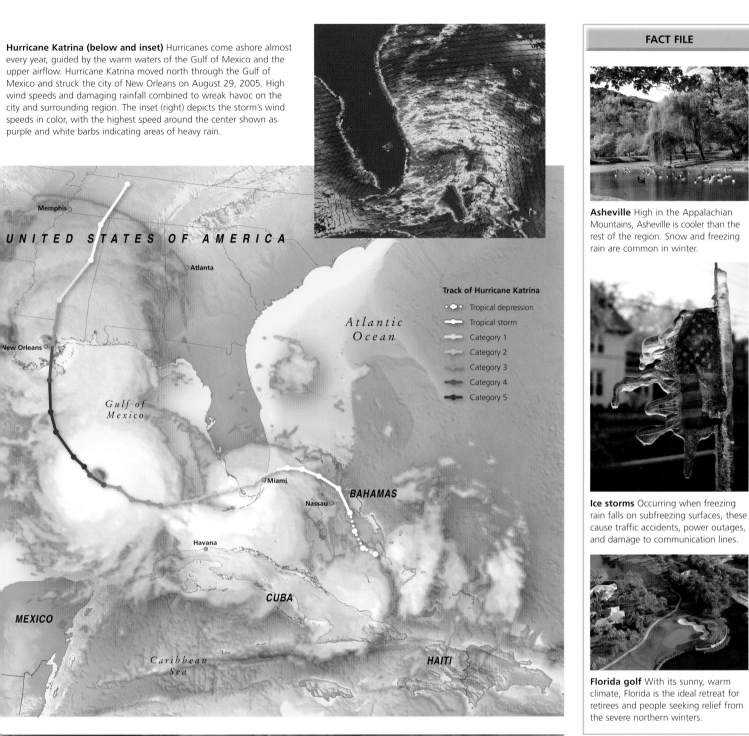

Memphis

UNITED STATES OF AMERICA

Atlanta

*Atlantic
Ocean*

New Orleans

Track of Hurricane Katrina

Tropical depression
Tropical storm
Category 1
Category 2
Category 3
Category 4
Category 5

*Gulf of
Mexico*

Miami

Nassau

BAHAMAS

Havana

MEXICO

CUBA

*Caribbean
Sea*

HAITI

1. Belize Mangroves and sea grasses thrive in Belize's tropical wet climate, protecting the coast from erosion.

Mangroves also filter pollutants, while sea grasses provide breeding grounds for turtles and many varieties of fish.

Belize

2. Baja California This dry region has less than 24 inches (610 mm) of rain each year. Influenced by subtropical

high pressure, summer days are hot; nights are cold. The cold California Current makes the north cooler.

Baja California, Mexico

3. Lake Petén Itzá This lowland lake is surrounded by dense tropical rainforest. High temperatures and

precipitation from a variety of sources, including the ITCZ, mean that the area is hot and wet year round.

Lake Petén Itzá, Guatemala

4. Yucatán brush forest This is in the northwest of the Yucatán Peninsula, where the climate is tropical seasonal. The long dry season lasts from November

to April. Annual rainfall, mostly from May to October, does not normally exceed 47 inches (1,200 mm).

Yucatán brush forest, Mexico

Mesoamerica

Extending from northern Mexico through Panama, this narrow landmass between the Caribbean and the Pacific has tropical and mountain climates, with a strong marine influence. Except at high altitudes, it is warm year round, and precipitation is high, due to the influence of hurricanes, the North American monsoon, and the Intertropical Convergence Zone (ITCZ).

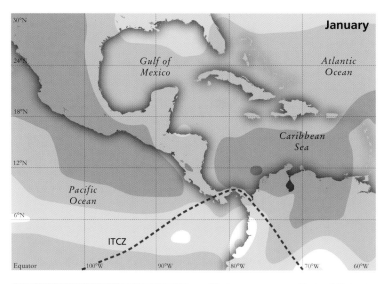

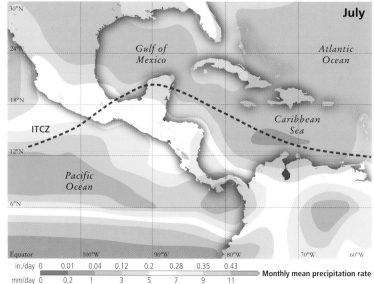

Monthly mean precipitation rate

in./day	0	0.01	0.04	0.12	0.2	0.28	0.35	0.43
mm/day	0	0.2	1	3	5	7	9	11

Intertropical Convergence Zone (above) The position of this belt of low pressure circling Earth near the Equator affects rainfall. When it moves south in January, there is less rain. In July it migrates as far north as the Yucatán Peninsula, bringing heavier rainfall to much of the region.

El Salvador (right) In the tropics, temperatures vary more with altitude than by season, so that valleys are hotter than higher up the mountains. While both have plenty of rain, it is usually wetter on the mountain slopes than in the lowlands.

Cloud forest (left) The year-round clouds over the forested Costa Rican Cordilleras form when the northeast trade winds are forced upslope. Many species of birds and animals thrive in this unique environment.

Tropical seasonal (below) In Nicaragua the wet season is from late May to January and the dry season from January to mid-May. The warm temperatures of the tropics are moderated in mountainous regions.

Hurricane frequency (above) Hurricanes occur in this region every year in the warm waters of the eastern Pacific as well as in the Caribbean. In this image, Hurricane Dean is seen crossing the Yucatán Peninsula in August 2008.

FACT FILE

Climograph These cities have well-defined rainy seasons; they are dry from November to May. Acapulco and San José are wetter. Temperatures vary little throughout the year, but Mexico City is coolest because of its altitude.

Average precipitation
Max. temperature
Min. temperature

1. ACAPULCO, MEXICO

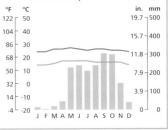

2. MEXICO CITY, MEXICO

3. MANAGUA, NICARAGUA

4. SAN JOSÉ, COSTA RICA

Islands of the Caribbean

Surrounded by the warm waters of the Atlantic and the Caribbean, these islands are humid all year, with tropical wet and seasonal climates. Total sunshine hours and solar radiation intensity are high, but the constant trade winds moderate temperature. Rain comes from various sources, including easterly waves in the atmosphere and the hurricanes they create.

Floods (below) These satellite views show how, in 2004, Hurricane Jeanne's torrential rains filled a large, previously dry lake basin (pink, left image) near Gonaïves in Haiti. The flooded lake extended well beyond its normal shores (dark blue, right image).

FACT FILE

Climographs In these Caribbean cities, it is uniformly warm all year, with little annual or daily variation. Precipitation, however, is variable. The extremes are St. Croix, which is arid, and Port of Spain, which is wet from June through December.

- Average precipitation
- Max. temperature
- Min. temperature

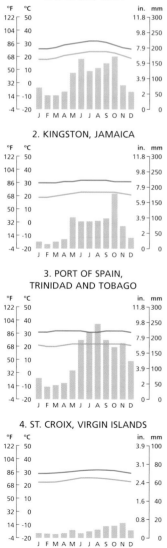

1. HAVANA, CUBA

2. KINGSTON, JAMAICA

3. PORT OF SPAIN, TRINIDAD AND TOBAGO

4. ST. CROIX, VIRGIN ISLANDS

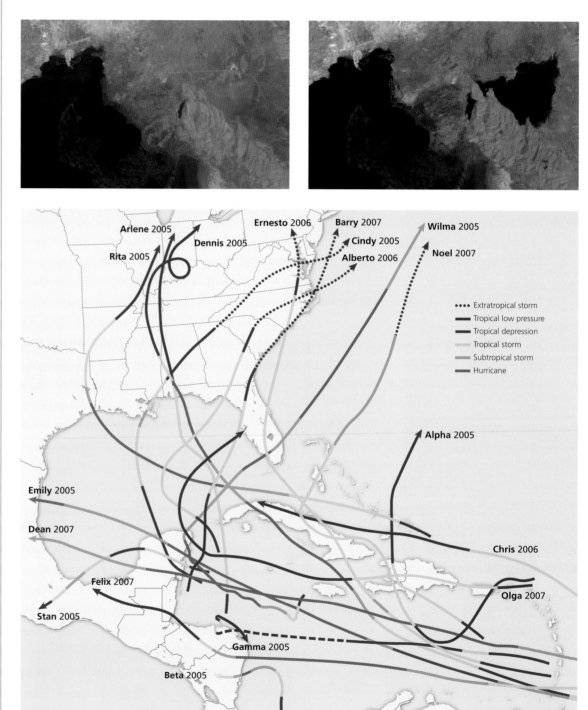

Arlene 2005
Rita 2005
Dennis 2005
Ernesto 2006
Barry 2007
Cindy 2005
Alberto 2006
Wilma 2005
Noel 2007

Emily 2005
Dean 2007
Felix 2007
Stan 2005
Gamma 2005
Beta 2005

Alpha 2005
Chris 2006
Olga 2007

•••• Extratropical storm
— Tropical low pressure
— Tropical depression
— Tropical storm
— Subtropical storm
— Hurricane

St. Lucia (above) In the eastern Caribbean, lush tropical vegetation around the two volcanic plugs, Gros Piton and Petit Piton, thrives in St. Lucia's heavy rainfall and constant warm temperatures.

Hurricane tracks, 2005–2007 (left) This region is particularly prone to hurricanes. Easterly waves form in the atmosphere off northwestern Africa and are carried west by the trade winds to the Caribbean, where the very warm water provides ideal surface conditions for hurricanes.

Trinidad and Tobago (above) Scarlet ibises and many other bird species thrive in this island's Caroni Swamp mangroves. The climate is seasonal tropical, but enough rain falls to maintain the mangroves.

Tropical paradise (right) Coconut palms, like these in Barbados, grow in the hot, wet climates and sandy coastal soils found in most Caribbean and neighboring Atlantic islands.

SOUTH AMERICA

The continent of South America extends from about 12°N of the Equator to subpolar latitudes at 55°S. It is almost completely surrounded by ocean and in the west is ridged for more than 4,500 miles (7,200 km) from north to south by the Andes mountain chain, which soars to an average height of 13,000 feet (4,000 m). Together, these features create nine climates, ranging from tropical seasonal in the north to subpolar in the south, and including the tropical wet climate of the Amazon and the alpine climate high in the Andes Mountains, where glaciers exist all year.

CLIMATIC INFLUENCES

The Intertropical Convergence Zone (ITCZ) is a belt of heavy rainfall at the convergence of the trade winds. As the seasons change, the ITCZ moves north of the Equator in July and south in January, contributing to South America's tropical wet and tropical seasonal climates. The very wet conditions produced by the ITCZ are increased by the Amazon rain forests, which help to keep that region moist when they transpire. South of the tropics, at midlatitudes, the moist westerlies produce enough rainfall to support lush temperate forests.

The latitudinal and seasonal variation of temperature is determined chiefly by the position of the sun. At low latitudes, such as in the Amazon, the sun is overhead for most of the year. It is warmer there than at middle latitudes or high latitudes such as in Patagonia, where the sun is never overhead. This is why temperatures range from hot in the tropics to frigidly cold at the subpolar latitudes of Tierra del Fuego.

The Andes mountain range also affects temperatures and moisture. Cooler temperatures and glaciers occur high in the tropical Andes, where at lower altitudes it would be too warm for snow to fall and ice to form.

The Andes also obstruct the eastward flow of moist air from the Pacific Ocean. In midlatitudes they cause heavy orographic rainfall along the west of the continent and drier climates in the east, in the rain shadow of the Andes.

Cold ocean currents play a significant role in keeping both temperature and precipitation along the west coast lower than might be expected. The west coast in the subtropics and equatorial regions is dry because the air that moves in from the Pacific is cool and stable. The clouds that form do not bring rain and some of those areas, such as the Atacama Desert, remain dry for years. Often, when precipitation occurs, it is brought by periodic El Niño events. At these times, the normally cold ocean water warms, allowing the air above it to become moist, which leads to thunderstorms and torrential rain. Along the east coast, ocean currents are warmer. As a result, the atmosphere is warmer and more moist, so that the breezes bring heavy rainfall to coastal regions such as Brazil's Atlantic Forest.

The cool ocean off the subtropical west coast contributes to the Mediterranean climate there, keeping winters cool and moist. Summers are long and dry due to the intensification of the subtropical high pressure system at this time.

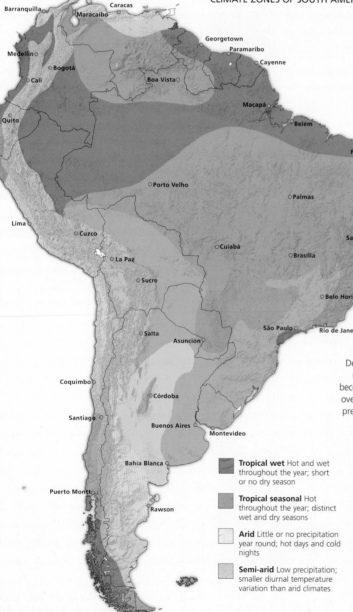

CLIMATE ZONES OF SOUTH AMERICA

Tropical wet Hot and wet throughout the year; short or no dry season

Tropical seasonal Hot throughout the year; distinct wet and dry seasons

Arid Little or no precipitation year round; hot days and cold nights

Semi-arid Low precipitation; smaller diurnal temperature variation than arid climates

Mediterranean Hot, dry summers; mild, moist winters, occasionally below freezing

Subtropical Warm and moist; hot summers and cooler, drier winters

Temperate Four distinct seasons; precipitation year round; warm summers and cold winters

Subpolar Very cold throughout the year; no true summer; tundra vegetation

Mountain Colder than low-level locations found at the same latitude

Winds and currents (right) When westerly winds pass over cold ocean currents, they become cold and dry, creating the cold, dry climate of the Atacama Desert along the west coast. On the east, the southeast trade winds become warm and moist as they pass over warm ocean currents, bringing precipitation when they reach land.

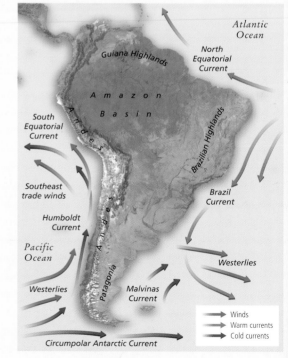

Tierra del Fuego (above) This group of islands at the southernmost tip of the South American continent has an inhospitable subpolar climate that is extremely cold, windy, and rainy.

Andes (right) Climate in this long mountain range varies from warm and wet in the north, at tropical latitudes, to much colder in the south. Near the peaks, the climate is cold alpine.

Atacama Desert (left) This is one of the driest regions on Earth, its climate influenced by the cold Humboldt Current that flows off the Chilean coast. High in the Andes, it is a cold desert, and when precipitation occurs it falls as snow.

Tropical rain forest (below) In South America's equatorial region, the extensive rain forests of the Amazon Basin, such as these in Venezuela, have a very warm, moist tropical climate.

Tropical North and East

Straddling the Equator, the Amazon Basin has a tropical wet climate with high average rainfall and consistently high temperatures. North and south of the basin, the seasonal migration of the Intertropical Convergence Zone (ITCZ) brings clearly defined wet and dry seasons. Northern Venezuela's semi-arid climate is influenced by the cool currents of the Caribbean Sea.

Iguaçu Falls (below) In southern Brazil's tropical seasonal climate, the flow of water over these spectacular falls is greatest in the rainy season, which lasts from January to March, and less during the rest of the year, when it is drier.

Seasonal shift (below right) The trade winds, and the ITCZ into which they converge over South America, shift north in July and south in January. Because heavy rains are associated with the ITCZ, the rainy season follows this shift, moving north and south of the Equator and creating distinct wet and dry seasons.

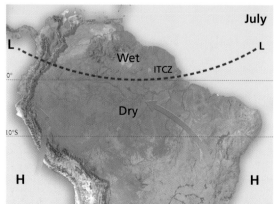

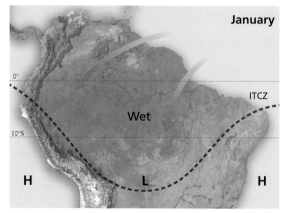

Cerrado (right) This extensive region in central Brazil experiences a seasonal tropical climate. Enough rain falls during the year to support savanna vegetation, but the long dry season discourages the growth of trees, which require more water.

Amazon storms (left) Rain-forest trees produce vast amounts of water vapor, which makes the atmosphere buoyant and rises to form clouds. These then develop into the thunderstorms that occur freqently in the Amazon's tropical wet climate.

FACT FILE

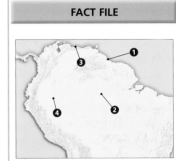

Climographs Temperatures are high year round. Except in drier Caracas, rainfall is ample. In Georgetown it peaks twice, while Manaus is wetter during the first half of the year. Rainfall in Iquitos varies little throughout the year.

Average precipitation
Max. temperature
Min. temperature

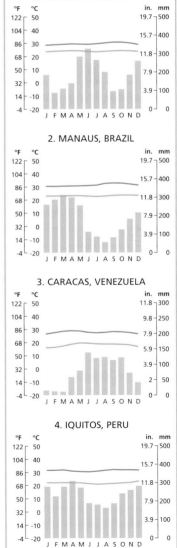

1. GEORGETOWN, GUYANA

2. MANAUS, BRAZIL

3. CARACAS, VENEZUELA

4. IQUITOS, PERU

Arid West Coast

This western coastal strip stretching from Peru to northern Chile has one of the world's most unusual climates. It experiences extreme aridity, because the prevailing southeasterly winds force the upwelling of the cold, deeper waters that form part of the Humboldt Current. This keeps coastal air temperatures cool, and the clouds that form do not produce rain.

Cold desert (below) Water from melting snow and glaciers in the Andes flows into depressions in the Atacama Desert. In this cold, arid high desert it quickly evaporates, forming extensive salt pans that attract birds like these pink flamingoes.

FACT FILE

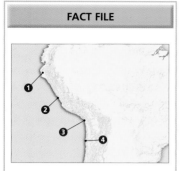

Climographs These places are hot and very dry. Chiclayo has the most precipitation, with a clear maximum in the first half of the year. Arica and Antofagasta are driest. Temperatures are high all year, but slightly lower in winter.

— Average precipitation
— Max. temperature
— Min. temperature

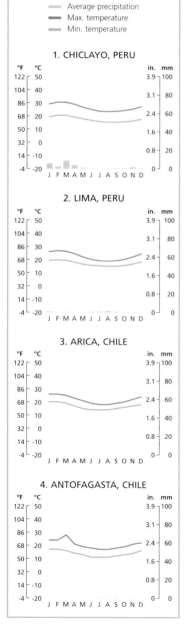

1. CHICLAYO, PERU

2. LIMA, PERU

3. ARICA, CHILE

4. ANTOFAGASTA, CHILE

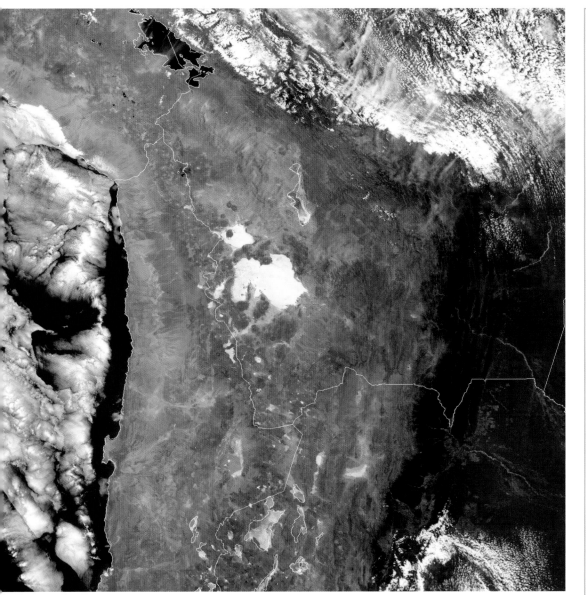

Wildlife Relatively few birds and animal species have adapted to the aridity of the Atacama. Scorpions and insects make their home in the lichen and are preyed upon by lizards. Occasionally foxes or mice may be seen.

Alpaca This herbivore lives in the high desert and mountains, protected from the cold by thick fur. Its stomach can extract the maximum nutrients from the poor grasses found there.

Viscacha Living at cold high altitudes, this desert rabbit has dense fur to keep it warm. It eats mosses and grasses that grow on the rocks where it often spends most of the day.

Desert fox This animal can survive in a wide range of habitats, from the cold, high Andes to the dry Pacific coast. Although hunted and trapped for its fur, its population remains stable.

Moisture block (above) The Andes Mountains (darker brown, center, in this satellite view) also contribute to the aridity of the Atacama Desert (lighter brown, left). They block the westward movement of moist air from wetter regions in the east (dark green, right).

Dunes (right) In this arid climate, very few rivers flow to the ocean transporting sediment. Instead, currents move sand from the south to form sand dunes along the Peruvian coast in the north.

Cloudy but dry (left) Along the coast the Atacama Desert can be cool and cloudy but extremely dry. Winds that blow over the cold Humboldt Current keep it cool, so that although clouds form there is no rain.

WEATHER WATCH

Wetter or warmer? Periodic warming associated with El Niño events can bring torrential rain. In some scenarios these events are predicted to become more frequent, which would change the arid nature of this region. Other scenarios suggest that it will get warmer, with an increase in aridity.

Andes Mountain Chain

In this long north–south chain, climates vary from very rainy to very dry, and from very hot to very cold. In the north it is warm and wet at lower levels, but above 15,000 feet (4,500 m) snow exists all year, while cities at mid-elevations, such as Bogotá, have moderate climates. In the southwest it is wet, but in the southeast dry rain-shadow climates are found.

La Paz (above) Located at an elevation of 12,000 feet (3,660 m), the Bolivian city of La Paz is the world's highest capital. Even though it is only about 15° south of the Equator, it can be quite cold and experiences occasional snow in the winter months.

Chimborazo, Ecuador (left) This volcano in the north is 20,700 feet (6,310 m) high. Although it is near the Equator, most of its precipitation falls as snow. Winters are cool and dry; summers are warmer and wet.

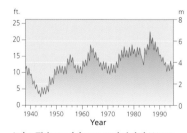

Lake Titicaca (above and right) Water levels in this lake high in the Andes, on the border betweeen Peru and Bolivia, are high during the wet season and low in the dry season. Abrupt drops in level in the 1940s and 1983–84 mark El Niño events, when very little precipitation occurred.

Alpine cold (above) The peaks of this vast mountain system remain so cold that they are snow-covered year round and have extensive glaciers. Valleys are often warmer and can be snow-free.

Peruvian Andes (right) Along the Inca Trail, in the valleys of the Peruvian Andes, temperatures are generally moderate. Heavy rain characterizes the summer wet season, but it is sunny and dry in winter.

FACT FILE

Climographs Moderate temperatures vary little, except for a clear winter minimum in Santiago. Precipitation is low, but varies: in winter it is higher in Santiago and lower in Cochabamba; in Bogotá it is higher in spring and fall.

■ Average precipitation
■ Max. temperature
■ Min. temperature

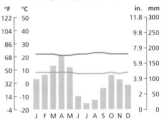

1. BOGOTÁ, COLOMBIA

°F	°C		in.	mm
122	50		11.8	300
104	40		9.8	250
86	30		7.9	200
68	20		5.9	150
50	10		3.9	100
32	0		2	50
14	-10			
-4	-20	J F M A M J J A S O N D	0	0

2. QUITO, ECUADOR

°F	°C		in.	mm
122	50		11.8	300
104	40		9.8	250
86	30		7.9	200
68	20		5.9	150
50	10		3.9	100
32	0		2	50
14	-10			
-4	-20	J F M A M J J A S O N D	0	0

3. COCHABAMBA, BOLIVIA

°F	°C		in.	mm
122	50		3.9	100
104	40			
86	30		3.1	80
68	20		2.4	60
50	10		1.6	40
32	0		0.8	20
14	-10			
-4	-20	J F M A M J J A S O N D	0	0

4. SANTIAGO, CHILE

°F	°C		in.	mm
122	50		3.9	100
104	40			
86	30		3.1	80
68	20		2.4	60
50	10		1.6	40
32	0		0.8	20
14	-10			
-4	-20	J F M A M J J A S O N D	0	0

Pampas and Subtropical Regions

Extending east of the Andes, this central and eastern area has subtropical and semi-arid climates. The semi-permanent subtropical high pressure over the area limits rainfall, although heavy rain may fall in summer. In the west, the rain-shadow effect of the Andes reduces rainfall; in the east cool, moist Atlantic breezes keep temperatures mild and precipitation moderate.

Pampas (right) Argentina's extensive, flat grasslands are ideal for raising cattle and are typical of a semi-arid climate, where precipitation is limited. Rain falls mostly in winter, while the rest of the year it is relatively dry.

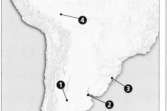
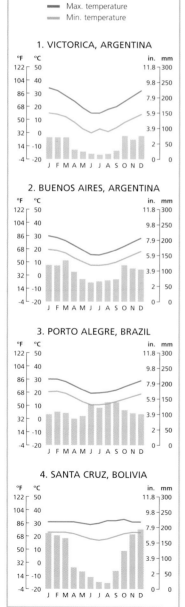

Montevideo, Uruguay (above) This port's subtropical climate is tempered by the cool, moist marine air of the Rio de la Plata estuary. Rain is evenly spread through the year; temperatures are moderate in summer and cool in winter.

Argentinian vineyards (left) Grapes thrive in the largely sunny, dry climate of the Andes foothills. In winter snow falls at higher elevations, but lower down, where temperatures are moderate, precipitation falls as rain.

Aridity (below) Located in western Argentina, in the rain shadow of the Andes, Mendoza is very dry. Droughts are common and the vegetation consists mainly of grasses, upon which local horses thrive.

Montevideo
Buenos Aires
Brazil Current
Bahia Blanca

Falkland Islands
Malvinas Current

| 40 | 46 | 53 | 61 | 68 | 75 | 82 °F |
| 2 | 4 | 8 | 12 | 16 | 20 | 24 | 28 °C |

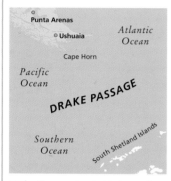

Drake Passage Located between Cape Horn and the South Shetland Islands of Antarctica, this body of water connects the southwestern Atlantic and the southeastern Pacific oceans.

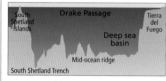

Depth profile Drake Passage varies in depth from 3 miles (5,000 m) in the South Shetland Trench to very shallow just south of Tierra del Fuego, where waves are shorter, steeper, and more hazardous for shipping.

Roaring forties These almost constant gale-force winds blow between 40°S and 50°S, producing enormous waves. Easterly winds blowing counter to the current make the waves even larger.

WEATHER WATCH

Retreating glaciers Increasing temperatures and reduced precipitation at higher altitudes are causing Patagonian glaciers to become thinner and retreat. The melting of these glaciers not only contributes to the predicted rise in sea levels but also threatens the water supplies of many people.

Patagonia

This southern part of Argentina and Chile extends down to the southern tip of South America, has both Atlantic and Pacific coasts, and includes the Falkland Islands. Its climates—arid, semi-arid, temperate, and subpolar—are as varied as its landscape, ranging from the moist Pacific coast to the dry southern plains of Argentina, and to frigidly cold Tierra del Fuego.

Dry habitat (above) Patagonia's dry east coast is a favored habitat of Magellan penguins, which nest in the small shrubs and tussock grass that grow in the sandy soil.

Temperate west (below) The cool, moist westerlies from the Pacific give Villarrica, southern Chile, a temperate climate. Lush forests flourish at lower altitudes and volcanic peaks are snow-capped.

Steppe (below) Lying in the Andes' rain shadow, the Argentinian steppe is dry. Enough rain falls to support small shrubs and grasses, which provide grazing for extensive sheep and cattle ranches.

Temperatures (below) Patagonia's mean annual temperatures are affected by latitude, altitude, and the ocean. They are highest in the north and on the coast. Inland, at higher altitudes, it is colder.

Perito Moreno Glacier, Argentina (above) This is one of many glaciers that exist in Patagonia at high altitudes, where there are cool-to-cold temperatures year round and ample snowfall.

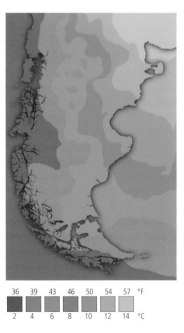

36 39 43 46 50 54 57 °F
2   4   6   8   10 12 14 °C

Precipitation (right) In the west, the westerly winds and cold Pacific Ocean make the moorland cool and wet. Higher up, where precipitation is ample but temperatures are much lower, evergreen and deciduous forests grow. In the east, steppe grassland prevails in the dry conditions.

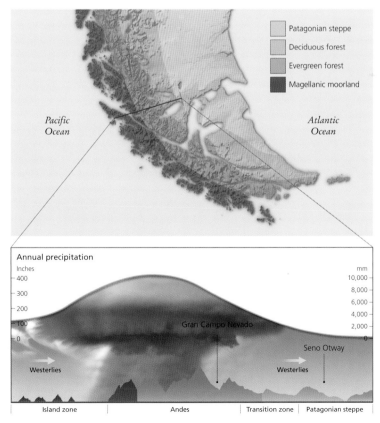

Pacific Ocean

Atlantic Ocean

Patagonian steppe

Deciduous forest

Evergreen forest

Magellanic moorland

Annual precipitation

Inches   mm
400   10,000
300   8,000
200   6,000
                4,000
100   2,000
0   0

Gran Campo Nevado

Seno Otway

Westerlies   Westerlies

Island zone | Andes | Transition zone | Patagonian steppe

FACT FILE

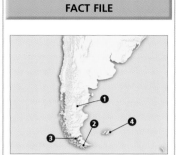

Climographs Temperatures range from moderate to cool in summer to cold in winter. All locations below show a winter minimum. Precipitation is low: Sarmiento is driest, Stanley is wettest, and Ushuaia is slightly wetter in fall.

— Average precipitation
— Max. temperature
— Min. temperature

1. SARMIENTO, ARGENTINA

°F   °C   in.   mm
122   50   3.9   100
104   40   3.1   80
86   30   2.4   60
68   20
50   10   1.6   40
32   0   0.8   20
14   -10
-4   -20   0   0
      J F M A M J J A S O N D

2. USHUAIA, ARGENTINA

°F   °C   in.   mm
122   50   3.9   100
104   40   3.1   80
86   30   2.4   60
68   20
50   10   1.6   40
32   0   0.8   20
14   -10
-4   -20   0   0
      J F M A M J J A S O N D

3. PUNTA ARENAS, CHILE

°F   °C   in.   mm
122   50   3.9   100
104   40   3.1   80
86   30   2.4   60
68   20
50   10   1.6   40
32   0   0.8   20
14   -10
-4   -20   0   0
      J F M A M J J A S O N D

4. STANLEY, FALKLAND ISLANDS

°F   °C   in.   mm
122   50   3.9   100
104   40   3.1   80
86   30   2.4   60
68   20
50   10   1.6   40
32   0   0.8   20
14   -10
-4   -20   0   0
      J F M A M J J A S O N D

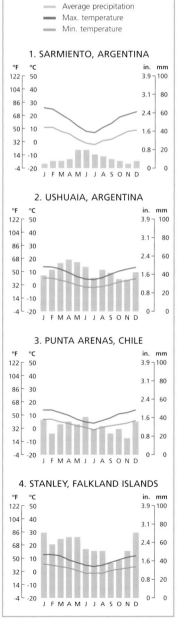

EUROPE

Extending from the Atlantic Ocean to the Ural Mountains in Russia, and from the Arctic to the Mediterranean Sea, Europe is Earth's second smallest continent. Its climates are mainly influenced by latitude, proximity to large bodies of water, and altitude. Far from the Equator, Europe has no tropical regions; its eight climate zones range from polar, where cold air masses dominate, to subtropical and Mediterranean, which are affected by the seasonal expansion and contraction of the subtropical high-pressure system. In between lie the boreal, continental, and temperate climates, and the high-altitude climates of the west–east mountain chains.

REGIONAL FACTORS

While latitude, elevation, and closeness to oceans are the major determinants of climate in Europe, as elsewhere, many regional factors are also important.

In northern Europe, cold polar air masses affect temperatures and precipitation throughout the year, but especially in winter as they move south. Frigid temperatures and the lack of sunlight make winter very long and severe. The frontal systems that form along the boundaries of the cold air masses provide the instability that favors the frequent development of gales and storms. These are responsible for much of the precipitation that occurs in the winter months.

In western Europe, the Atlantic Ocean and the midlatitude westerly winds are responsible for the milder temperate climate of that region. The warm Gulf Stream ocean current, which becomes the North Atlantic Drift as it moves northeast in the Atlantic Ocean, transports heat and moisture from the southern latitudes to areas north of the Arctic Circle. During winter this keeps the coasts there ice-free and warmer than coasts in similar latitudes in the Pacific Ocean.

The westerly winds steer numerous low-pressure systems into the coastal regions of western Europe, especially in winter, but also in the other seasons. As a result, annual precipitation is high, and there is ample rainfall in every season.

The Alps physically separate northern from southern Europe. This extensive, very high mountain range prevents the northern, cold air from penetrating south into Mediterranean Europe. The mountains also create their own climates, ranging from temperate in the foothills to the cold alpine climate of the high peaks, with their perennial snow cover.

In southern Europe, summers are long, hot, and dry, influenced by the expanding subtropical high pressure and hot, dry winds from northern Africa. Winters are cool, moderated by the Mediterranean Sea. They are also rainy as a result of the frequent storms that move with the general flow of the midlatitude westerlies.

Eastern Europe lies farthest from any marine influence, and so much of it experiences climates that characterize inland continental regions. Winters, influenced by cold polar and subpolar air masses, are long and severely cold; summers are short and cool. Precipitation is plentiful all year, but highest in summer months. It is the result of unstable air that occurs along the frontal zones when the warmer tropical air masses from the south meet the cold polar and subpolar air masses from the north.

Russian tundra (above) In the subpolar climate of northern Russia, only the grasses and mosses of the tundra can survive the long, severe winters and short, cool summers, when most of the rainfall occurs.

CLIMATE ZONES OF EUROPE

Semi-arid Low precipitation; smaller diurnal temperature variation than arid climates

Mediterranean Hot, dry summers; mild, moist winters, occasionally below freezing

Subtropical Warm and moist; hot summers and cooler, drier winters

Temperate Four distinct seasons; precipitation year round; warm summers and cold winters

Continental Cool and moist; warm summers and severe winters

Boreal Cool summers and very severe winters with snow; evergreen forest vegetation

Subpolar Very cold throughout the year; no true summer; tundra vegetation

Mountain Colder than low-level locations found at the same latitude

Reykjavik

Oslo · Helsinki
Stockholm · Tallinn
Dublin · Riga · Moscow
Copenhagen · Vilnius
London · Minsk
Amsterdam · Berlin · Warsaw
Brussels · Kiev
Paris · Prague
Bern · Vienna · Bratislava · Chişinău
Budapest
Ljubljana · Zagreb · Bucharest
Lisbon · Madrid · Belgrade
Sarajevo · Sofia
Rome · Tirana · Skopje · Istanbul
Athens
Valletta

Lindos, Greece (right) In southern Europe's Mediterranean climate, summers are long, hot, and dry. The low annual rainfall, mostly in winter, can support only short grasses, small trees, and shrubs.

Storms (below) Many storms move from North America's east coast northward and eastward into Europe. Their track shifts north or south with pressure differences, called the North Atlantic Oscillation (NAO). Areas where the worst storms cluster are shown in darker blue and brown.

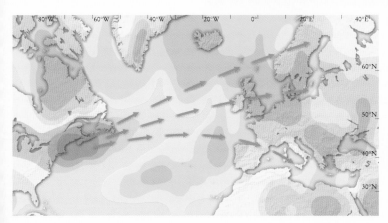

Continental winter (above) Blizzards and snowstorms are common in Moscow's very long, cold winter. Snowfalls begin as early as October and do not come to an end until late spring.

Ice-free coast (right) Although parts of the Norwegian coast lie north of the Arctic Circle, they remain free of ice in the winter months, thanks to the warm waters of the Gulf Stream/North Atlantic Drift.

European Arctic

This region includes far northern Scandinavia and Russia west of the Ural Mountains, as well as Iceland. Dominated by the polar air masses, winters are long and frigidly cold. At least three months are dark in winter, when the North Pole is tilted away from the Sun. Summers are short and relatively cool, with a minimum of three months of constant daylight.

Lofoten Islands (right) Although they are north of the Arctic Circle, these islands experience a temperate summer climate, thanks to the warm Norwegian Current. Winters are quite mild and summers cool.

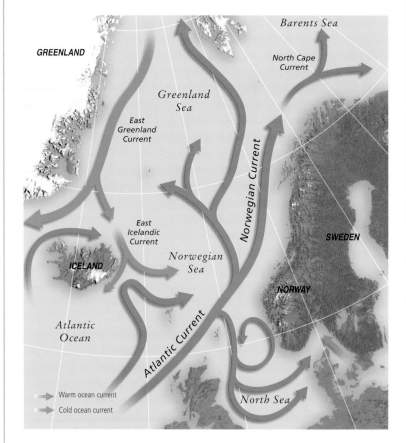

Map labels: GREENLAND · Barents Sea · North Cape Current · Greenland Sea · East Greenland Current · East Icelandic Current · Norwegian Current · SWEDEN · ICELAND · Norwegian Sea · NORWAY · Atlantic Ocean · Atlantic Current · North Sea

→ Warm ocean current
→ Cold ocean current

Spitzbergen Islands, Norway (above) The far northern Arctic, beyond the influence of warm ocean currents, is extremely cold year round, with sea ice and permanent glaciers on land.

Heat transfer (left) The warm Norwegian Current, fed by the warm southern Gulf Stream, keeps Scandinavia's northern coastal regions warmer than they would otherwise be at such high latitudes.

Midnight sun (above) For at least three months in summer the sun never goes down in regions north of the Arctic Circle. It can still be seen at midnight, as pictured here at North Cape, Norway.

Iceland (right) When frontal low-pressure systems develop into storms and gales over the North Atlantic in winter, the exposed coasts of Iceland are pounded by high waves driven by the strong winds.

Ice-free port (below) The warm ocean currents reach as far north as Murmansk, Russia, the largest city north of the Arctic Circle, ensuring that its harbor never freezes in winter.

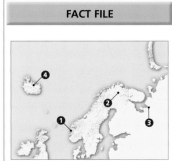

FACT FILE

Climographs All have cold winters and mild summers; Archangel has the greatest temperature range. Akureyri has the coldest summers and the least precipitation. All have year-round precipitation, tending to be higher in fall.

- Average precipitation
- Max. temperature
- Min. temperature

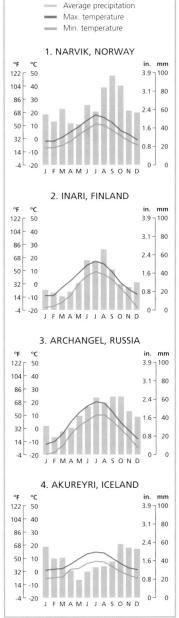

1. NARVIK, NORWAY

2. INARI, FINLAND

3. ARCHANGEL, RUSSIA

4. AKUREYRI, ICELAND

Norway spruce This conifer is well adapted to boreal winters. It is able to maintain high rates of photosynthesis in cold temperatures and survive below-freezing weather for extensive periods without suffering any damage.

Shape The spruce's conical shape allows it to shed snow very easily, so that its branches do not break under the weight of the snow.

Leaves The cells of the needle-like leaves have large spaces between them. This prevents damage during freezing and thawing.

Nutrients Evergreen leaves begin photosynthesizing right after the spring thaw, instead of waiting for new summer growth to produce nutrients.

Northern Forests

Vast stands of evergreen trees flourish in the boreal climate of northern Europe between the Arctic Circle and 60°N. During its severely cold winters, storms from the North Atlantic bring heavy snow. Summers are mild, but changeable as cold and warm fronts move through. This instability is expected to increase as temperatures continue to warm.

Snow cover (below) While coasts remain relatively warm and snow-free, winters in the interior are long, with frequent snowfalls and little melt. Thick snow cover lasts for up to six months in winter, as seen here in this satellite view of Scandinavia.

Severe winters (left) When moist marine air meets colder continental air, a frontal boundary forms. This contributes to the heavy snowfalls and blizzards that are common in this region in winter.

Mild summers (above) Helsinki, Finland, enjoys short but warm summers with up to 19 hours of daylight. Precipitation is low, but so is evaporation, so summer is more humid than might be expected.

Finnish marshland (above) In April, as temperatures rise and stay above freezing, spring comes late to Europe's northern forests. The frozen ground thaws, but poor drainage leaves the soil waterlogged, and vast areas become swampy.

Forested landscape (left and above) Extensive forests of spruce and pine cover this subarctic region. In winter, thick snow cover lasts many months, as in the Swedish forest shown on the left. Repovesi National Park, Finland, is pictured above in summer, when the conifers produce new growth in the warmer temperatures, higher rainfall, and longer hours of sunlight.

FACT FILE

Climographs Winters are long and cold; summers are short and mild. Precipitation is constant, but usually higher in summer than in winter. In Härnösand, higher precipitation lasts from late summer through fall and early winter.

— Average precipitation
— Max. temperature
— Min. temperature

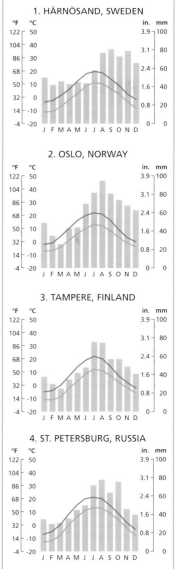

1. HÄRNÖSAND, SWEDEN

2. OSLO, NORWAY

3. TAMPERE, FINLAND

4. ST. PETERSBURG, RUSSIA

Climographs In all these cities, winters are cool and summers are mild to moderate. Utrecht has the widest temperature range. Moderate year-round precipitation varies little, but more tends to occur in fall and early winter.

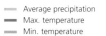
- Average precipitation
- Max. temperature
- Min. temperature

1. LONDON, U.K.

2. DUBLIN, IRELAND

3. UTRECHT, NETHERLANDS

4. HAMBURG, GERMANY

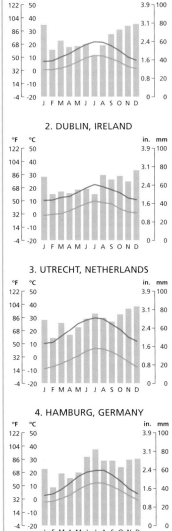

Temperate Western Europe

Lying in the midlatitudes, this region extends from the British Isles to eastern Germany and from parts of Scandinavia to northern Spain. The Atlantic Ocean moderates coastal temperatures and is the source of moisture in the many low-pressure systems that are carried by the westerlies and make the weather very changeable. Rain falls almost daily.

Heat waves (right) In July 2006 temperatures in Paris reached 97°F (36°C). While summers are generally moderate, especially in coastal areas, several consecutive days of much higher than normal temperatures can occur inland.

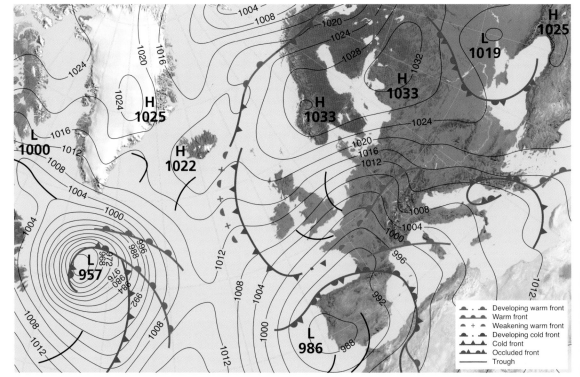

- Developing warm front
- Warm front
- Weakening warm front
- Developing cold front
- Cold front
- Occluded front
- Trough

Changeable weather (above) This typical winter synoptic chart shows eastward-moving low-pressure systems with embedded cold, warm, and occluded fronts. As a cold front passes over an area, temperatures drop, but the opposite happens with a warm front. An occluded front occurs when a warm front is overtaken by a cold front. Precipitation may accompany all three frontal types.

Rural fall (left) This is the most colorful of the four distinct seasons in temperate regions such as the U.K. Summers are moderate and winters are seldom below freezing for extended periods.

Rare snow (right) Winter precipitation is usually rain. When snowstorms do occur, they can be quite severe and make driving conditions hazardous, as on this major highway in southern England.

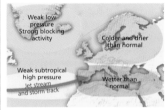

Mediterranean Europe

This region covers southern Spain, France, and Italy, as well as coastal Slovenia, Croatia, and Bosnia-Herzegovina, together with Greece and the islands of the Mediterranean and Aegean seas. Influenced by the expanding and intensifying subtropical high pressure in the south, summers are long, hot, and dry. Winters are mild and moist, and cyclones are frequent.

Crete (below) High temperatures, clear blue skies, and very little precipitation are typical of summer in this Greek island and other parts of the southern Mediterranean, making this region an ideal destination for sunseeking holidaymakers.

Epirus In this mountainous region of northern Greece, the summer heat and humidity are tempered by the meltemi winds, but in the winter the mountains are snow-covered.

Olives Native to the Mediterranean, olive trees prefer its sunny, hot, dry summers and moderate winters. Extensive groves like these in Portugal are cultivated in the region.

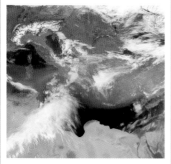

African dust Hot, dry winds from northern Africa often carry dust across the southern Mediterranean, as shown in this satellite image, and farther north into Austria and Hungary.

WEATHER WATCH

Warmer and drier As temperatures rise, more frequent droughts and forest fires are predicted for the European Mediterranean region. These changes will affect the region's culture and lifestyle. As temperatures rise and rainfall decreases, viticulture and wine production may decline.

Tuscany, Italy (above) While the southern Mediterranean is quite dry in summer, rain is more frequent farther north, where easterly winds that blow inland off the sea bring moisture and cooler temperatures.

Kalamata, Greece (left) In the southern Mediterranean, short thunderstorms may occur in the late afternoon, when the warm, moist air becomes convectively unstable. They usually dissipate by evening.

Lavender fields (right) The south of France has very sunny, warm, relatively dry summers, while winters are mild and generally frost-free. This climate is perfect for the cultivation of lavender, a perennial native of the Mediterranean region.

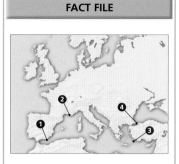

Cyclone paths (right) Storms from the far west tend to move northeast in December (**1**), shifting to an easterly path in February (**2**). Storms in the Aegean occur more to the north (**3** and **4**) in December and shift farther south (**5**) in February.

Local winds (below) While the prevailing winds are westerly, the region has many local winds. These develop chiefly in response to the varied topography and land–sea contrasts in temperature.

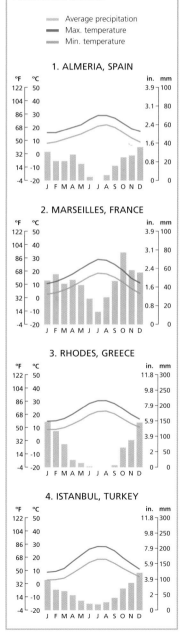

Climographs These cities have hot, dry summers and mild, wet winters. Almeria and Rhodes are the driest in summer, with no rainfall in some months. In winter, Marseilles and Istanbul are much wetter than Almeria and Rhodes.

Average precipitation
Max. temperature
Min. temperature

1. ALMERIA, SPAIN

2. MARSEILLES, FRANCE

3. RHODES, GREECE

4. ISTANBUL, TURKEY

Foehn

Bora

Mistral

Tramontana

Bora

Atlantic Ocean

Warm winds
Cold winds
Other winds

Marin

Zephyr

Etesian

Libeccio

Levantades

Skysweeper

Gregale

Ponente

Meltemi

Datoo

Imbat

Mbatis

Levanter

Sirocco

Leveche

Xlokk

Mediterranean Sea

Gharbi

Ghibli

Khamsin

FACT FILE

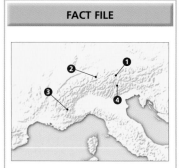

Climographs Winters are cool but coldest at Mt. Säntis summit, where summers are also coolest. Innsbruck and Bolzano are warmest in summer. Precipitation occurs all year; it is higher in summer, except in Embrun.

Average precipitation
Max. temperature
Min. temperature

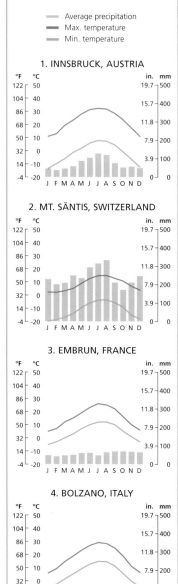

1. INNSBRUCK, AUSTRIA

2. MT. SÄNTIS, SWITZERLAND

3. EMBRUN, FRANCE

4. BOLZANO, ITALY

European Alps

These mountains stretch in an arc from France across Switzerland and Austria to northern Italy. Not only do variations in elevation and topography create unique local microclimates, but climate is also influenced by air masses that move into the Alps from the west, north, east, or south at different times of year, producing different kinds of weather.

Alpine summer (below) In much of the French Alps, pictured here, temperatures rise above freezing in summer, unlike in the higher Swiss Alps. As a result, most of the snow deposited by winter snowstorms melts, exposing the jagged rocky peaks.

Mild microclimate (above) Located in the lee of Mount Pilatus, the Swiss city of Lucerne enjoys a milder climate than other parts of the Alps, thanks to the effect of the dry foehn wind, which warms as it descends the leeward slopes.

Trentino–Alto Adige, Italy (below) On the southern side of the Alps, this small region has a warm, sunny climate that is well suited to the cultivation of a large variety of grapes at a range of altitudes.

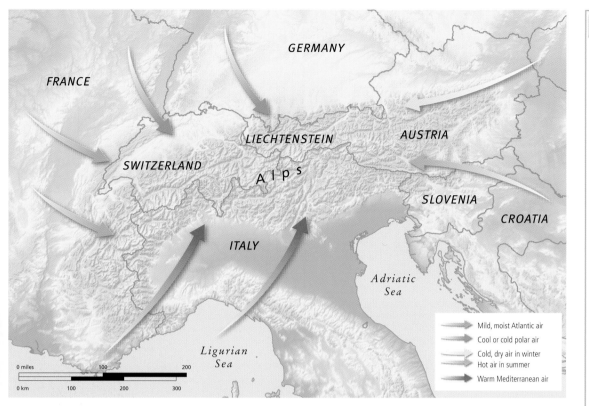

Air masses (above) Northern polar air lowers temperatures. Mild, moist air from the Atlantic brings heavy precipitation in winter. Air from the east is cold in winter and hot in summer, while air from the south brings warmer weather.

Valley fog (left) In winter, fog often shrouds valleys but the mountaintops stay sunny, as in this view of the Bavarian Alps. Cool air sinks to the valley bottom, and the warm air passing above it traps the fog.

Grindelwald, Switzerland (below) The highest peaks receive precipitation as snow all year and they remain snow-capped in summer. In the warmer lower valleys, ample rainfall supports pasture.

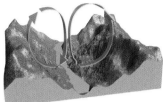

Eastern Europe

Most of this inland region experiences a continental climate. Far from marine influences, winters are extremely cold due to the southward movement of cold polar air. Summers are short and mild, and precipitation is constant, but not heavy, throughout the year. Extreme weather, such as droughts and heat waves in summer and heavy winter snowfalls, can occur.

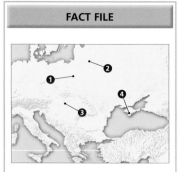

Climographs Vilnius has the coldest winters, while Simferopol has the warmest summers. Precipitation occurs all year, and is highest in the summer months. Vilnius is the wettest of these cities and Simferopol is the driest.

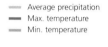
- Average precipitation
- Max. temperature
- Min. temperature

1. WARSAW, POLAND

2. VILNIUS, LITHUANIA

3. BUDAPEST, HUNGARY

4. SIMFEROPOL, UKRAINE

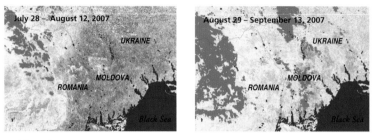
July 28 – August 12, 2007

August 29 – September 13, 2007

Moldova drought Some parts of eastern Europe are prone to drought. In summer 2007, the drought in Moldova was the worst of nine dry periods since 1990. Brown areas **(above left)** represent devastated vegetation when the drought was at its worst. Rains in early September had begun to restore some of the vegetation (light green, **above right**), but it was too late to save crops, and yields were dramatically reduced.

Mild summers (below) In Ukraine, year-round precipitation, with a slight increase in summer, and mild summer temperatures provide ideal conditions for the cultivation of rapeseed, which is used to produce biodiesel.

Thunderstorms (left) In regions with continental climates, the atmosphere warms up and becomes unstable in summer. The afternoon thunderstorms that result can produce dangerous lightning, such as seen here in Kiev, Ukraine.

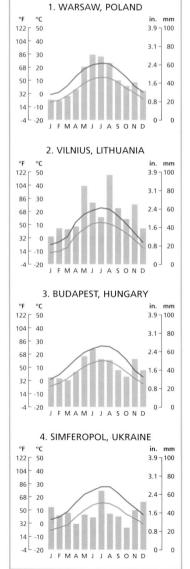

Frozen river (right) In eastern Europe's long and severely cold winters, rivers like the Moskva in Moscow freeze over from as early as November and do not thaw until around the end of March.

Tatra Mountains, Slovakia (above)
When polar and arctic air masses move south in winter, there are heavy snowfalls along the frontal zones that form, providing good conditions for skiing.

Crimea (right) Unlike the rest of eastern Europe, the southern Crimea has a dry climate. Because this region is protected from the cold north winds, it is cool, rather than cold, in winter.

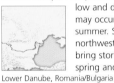

ASIA

Stretching from northern Arctic latitudes to the islands of the Indian Ocean just south of the Equator, this vast region contains most of the world's climates. In the north, the Siberian high dominates in winter; the south is strongly influenced by monsoons and the shifting Intertropical Convergence Zone (ITCZ). Tropical cyclones augment rainfall in summer, while extensive mountain ranges such as the Himalayas moderate the distribution of precipitation. The subtropical high-pressure system near 30°N is largely responsible for the existence of deserts around that latitude.

ITCZ (right) This belt of strong convection and heavy rainfall occurs where the trade winds converge. In summer it shifts north of the Equator into southern Asia, making the summer monsoonal rains even heavier.

CLIMATIC INFLUENCES

Continentality—the effect of large landmasses on temperature—is an important determinant of climate in Asia, along with latitude and altitude. Temperatures range from extremely hot in southeastern equatorial regions and the southwestern deserts to frigidly cold in high subpolar latitudes. Intense solar radiation keeps temperatures high at low latitudes, but permanent snow and ice also exist wherever there are very high mountain ranges, such as the Himalayas. In the subtropics, the clear skies and dry conditions that accompany high atmospheric pressure allow maximum amounts of solar radiation to be absorbed by Earth's surface; temperatures soar by day, but drop sharply at night in the absence of the sun. In high subpolar latitudes, the Siberian high keeps skies clear in winter, but temperatures remain very low because of the very short winter days and the cold winds associated with the high pressure.

Like temperature, precipitation also decreases poleward, but there is greater regional variation. In equatorial areas, the effects of the summer monsoonal rains, the ITCZ, tropical cyclones, and daily thunderstorms contribute to very heavy rainfall. It is continentality that causes the seasonal shift in wind direction known as the monsoon. In summer, southwesterly winds from the Indian Ocean converge on southern Asia, bringing a period of heavy, prolonged rainfall that can last from April to October. In winter, the reverse occurs: cold, dry northeasterly winds blow from the continent's interior, bringing dry weather to coastal areas. As a result, regions at latitudes near 15°N experience a tropical seasonal climate, but in the deeper tropics closer to the Equator it is always hot and humid. Farther north and

farther inland, at a distance from the ocean, precipitation is low and seasonal. Some moisture may be carried on the westerly winds, producing summer thunderstorms with heavy rain.

While the subtropical high-pressure system is the main cause of the aridity of the southwestern deserts, vast dry regions in central Asia like the Gobi desert owe their existence to the Himalayas. This extensive mountain system prevents central Asian regions from receiving any summer monsoonal precipitation. In winter they are subject to cold, dry, northeasterly winds that originate in the Siberian high-pressure system, so annual precipitation is very low.

CLIMATE ZONES OF ASIA

■	**Tropical wet** Hot and wet throughout the year; short or no dry season		**Subtropical** Warm and moist; hot summers and cooler, drier winters
■	**Tropical seasonal** Hot throughout the year; distinct wet and dry seasons		**Continental** Cool and moist; warm summers and severe winters
■	**Arid** Little or no precipitation year round; hot days and cold nights		**Boreal** Cool summers and very severe winters with snow; evergreen forest vegetation
■	**Semi-arid** Low precipitation; smaller diurnal temperature variation than arid climates		**Subpolar** Very cold throughout the year; no true summer; tundra vegetation
■	**Mediterranean** Hot, dry summers; mild, moist winters, occasionally below freezing		**Mountain** Colder than low-level locations found at the same latitude

Summer monsoon (left) Clouds looming over Agartala, in northeast India, announce the arrival of the summer monsoon, when moisture-laden winds bring heavy rains that often cause severe flooding.

Precipitation (above) Southern coastal regions and the southern slopes of the Himalayas have the highest precipitation. Farther inland it is drier as a result of the rain-shadow effect of the Himalayas and the distance from the ocean.

Himalayas (above) High in the Tibetan Himalayas, temperatures are always below freezing and peaks are permanently snow-covered. On the lower slopes, it is warmer and precipitation falls as rain.

Gobi desert, Mongolia (left) This high-altitude desert receives less than 8 inches (200 mm) of precipitation per year, so vegetation is sparse. Temperatures drop as low as –40°F (–40°C) in winter and soar to 104°F (40°C) in summer.

Siberian Arctic

The extreme cold of this region is attributable chiefly to the influence of the Siberian high-pressure system. Winters are dark, long, and constantly below freezing, with temperatures sometimes as low as −40°F (−40°C). In winter, precipitation is slight but frequent blizzards create white-out conditions. Summers are brief, warm, and wetter than winters.

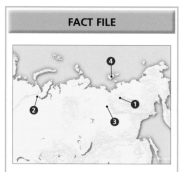

FACT FILE

Climographs Verkhoyansk, strongly influenced by the Siberian high, is coldest in winter and warmest in summer. At 75°N, Ostrov Kotel'nyy is below freezing all year. All have little precipitation, with most in summer. Salekhard is wettest.

▬▬▬ Average precipitation
▬▬▬ Average temperature

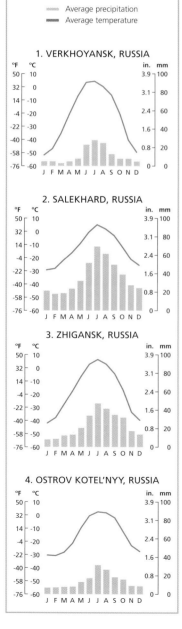

1. VERKHOYANSK, RUSSIA

2. SALEKHARD, RUSSIA

3. ZHIGANSK, RUSSIA

4. OSTROV KOTEL'NYY, RUSSIA

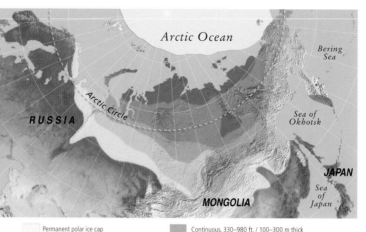

Permanent polar ice cap

Continuous, more than 1,640 ft. / 500 m thick

Continuous, 980–1,640 ft. / 300–500 m thick

Continuous, 330–980 ft. / 100–300 m thick

Discontinuous, maximum thickness 330 ft. / 100 m

Sporadic, maximum thickness 80 ft. / 25 m

Permafrost zones (above) Over much of the Siberian Arctic, as well as farther south, the soil is permanently frozen all year. Summers are too short to thaw the soil, except in the southern zones.

Spring floods (right) In this satellite view, ice shards in the Laptev Sea and the greening land mark the return of warmer weather in spring. Melting coastal ice can block waterways, such as the Lena River shown here, leading to flooding.

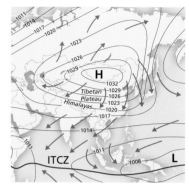

Siberian high (above) Caused by continentality, this anticyclone develops over Eurasia in late summer and persists through winter. Dry winds move out from its center, carrying very cold air far south.

Subpolar summer (right) In the short, warm Arctic summer, water from melting snow and light rainfall stays near the surface in this poorly drained region, providing enough moisture to support the tundra's low-growing vegetation.

Frozen landscape (left) In winter, snow and ice form a surface thick and stable enough to travel on relatively easily, and reindeer races are popular events among the indigenous Nenets people.

Polar winter (right) In the frigidly cold northern Siberian Arctic winter, frozen sea ice provides ideal conditions for polar bears stalking seals. Winter snowfalls increase the volume of glaciers (background).

Northern Asia

This region reaches from the Arctic Circle to the borders of Mongolia, and from eastern Russia across northeastern China to Japan. Its climates, mainly boreal and continental, vary little from west to east, but it is milder in the south than in the north, so that the boreal forest, or taiga, that covers most of the region changes to temperate forest in the south.

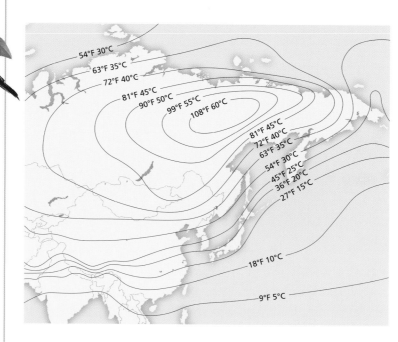

Temperature range (above) Maximum (summer) and minimum (winter) temperatures can differ by as much as 108°F (60°C) in northern areas, driven by the very low winter temperatures. The range decreases to the south.

Taiga (below) This dense belt of coniferous forest extends across northern Asia. The evergreen trees are adapted to the short, cool summer growing season and to the cold, dry conditions that prevail throughout the boreal winter.

Kamchatka (below) In northeast Russia, these erupting volcanoes melt glaciers, while cool temperatures aid the advance of glacial ice. The east and south receive heavy precipitation, mostly in summer.

ОЙМЯКОН – „ПОЛЮС ХОЛОДА"

Vladivostok (right) On the southeastern coast, the climate is milder. The cold, dry air from inland areas keeps winters cold, with clear skies. Summers can be rainy, but warm temperatures can continue into the autumn, when it is dry and sunny.

Anshan, China (right) The Siberian high-pressure system brings severely cold winters to northeastern China. While precipitation is usually low, occasional heavy snowstorms can be hazardous.

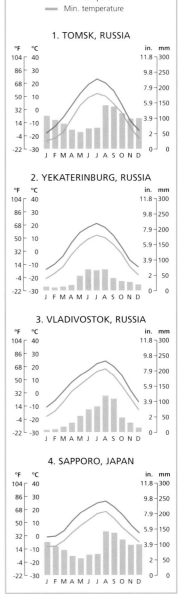

FACT FILE

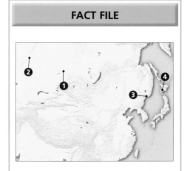

Climographs These cities have long, cool-to-very-cold winters and brief, mild summers. Sapporo is warmest. Precipitation occurs all year but is usually highest in summer. Tomsk and Sapporo are wettest, and Yekaterinburg is driest.

Average precipitation
Max. temperature
Min. temperature

1. TOMSK, RUSSIA

2. YEKATERINBURG, RUSSIA

3. VLADIVOSTOK, RUSSIA

4. SAPPORO, JAPAN

Cold Deserts and Steppe

These high-altitude arid and semi-arid regions extend from the eastern borders of Iran to northern China. The low precipitation they receive can be attributed to their great distance from oceans and/or the rain-shadow effect of mountains. The little winter snow that falls stays on the ground in the low temperatures. Summers are short and mild, with some heat waves.

FACT FILE

Climographs Winters are long, with temperatures ranging from cold in Dalandzadgad to cool in Ashgabat and Herat. Summers are hot: Ashgabat is hottest. All have low precipitation: Kashi has the least, Herat the most, peaking in winter.

Average precipitation
Max. temperature
Min. temperature

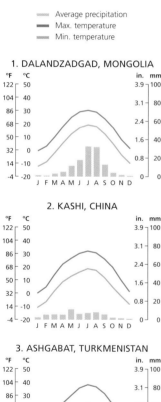

1. DALANDZADGAD, MONGOLIA

2. KASHI, CHINA

3. ASHGABAT, TURKMENISTAN

4. HERAT, AFGHANISTAN

Gobi desert (above) Stretching across Mongolia and China, the Gobi is intensely cold in winter: its northerly location on a high plateau leaves it exposed to winds from the north. While winters are dry, occasional blizzards may carry enough snow to leave a thin layer.

Rain-shadow effect (right) The Gobi and Takla Makan deserts lie in the lee of the Himalayas. Winds blowing northward from the Indian Ocean deposit their moisture on the southern slopes. As air descends the northern slopes, it is dry.

Takla Makan, China (left) The sand dunes of this vast desert are very high, some reaching over 980 feet (300 m). Constant strong winds move them an average of 490 feet (150 m) each year.

Mongolian steppe (below) This vast semi-arid region experiences long, cold winters and short, warm summers. Low summer rainfall supports open grassland and, with the aid of irrigation, agriculture.

FACT FILE

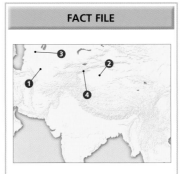

1. Karakum Desert This desert is hot and dry in summer, while cool average winter temperatures remain above freezing and there is little snow. Sparse rainfall occurs in winter and spring, and is supplemented by irrigation.
Karakum Desert, Turkmenistan

2. Takla Makan One of the world's largest sandy deserts, this region has very little precipitation: some eastern areas do not receive any. Summers are hot, but winters are quite cold, under the influence of cold air masses from Siberia.
Takla Makan Desert, China

3. Kazakh steppe This extensive semi-arid grassland region is severely cold and dry in winter. Precipitation occurs in spring and summer. The rains rejuvenate the frost-tolerant grasses, transforming the dry winter landscape into vast, green meadows.
Kazakh steppe, Kazakhstan and Russia

4. Tian Shan montane steppe This region occupies the lower slopes of an extensive mountain system in northwestern China. Lying between the drier desert valleys and the wetter high elevations, it receives enough rain to support grasses.
Tian Shan montane steppe, China

WEATHER WATCH

Longer growing season Mean annual temperatures have risen in these regions, mostly during the winter and spring, although warming has also occurred in other seasons. With the warmer weather, the growing season starts earlier and is longer. Agricultural practices will have to adapt to this change.

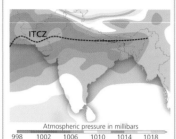

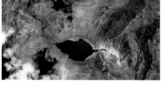

Himalayas

Altitude and the monsoons, along with latitude, are the main influences here. Climates vary with elevation from subtropical in the southern foothills to alpine on the highest peaks. Heavy rainfall on southern slopes swells the summer monsoon rains, but during the winter monsoon the Himalayas block cold air from the north, moderating India's climate.

Effects of altitude (below) The high elevation of the Himalayas forces moist air from the Indian Ocean to rise: very heavy rain falls on the lower southern slopes, decreasing with height. In the lee of the Himalayas, the Plateau of Tibet is dry.

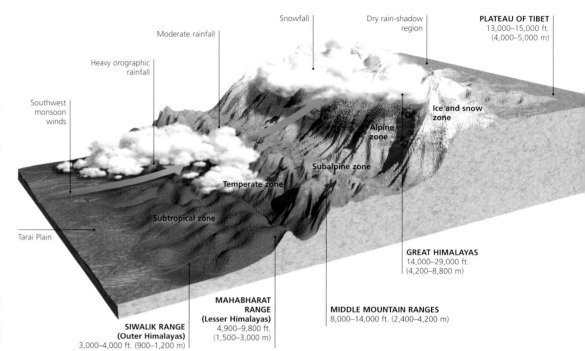

Snowfall

Moderate rainfall

Dry rain-shadow region

PLATEAU OF TIBET
13,000–15,000 ft.
(4,000–5,000 m)

Heavy orographic rainfall

Ice and snow zone

Southwest monsoon winds

Alpine zone

Subalpine zone

Temperate zone

Subtropical zone

Tarai Plain

GREAT HIMALAYAS
14,000–29,000 ft.
(4,200–8,800 m)

MAHABHARAT RANGE
(Lesser Himalayas)
4,900–9,800 ft.
(1,500–3,000 m)

MIDDLE MOUNTAIN RANGES
8,000–14,000 ft. (2,400–4,200 m)

SIWALIK RANGE
(Outer Himalayas)
3,000–4,000 ft. (900–1,200 m)

Great Himalayas In this southward view, the dry Plateau of Tibet contrasts with the highest Himalayas, which are perpetually covered in snow and ice in the very cold air of altitudes above 23,600 feet (7,200 m).

Mountain hazards (above) Even the most experienced mountaineers can find climbing above the snowline (16,000 ft./4,880 m) hazardous, with permanent snow and glaciers to be negotiated and sudden snowstorms in winter or summer.

Warmer valleys (right) In the lower valleys, the climate is warmer than in the higher Himalayas. Precipitation occurs as rain, supporting alpine meadows and dense stands of trees such as those shown here around the town of Punakha, Bhutan.

Avalanches (above) After heavy snowfalls, unstable snow on the steep slopes of the highest Himalayas can move downhill rapidly and unexpectedly. With global warming, avalanches are predicted to become more frequent in this region.

FACT FILE

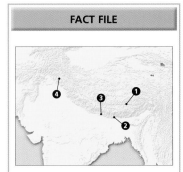

Climographs Winters are mild and summers cool to hot. Kathmandu and Srinagar are warmest. Winters are dry and summers are wet, except in Srinagar, which has more rain in winter–spring. Srinagar is driest and Darjiling wettest.

```
----  Average precipitation
----  Max. temperature
----  Min. temperature
```

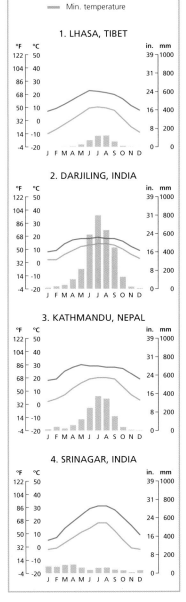

1. LHASA, TIBET

°F	°C		in.	mm
122	50		39	1000
104	40		31	800
86	30		24	600
68	20		16	400
50	10		8	200
32	0		0	0
14	-10			
-4	-20			

J F M A M J J A S O N D

2. DARJILING, INDIA

°F	°C		in.	mm
122	50		39	1000
104	40		31	800
86	30		24	600
68	20		16	400
50	10		8	200
32	0		0	0
14	-10			
-4	-20			

J F M A M J J A S O N D

3. KATHMANDU, NEPAL

°F	°C		in.	mm
122	50		39	1000
104	40		31	800
86	30		24	600
68	20		16	400
50	10		8	200
32	0		0	0
14	-10			
-4	-20			

J F M A M J J A S O N D

4. SRINAGAR, INDIA

°F	°C		in.	mm
122	50		39	1000
104	40		31	800
86	30		24	600
68	20		16	400
50	10		8	200
32	0		0	0
14	-10			
-4	-20			

J F M A M J J A S O N D

Southwestern Hot Deserts

Extensive hot deserts occupy most of the non-mountainous parts of the Middle East. Summers inland are hot and dry, with cooler winters, but coastal areas are cooler and more humid. By day, clear skies produced by the subtropical high pressure allow intense solar radiation and very high temperatures. At night, it becomes cooler as heat is lost to the atmosphere.

Aqaba, Jordan (below) Along the Red Sea coast, the strong marine influence produces a cooler, more moist climate than that of inland areas. Constant breezes also moderate the temperature. Low rainfall occurs in the cooler part of the year.

FACT FILE

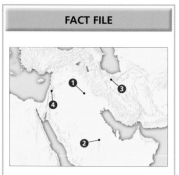

Climographs These desert cities all have mild winters and extremely hot summers, with very little rain. Negligible amounts fall in summer, but there is a slight increase in winter in all cities except Riyadh, which has a small increase in spring.

Average precipitation
Max. temperature
Min. temperature

1. BAGHDAD, IRAQ

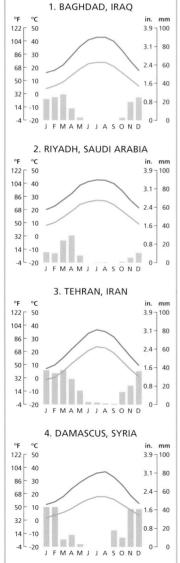

2. RIYADH, SAUDI ARABIA

3. TEHRAN, IRAN

4. DAMASCUS, SYRIA

Middle Eastern deserts (left) Summer temperatures routinely reach 113°F (45°C), by day. It is cooler at night and in winter, but it is still quite warm. Low rainfall often occurs as downpours in winter and spring.

Palmyra, Syria (below) In the Syrian Desert, winds sweep the ground clear of sand in some areas, leaving a surface of mostly flat, loose stones, called "reg," as seen around the ruins of this ancient city.

Negev desert, Israel (above) Here winds shape the barren, rocky landscape to create a desert type known as "hamada." In rare but torrential storms, normally dry valleys, or "wadis," may fill with water.

Rub' al-Khali, Arabian Peninsula (below) In this aerial view of the world's largest sand sea, or "erg," the rippled pattern is formed by longitudinal dunes shaped by the strong, constant wind.

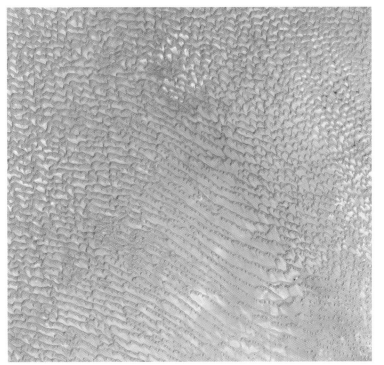

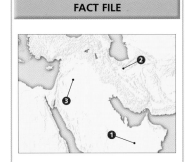

SANDSTORMS

Arid landscapes that have little vegetation to hold the soil and constant strong winds are prone to sandstorms. They occur frequently in the Arabian Desert, where there is an ample supply of sand. Whole dunes as high as 5,250 feet (1,600 m) may be transported as moving walls of sand. Visibility can drop to zero, and movement under these conditions is limited.

Monsoonal Subcontinent

Climates in this region vary from hot and wet in southern India to cooler and drier in the north. High temperatures are due to the intense solar radiation typical of low latitudes, while the Asian monsoon produces marked wet and dry seasons. Rainfall in the wet season is augmented by the Intertropical Convergence Zone (ITCZ) and tropical cyclones.

Hampi, Deccan Plateau (above) This mostly semi-arid central region of southern India receives its rain in the summer monsoon season, and is very hot and dry the rest of the year.

Bangladesh downpour (left) From April to September the summer monsoon winds bring heavy rain. Coming from the south, the winds pass over extensive warm ocean before depositing their moisture on land.

Thar Desert (below) In this northwestern desert, temperatures reach 106°F (41°C) in summer, dropping to around 82°F (28°C) in winter. Low rainfall varies annually, but occurs mainly in the monsoon season.

Monsoon onset (right) The rate at which the ITCZ shifts north influences the timing of the summer monsoon, as much of the rainfall is directly associated with the moisture-laden winds that converge on the ITCZ trough. The southern areas are the first to experience the rains, and the northwestern regions the last.

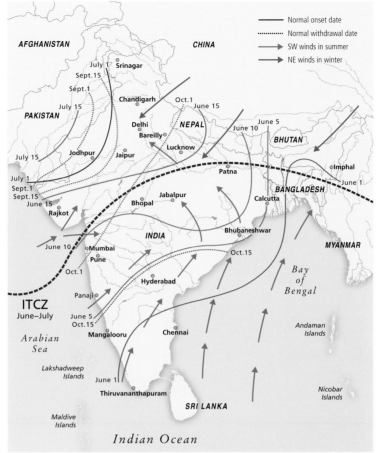

Normal onset date
Normal withdrawal date
SW winds in summer
NE winds in winter

AFGHANISTAN
CHINA
July 1 Srinagar
Sept.15
Sept.1
Chandigarh
Oct.1
June 15
PAKISTAN
July 15
Delhi
NEPAL
June 5
Bareilly
Lucknow
June 10
BHUTAN
July 15
Jodhpur
Jaipur
Patna
Imphal
July 1
BANGLADESH
June 1
Sept.1
Jabalpur
Sept.15
Bhopal
Calcutta
June 15
Rajkot
INDIA
Bhubaneshwar
MYANMAR
June 10
Mumbai
Pune
Oct.15
Oct.1
Bay
of
Bengal
ITCZ
Hyderabad
June–July
Andaman
Islands
Arabian
Sea
June 5
Oct.15
Panaji
Lakshadweep
Islands
Mangalooru
Chennai
Nicobar
Islands
June 1
Thiruvananthapuram
SRI LANKA
Maldive
Islands
Indian Ocean

Rice crops (above) Much of India's agriculture depends on the summer monsoon rains. If they arrive late, or not at all (during El Niño events), rice crops fail.

Summer and winter monsoons (right) Land heats up and cools down faster than water. In summer, air moves from the ocean to the land, where it rises, water vapor condenses, and heavy rain results. In winter, air moves in the opposite direction, leaving the land dry.

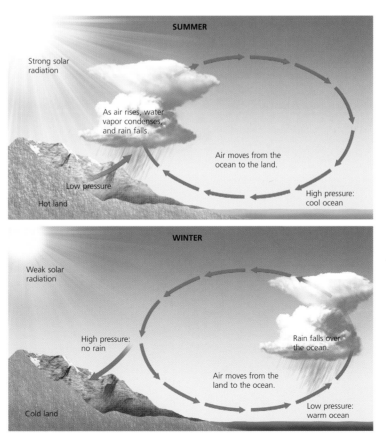

SUMMER
Strong solar radiation
As air rises, water vapor condenses, and rain falls.
Air moves from the ocean to the land.
Low pressure
Hot land
High pressure: cool ocean

WINTER
Weak solar radiation
High pressure: no rain
Rain falls over the ocean.
Air moves from the land to the ocean.
Cold land
Low pressure: warm ocean

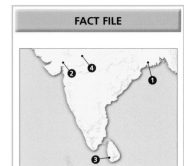

Climographs High temperatures vary little in Colombo, but spring and fall are hotter in Ahmadabad, Kota, and Calcutta. Most rain falls in summer, except for spring and fall peaks in Colombo, due to the onset and end of the monsoon.

Average precipitation
Max. temperature
Min. temperature

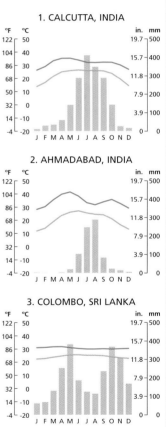

1. CALCUTTA, INDIA
°F °C in. mm
122 50 19.7 500
104 40 15.7 400
86 30 11.8 300
68 20 7.9 200
50 10 3.9 100
32 0
14 -10
-4 -20 0 0
J F M A M J J A S O N D

2. AHMADABAD, INDIA
°F °C in. mm
122 50 19.7 500
104 40 15.7 400
86 30 11.8 300
68 20 7.9 200
50 10 3.9 100
32 0
14 -10
-4 -20 0 0
J F M A M J J A S O N D

3. COLOMBO, SRI LANKA
°F °C in. mm
122 50 19.7 500
104 40 15.7 400
86 30 11.8 300
68 20 7.9 200
50 10 3.9 100
32 0
14 -10
-4 -20 0 0
J F M A M J J A S O N D

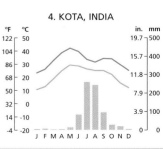

4. KOTA, INDIA
°F °C in. mm
122 50 19.7 500
104 40 15.7 400
86 30 11.8 300
68 20 7.9 200
50 10 3.9 100
32 0
14 -10
-4 -20 0 0
J F M A M J J A S O N D

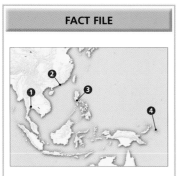
Climographs All four cities are hot year round, except Hong Kong, which has relatively mild winters. Winters are dry except in Kieta, where significant rain falls each month. Rainfall in Bangkok peaks in spring and fall.

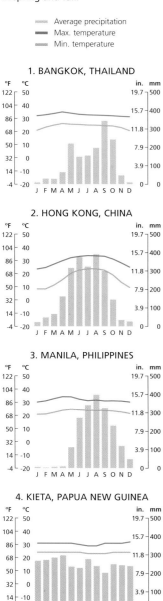

Average precipitation
Max. temperature
Min. temperature

1. BANGKOK, THAILAND

2. HONG KONG, CHINA

3. MANILA, PHILIPPINES

4. KIETA, PAPUA NEW GUINEA

Tropical Maritime Southeast

This region includes much of Southeast Asia, as well as Indonesia, Papua New Guinea, and the Philippines. Close to the Equator, the Intertropical Convergence Zone (ITCZ) creates hot, wet conditions all year. Farther north, the Asian monsoon determines rainfall, while southeast China has the wet summers and dry winters typical of a subtropical climate.

Vietnam (below) The tropical wet climate of Vietnam is ideal for rice cultivation. Much of the rainfall is associated with the ITCZ and local thunderstorms that form in the convectively unstable air.

Hong Kong (left) In this subtropical city, it is hot and wet in spring and summer, when solar intensity is high and the ITCZ moves into the region. Fall and winter are cooler and drier.

Squall lines (right) Between March and November, these lines of thunderstorms, known as sumatras, bring heavy rain to Peninsular Malaysia's west coast, which is sheltered from the monsoonal winds by high, eastern mountain ranges.

Imminent typhoon (above) Heavy clouds off the coast of Thailand announce the approach of a typhoon and heavy rain. The warm Andaman Sea sustains typhoons in this part of the region.

South Asian monsoon (right) In summer, air flows from the ocean toward low pressure over the hot Asian landmass, resulting in heavy rain. Attracted by the low pressure, the ITCZ shifts rapidly north, increasing this summer rainfall. The steady northward progression between April and June of this belt of heavy rain, shown by the red and orange areas in these charts, marks the onset of the monsoon. Less rain occurs in the areas shaded yellow and green, while blue indicates little or no rain.

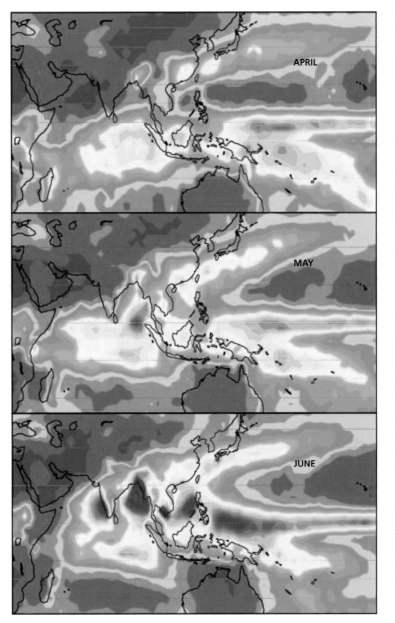

APRIL

MAY

JUNE

Irrawaddy River In the south, close to its delta, Myanmar's largest river has a tropical wet climate that is strongly influenced by the Asian summer monsoon rains. In its higher reaches, however, it flows over the dry Mandalay Plain.

Mandalay Plain Lying in the rain shadow of the Arakan Yoma Mountains, this dry area is protected from rain-bearing southwesterly winds.

Aridity Near Mandalay, the Irrawaddy River's banks are very arid. In summer it can be very hot, but in winter temperatures are moderately cool.

Amarapura Annual precipitation in the upper Irrawaddy is less than 30 inches (760 mm), and temperatures are cooler than in the delta region.

Extreme poverty More frequent El Niño events will cause the monsoon, and crops, to fail more often. As sea levels rise, coastal and delta farmlands will become more saline and less productive, and freshwater fish will not survive. Food shortages and extreme poverty will follow these changes.

AFRICA

Most of the African continent lies within the tropics, but there are also regions in the subtropics, especially in the north. As a result, climates range from tropical wet and tropical seasonal through to subtropical and Mediterranean, and tend to fall symmetrically on either side of the Equator. More temperate climates are found in higher-altitude regions, where temperatures are cooler. While Madagascar is a separate island, it is subject to the same climatic influences as the rest of Africa and experiences similar climate types.

CLIMATIC INFLUENCES

Africa's climates are influenced by a variety of factors, including latitude, altitude, ocean currents, and continentality. High solar angles due to Africa's location in the tropics subject the continent to intense solar radiation, which keeps temperatures high year round. In the subtropics, the subtropical high reduces cloud cover, so that more solar radiation reaches the surface there than near the Equator, producing the highest temperatures. These conditions, plus the lack of rainfall, are the reasons why Africa's vast deserts—the Sahara, Namib, and Kalahari—have formed in the subtropics.

In Africa's highland regions, temperatures often remain below freezing all year, producing alpine climates and supporting permanent, though now decreasing, glaciers and snow fields like those of Mount Kilimanjaro. In the extensive high plateaus of East Africa, temperatures are cool, but usually stay above freezing, creating more temperate climates. Along the coast, the ocean moderates temperature. In the southwest, the air flowing inland off the cold Benguela Current keeps the Namib Desert cooler than might be expected at these latitudes. On the eastern side of the continent, the warm Agulhas Current is the source of moisture carried inland by the trade winds.

There is little seasonal variation in the amounts of solar radiation Africa receives. As a result, although winters are cooler than summers, the seasons tend to be marked by rainfall variation rather than by changes in temperature. The main source of rainfall is the Intertropical Convergence Zone (ITCZ), the belt along which the trade winds meet. The equatorial region has the wettest climate; its rainy season can last from 10 to 12 months. At higher latitudes, rainfall becomes more seasonal as the ITCZ migrates north or south of the Equator. Regions close to the subtropics have wet summers and dry winters, but those lying between the Equator and the subtropics have two wet seasons as the ITCZ moves north and then south over them. In subtropical regions, the extensive high-pressure system is the dominant influence, especially in the heart of northern Africa. There is very little rain, and what little falls is brought by isolated thunderstorms or midlatitude systems that approach closer to the Equator than normal. The combined effect of the ITCZ's migration and the subtropical high is to create a symmetrical pattern of rainfall bisected by the Equator—with the highest rainfall occurring near the Equator and lowest in the subtropics.

Land of contrasts (right) The hot, dry climate of the savanna in Amboseli National Park, Kenya, is produced by the rain-shadow effect of Mount Kilimanjaro, a volcanic massif 19,340 feet (5,895 m) high that blocks the prevailing winds.

Sahara (right) Clear skies produced by constant subtropical high pressure combine with intense solar radiation to create the arid climates of Africa's vast deserts.

CLIMATE ZONES OF AFRICA

Tropical wet Hot and wet throughout the year; short or no dry season

Tropical seasonal Hot throughout the year; distinct wet and dry seasons

Arid Little or no precipitation year round; hot days and cold nights

Semi-arid Low precipitation; smaller diurnal temperature variation than arid climates

Mediterranean Hot, dry summers; mild, moist winters, occasionally below freezing

Subtropical Warm and moist; hot summers and cooler, drier winters

Temperate Four distinct seasons; precipitation year round; warm summers and cold winters

Mountain Colder than low-level locations found at the same latitude

Dust storms (below) In this satellite view, dust plumes blow over the Mediterranean Sea from Libya. When drought destroys vegetation, the loosened, dry soil can be carried long distances by strong winds.

Tropical wet climate (above) Dense rain forest flourishes in regions like Cameroon, where it is hot all year because the sun is always close to the zenith, and constantly wet owing to the ever-present ITCZ.

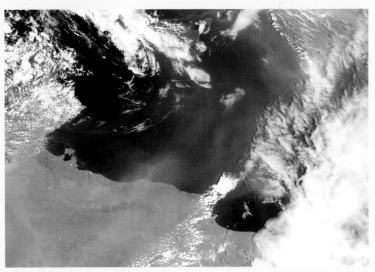

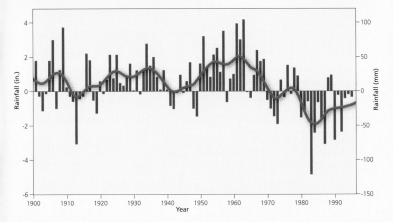

Rainfall, 1900–2000 (above) Africa is subject to fluctuating rainfall, but since 1968 the mean (red line) has been noticeably lower, with marked droughts in 1973, 1984, and 1992.

Influences (right) The ITCZ migrates south in January and north in July, bringing with it heavy rainfall. Climates are also affected by cold (blue) and warm (red) ocean currents, and the trade winds.

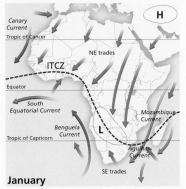

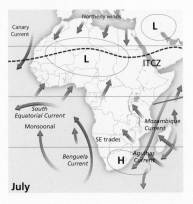

North Africa

Along North Africa's coast, cities like Algiers, Tunis, and Alexandria enjoy pleasant climates with moderate temperatures, generally dry summers, and moist winters. Moving south to the fringes of the Sahara desert, the climate becomes progressively hotter and drier. In the northwest, the Atlas Mountains have a significant impact on temperature and precipitation.

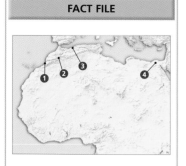

Monastir, Tunisia (above) North Africa's coastal climate is moderated by the Mediterranean Sea. Hot, dry summers and mild winters with moderate rainfall make the Tunisian coast attractive to tourists.

Sirocco (below) This hot, dry wind, locally known as the chili, ghibli, or khamsin, blows when air from the south is drawn into low-pressure areas (depressions) that cross the Mediterranean in spring and fall.

Track of depression

Chili

L

Mediterranean Sea

Ghibli

Khamsin

Atlas Mountains (above) Cold polar air brings more rain to the northern slopes of these mountains than to the south, where hot, dry tropical air prevails. In the High Atlas region, precipitation occurs as snow.

FACT FILE

Climographs These cities have strongly seasonal rainfall: winters are moist, summers are dry. Algiers is wettest and Alexandria driest. Winters are cool, while summers are warm to hot. Tripoli is hottest; Rabat is coolest.

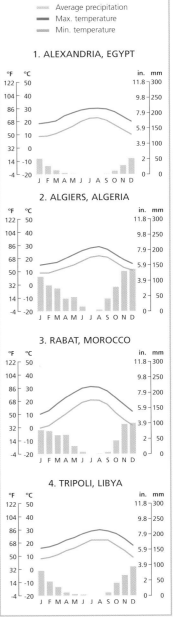

- Average precipitation
- Max. temperature
- Min. temperature

1. ALEXANDRIA, EGYPT

2. ALGIERS, ALGERIA

3. RABAT, MOROCCO

4. TRIPOLI, LIBYA

Morocco (below) Influenced by hot, dry air from the south and cool, northern sea breezes, Morocco has hot, dry summers and cooler winters, with coastal rain. These are ideal conditions for olive cultivation.

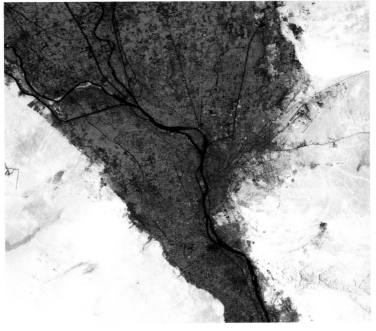

Cairo (above) Flanked by the desert, Egypt's capital (mainly red) on the banks of the Nile has an arid climate with an average annual rainfall of only 0.4 inches (10 mm). In summer, days are hot and dry; nights are much cooler. Winters are cool, with a little rain.

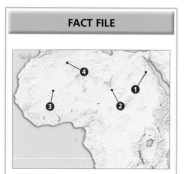

Climographs There is very little rainfall or seasonal temperature variation in these desert cities; summer tends to be hotter than winter, and it is much cooler at night. Timbuktu is slightly cooler in summer, when it receives some rain.

Average precipitation
Max. temperature
Min. temperature

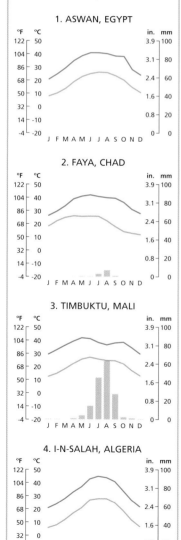

1. ASWAN, EGYPT

2. FAYA, CHAD

3. TIMBUKTU, MALI

4. I-N-SALAH, ALGERIA

Sahara

The world's largest desert extends over most of the widest part of northern Africa. The permanent subtropical high-pressure system at these latitudes ensures that skies remain clear all year, exposing the surface to intense solar radiation and resulting in extremely high temperatures. The desert's great distance from moisture sources increases its heat and aridity.

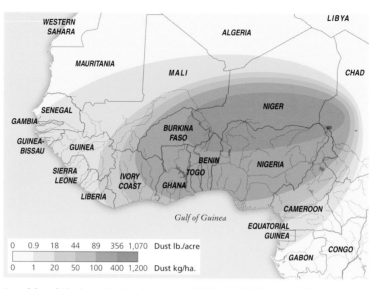

0   0.9   18   44   89   356 1,070   Dust lb./acre

0   1   20   50   100   400 1,200   Dust kg/ha.

Dust (above) The harmattan is a dry, northeasterly wind originating in the sub-tropical high. Strongest in winter, it carries dust from the central Sahara to the Gulf of Guinea, and sometimes across the Atlantic.

Aridity (right) The vast, arid Sahara contrasts with wetter central Africa in this satellite image. Aridity inland is due to the subtropical high, but along the Atlantic coast is caused by the cold Canary Current.

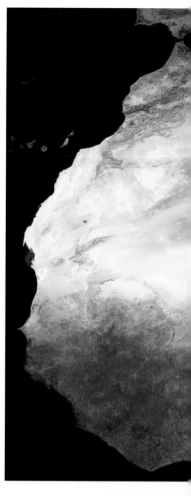

Great Western Erg, Algeria (above) Rainfall is so sporadic that some years it is not measurable. The consequent lack of vegetation allows the wind to shape dunes into vast sand seas, known as ergs.

Akakus, Libya (right) In this very arid region in southern Libya, dramatic rock arches are common among the dunes. Strong winds transport sand and abrade the rocks into unusual formations.

Shifting dunes The wind transports dust and sand grains, moving sand dunes as much as 330 feet (100 m) per year. The dust is carried by the air, but the sand grains are dragged and rolled along the surface in a series of short hops.

Wind direction | Wind deposits sand grains.

1 The wind drives the sand along and deposits it on the exposed side of a low obstacle, such as a rock or shrub. A gently sloping surface forms.

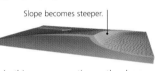

Slope becomes steeper.

2 As this process continues, the slope upon which the sand is being deposited becomes steeper and longer, building up the dune.

Loose sand slips down.

3 Eventually the leeward slope (the side sheltered from the wind) is so steep that any additional grains of sand slip down the slope.

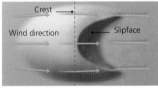

Crest
Wind direction | Slipface

4 Viewed from above, sand grains accumulate at the crest, then fall down the slipface, and the whole dune begins to move forward.

Aïr Mountains, Niger (left) Much of the Sahara consists of high, stony plateaus, called hamadas. This part of the Aïr Mountains experiences a very hot, extremely dry climate, unlike the wetter southern part of the range.

One future scenario Scientists think that global warming might bring increased rainfall to the southern edge of the Sahara, ending severe droughts. Differential heating of the land and ocean might produce changes in atmospheric circulation, and these might bring more moisture into the region.

Sahel

Forming a narrow transition zone between the Sahara and equatorial Africa, the Sahel experiences seasonal precipitation, mainly due to the shifting of the Intertropical Convergence Zone (ITCZ). Summers are hot and wet, and winters cooler and dry. The low levels of rainfall support only grasslands and savanna. Droughts are frequent and may be extensive.

Seasonal rainfall (below) When the rain-bearing ITCZ shifts to the north in August, the Sahel and regions to its south receive the most rain (red and yellow areas). When the ITCZ shifts south in winter, it is the dry season in the Sahel.

Drought Even during the rainy season, water in the Sahel is limited. Since the late 1960s, this region has experienced extended periods of severe drought, which have had a devastating impact on the land and local economies.

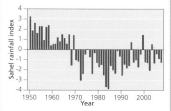

Annual rainfall From 1950 to 1970, rainfall was generally higher than average. From 1970 to 1990, however, very few years experienced above-average rainfall and drought prevailed.

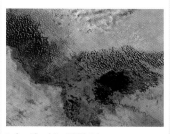

Lake Chad By 2001 this once-extensive expanse of water had shrunk to a small, shallow lake, owing to the demand for irrigation and potable supplies during long drought periods.

Herding The severe 1983 drought in the Sahel devastated herds of longhorn cattle in Mali. Herding activities have gradually been moving south, where there is more water and pasture.

Uncertain future The impact of climate change on the Sahel is uncertain. Some models predict drier conditions; others suggest it will be wetter. The outcome seems to depend on whether or not the ocean response to climate change is greater than the land response. Currently, the Sahel is wetter.

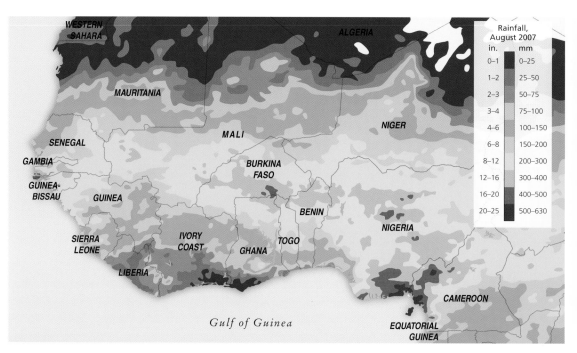

Rainfall, August 2007	
in.	mm
0–1	0–25
1–2	25–50
2–3	50–75
3–4	75–100
4–6	100–150
6–8	150–200
8–12	200–300
12–16	300–400
16–20	400–500
20–25	500–630

Gurara Falls, Nigeria (above) The torrential rain that occurs in the summer wet season swells the flow of water over these falls, which in the dry season have very little water.

Dust storm, Mali (right) The harmattan wind carries dust and sand into the Sahel from the Sahara to its north. These storms can make everyday chores such as drawing well water very difficult.

Long dry season (right) From October to May, there is very little precipitation. In the Sahelian savanna, giraffes obtain much of their water—and food—from the leaves of acacia (or thorn) trees, which are adapted to retain moisture.

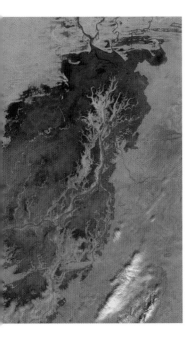

Niger Delta, Mali (above) At the end of the long wet season, this fertile inland delta provides much-needed resources. Fed by the floodwaters of the Niger and Bani rivers, the delta nearly doubles in extent.

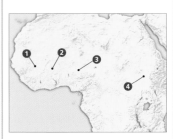

FACT FILE

Climographs These Sahel cities are very hot all year, with a slight cooling in summer that is associated with the rainy season. In winter, all four are extremely dry. In summer, Ouagadougou is wettest, and Malakal is driest.

Average precipitation
Max. temperature
Min. temperature

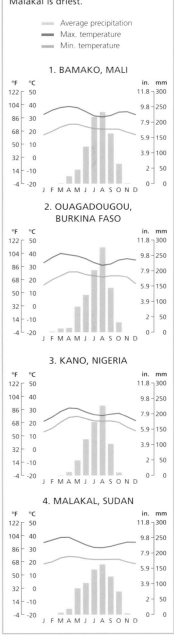

1. BAMAKO, MALI

2. OUAGADOUGOU, BURKINA FASO

3. KANO, NIGERIA

4. MALAKAL, SUDAN

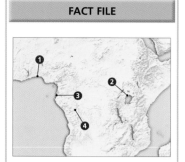
Climographs These equatorial cities are very warm all year, with little seasonal variation in temperature, but nights are cooler. Rainfall amounts are ample: all except Lagos have two rainy seasons. Libreville is wettest.

------ Average precipitation
—— Max. temperature
—— Min. temperature

1. LAGOS, NIGERIA

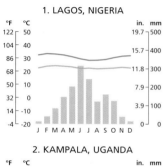

2. KAMPALA, UGANDA

3. LIBREVILLE, GABON

4. KINSHASA, DEMOCRATIC REPUBLIC OF CONGO

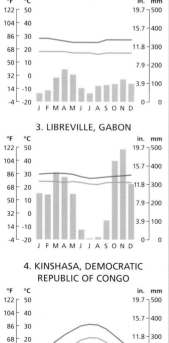

Equatorial Belt

Covering most of West Africa, this region straddles the Equator between 10°N and 10°S. It is hot and wet year round close to the Equator. Farther away it is still hot, but the double passage of the Intertropical Convergence Zone (ITCZ) may bring a rainy season in both May and November. At the area's fringes, there tends to be a wet summer, dry winter pattern.

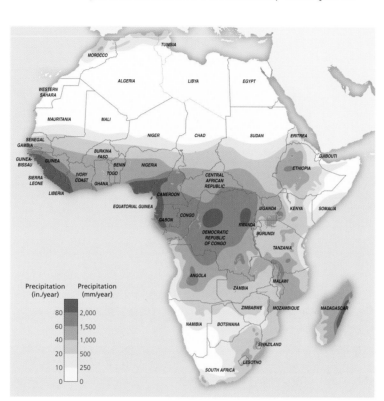

Precipitation (in./year) | Precipitation (mm/year)
80 | 2,000
60 | 1,500
40 | 1,000
20 | 500
10 | 250
0 | 0

Average annual rainfall (above) Inland and coastal areas of central Africa receive the most rain, influenced by the ITCZ. At this convergence of the trade winds, air rises and cools to produce heavy rain.

West African monsoon (below) When the ITCZ is over the Gulf of Guinea (top), the dusty northeast trade winds prevail. When it shifts north over West Africa (bottom), warm, moist air brings heavy monsoon rainfall to the region.

Northern winter

Harmattan wind: warm, dry, dusty

ITCZ

Gulf of Guinea

Northern summer

ITCZ

Monsoon flow: wet

Alpine climate (left) Uganda's Rwenzori Mountains soar above 16,000 feet (5,000 m) and, although near the Equator, have alpine vegetation, glaciers, snow fields, and cold alpine lakes.

Niger Delta, Nigeria (above) Near the mouth of the Niger River, this hot, tropical area is one of the wettest parts of Africa. The rainy season lasts eight months, and even in the dry season substantial rain falls.

Odzala National Park (above) This rain forest in the Congo thrives in the long rainy season of the tropical wet climate, when high humidity often shrouds its trees in mist.

Garamba National Park (below) The tropical seasonal climate in northeast Democratic Republic of Congo supports grasslands and savanna, where buffaloes and other large mammals roam.

FACT FILE

1. Simien and Bale mountains
Although in an equatorial region, these mountains have an alpine climate. Higher than 13,000 feet (4,000 m) in places, they have cool year-round temperatures. Rainfall feeds the many cold streams and alpine lakes.

Simien and Bale mountains, Ethiopia

2. Asmara At an elevation of 7,628 feet (2,325 m), this region has a temperate climate. Temperatures are pleasantly warm in the daytime and change little through the year. Rainfall is low, but there is more rain in summer.

Asmara, Ethiopia

3. Hargeysa Located below 4,900 feet (1,500 m), this city has a tropical wet climate, with very high temperatures in the warmer season, and moderate-to-cool weather in the cooler season. Precipitation is quite high.

Hargeysa, Somalia

4. Afar Depression This low-lying part of the Horn of Africa is extremely hot and dry, with an average annual rainfall of less than 7 inches (178 mm) and no rain for most of the year. Droughts are common. In the Danakil Desert, extreme aridity forms large salt pans, on which a salt-mining industry is based.

Afar Depression, Eritrea/Ethiopia/Djibouti/Somalia

WEATHER WATCH

Decreasing water supplies, higher temperatures The Rwenzori glaciers and snow fields are expected to disappear within 20 years, curtailing irrigation and the generation of hydroelectric power. With rising temperatures, the habitats of rare plants and endangered animals will be under severe threat.

African elephant This huge mammal is particularly suited to life in the tropical savanna because it can survive for long periods without water.

Giraffe Predators are easily spotted by this lofty animal. It obtains water from the acacia leaves it eats and needs to drink only once every few days.

African lions In the savanna these carnivores are sure to find a large number of prey and adequate cover. The adults have no natural predators.

Tropical Seasonal East Africa

This region takes in southern Kenya, most of Tanzania, Mozambique, and Madagascar. Some areas have only one rainy season, from about October to March. In others, the twice-yearly passage of the ITCZ produces two rainy seasons—approximately March–May and October–December. Otherwise, it is very dry, in some parts for so long that only grasses grow.

Tropical cyclones (below) These occur most frequently in the latter part of the summer rainy season in southeast Africa and bring large amounts of rainfall. In this satellite view, Tropical Cyclone Ivan is seen approaching Madagascar in February 2008.

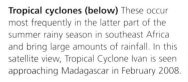

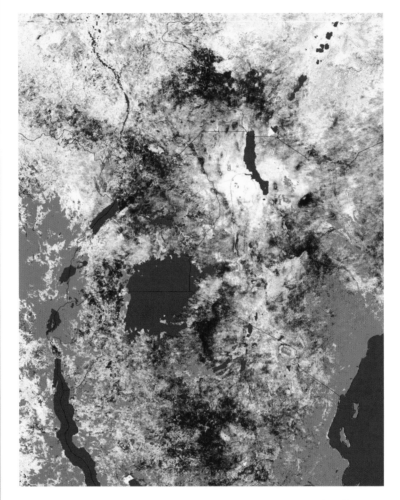

Drought (left) Rainfall can vary significantly from year to year. In 2005 and 2006, the failure of the seasonal rains led to severe drought. The brown colors in this satellite image of the area around Lake Victoria indicate stressed vegetation. Gray patches represent cloud cover.

Imminent storm (below) During the wet season, thunderstorms frequently happen in the late afternoon, when the temperature is highest and the atmosphere most convective.

Two dry seasons (above) Masai herders have difficulty finding pasture for their cattle during Tanzania's two dry seasons, one short and hot, and the other long and cooler.

Mount Kilimanjaro (below) While savanna grasslands are typical of lower altitudes in East Africa, this mountain in the same region has an alpine climate, in which precipitation falls as snow.

FACT FILE

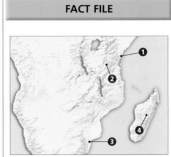

Climographs These places are warm to hot all year. Winters are slightly cooler, with the least rainfall, and are driest in Dodoma. Antananarivo has the wettest summers, while Mombasa has higher rainfall in spring and fall.

— Average precipitation
— Max. temperature
— Min. temperature

1. MOMBASA, KENYA

2. DODOMA, TANZANIA

3. MAPUTO, MOZAMBIQUE

4. ANTANANARIVO, MADAGASCAR

Southwestern Deserts

The Namib and Kalahari deserts cover an area extending from the western coast of Namibia to Botswana and up to southwest Angola. The southern subtropical high-pressure system is responsible for the clear skies and lack of rainfall over both deserts. Bordering the coast, the Namib is cool and subject to frequent advective fog. Inland, the Kalahari is warmer.

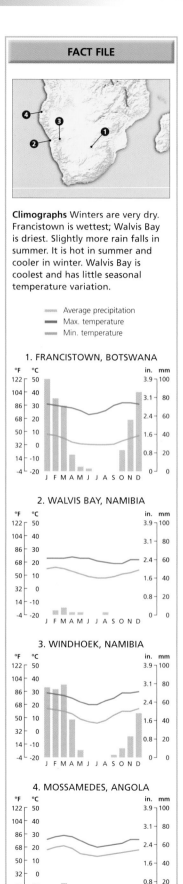

FACT FILE

Climographs Winters are very dry. Francistown is wettest; Walvis Bay is driest. Slightly more rain falls in summer. It is hot in summer and cooler in winter. Walvis Bay is coolest and has little seasonal temperature variation.

- Average precipitation
- Max. temperature
- Min. temperature

1. FRANCISTOWN, BOTSWANA

2. WALVIS BAY, NAMIBIA

3. WINDHOEK, NAMIBIA

4. MOSSAMEDES, ANGOLA

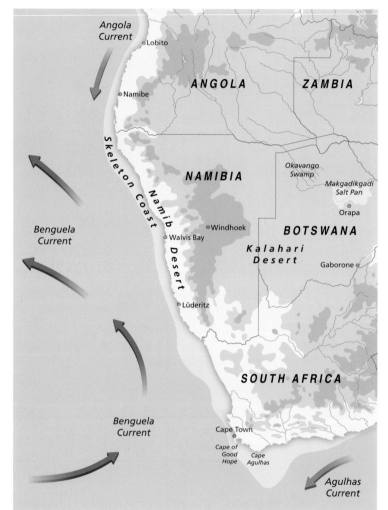

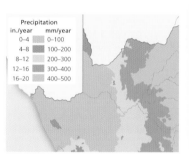

Precipitation

in./year	mm/year
0–4	0–100
4–8	100–200
8–12	200–300
12–16	300–400
16–20	400–500

Dust storm, Botswana (right) At the height of the dry season, intense solar radiation coupled with the lack of moisture parches the vegetation, and strong winds raise dust storms. Only drought-resistant trees like the acacia remain green.

Cold versus hot desert (left) Winds blowing off the Atlantic Ocean over the cold Benguela Current keep the Namib Desert cool. Away from the coast, the Kalahari Desert is hotter, except at the higher altitudes of its inner tablelands.

Kalahari rainfall (below) While this desert is very dry in the west and southwest, it receives more rainfall in the east, where there is savanna vegetation.

Extreme aridity (right) Rainfall in the Namib is less than 2.5 inches (64 mm) per year; moisture is derived only from the fog blown ashore. Strong winds build the world's tallest dunes here, which turn deep orange as the iron they contain is oxidized.

Skeleton Coast (below) Along the northern coast of the Namib Desert, shipwrecks have been half-buried by the ever-shifting dunes, bearing witness to the hazardous conditions created by the dense coastal fog and strong ocean currents.

FACT FILE

Benguela Current This current is cold because it originates in the south, and because water upwells along the Atlantic coast. The cold temperatures cause fog to form, and make the Namib Desert cooler than the Kalahari, which lies at the same latitudes, but inland.

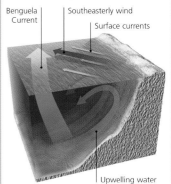

Benguela Current   Southeasterly wind

Surface currents

Upwelling water

Upwelling Winds drive the surface water away from the land, allowing cold, nutrient-rich water from deeper in the ocean to rise to the surface.

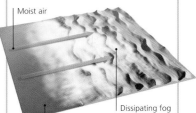

Moist air

Dissipating fog

Cold water surface

Fog Moist air cools as it passes over the cold water surface, water vapor condenses, and fog forms. Winds blow the fog inland, where it dissipates.

WELWITSCHIA MIRABILIS

This ancient plant can survive in extremely arid conditions and is endemic to the Namib Desert. The stomata of its leaves open when it is foggy to allow the plant to absorb moisture. When it is sunny, the stomata close, preventing moisture loss. The leaves lie close to the ground, which keeps the soil moist and cool. In this way, the plant can cope with temperatures as high as 149°F (65°C).

Southern Africa

In this region, altitude and latitude—mostly south of 10°S—produce climates ranging from temperate to subtropical. The temperate climates of southern coastal areas are influenced by the cold westerlies in the west and tropical cyclones in the east. In the interior, extensive high plateaus redistribute rainfall and promote cooler temperatures than at sea level.

Drakensberg Mountains (right) Soaring above 11,400 feet (3,475 m), this range parallel to the southeastern coast has an annual rainfall of 39.4 inches (1,000 mm). It can be hot and sunny by day, but temperatures can drop to freezing at night.

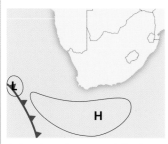

1. Forming When warm air flowing counterclockwise out of high pressure meets cooler southern air, a cold front and a low-pressure system form.

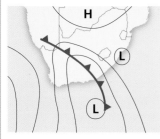

2. Rain The warm air ahead of a cold front rises to form clouds, bringing heavy rain and strong winds to regions along the frontal boundary.

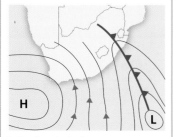

3. Passing As the low-pressure system, cold front, and rainfall move away, a high-presssure system and colder, drier air from the south replace them.

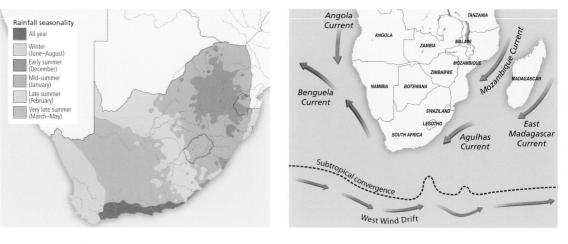

Rainfall seasonality
- All year
- Winter (June–August)
- Early summer (December)
- Mid-summer (January)
- Late summer (February)
- Very late summer (March–May)

Rainfall in South Africa (above) The far south receives rain all year. The rainy season occurs in winter in the southwest and gradually shifts from late to early summer toward the northeast.

Ocean currents (above right) A variety of currents influence air temperatures. Southward-flowing currents in the southeast are warm, but the northward-flowing Benguela in the west is cold.

Western Cape, South Africa (below) The shrub and heath vegetation known as fynbos is adapted to the long, dry summers and cool, moist winters of the Mediterranean climate in the southwest.

Semi-arid Karoo (above) This high plateau in central South Africa is protected from the prevailing winds by the coastal escarpments. It receives as little as 16 inches (400 mm) of rain per year, most of which falls in summer.

Matopos National Park, Zimbabwe (above) Belonging to the savannas that cover much of southern Africa, this region has hot, wet summers and cool, dry winters. High plateaus to the south protect it from cold southerly air masses.

Cold front (left) In winter, the marine influence of the Indian and Atlantic oceans usually keeps temperatures in Cape Town mild. However, the city is also subject to the passage of cold fronts, which are heralded by an extensive cloud bank like the one pictured here behind the city's Victoria and Alfred Waterfront area.

FACT FILE

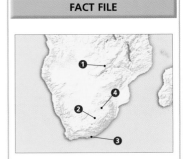

Climographs In Lusaka, spring and summer are warm; fall and winter are cooler. The other cities have warm summers and cool winters. Rainfall is mostly seasonal: Lusaka is wettest, Bloemfontein driest. Port Elizabeth has low rainfall all year.

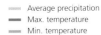

Average precipitation
Max. temperature
Min. temperature

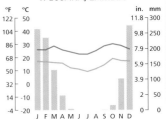

1. LUSAKA, ZAMBIA

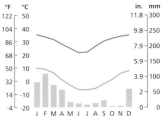

2. BLOEMFONTEIN, SOUTH AFRICA

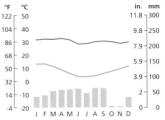

3. PORT ELIZABETH, SOUTH AFRICA

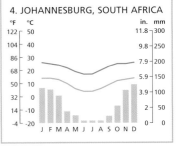

4. JOHANNESBURG, SOUTH AFRICA

OCEANIA AND ANTARCTICA

This vast area of ocean and land reaches from the Hawaiian Islands at 20°N to the Southern Ocean and Antarctica in the far south. Its land masses vary from the hot, dry Australian continent to the frozen expanses of Antarctica, and from New Zealand's two large islands to some 10,000 smaller Pacific islands and tiny atolls. The different latitudes, topographies, and orientations combine with the prevailing atmospheric circulation to give rise to climates that range from the hot tropical wet or hot arid to the cool temperate or extremely cold, dry polar.

CLIMATE ZONES OF OCEANIA AND ANTARCTICA

Tropical wet Hot and wet throughout the year; short or no dry season

Tropical seasonal Hot throughout the year; distinct wet and dry seasons

Arid Little or no precipitation year round; hot days and cold nights

Semi-arid Low precipitation; smaller diurnal temperature variation than arid climates

Mediterranean Hot, dry summers; mild, moist winters, occasionally below freezing

Subtropical Warm and moist; hot summers and cooler, drier winters

Temperate Four distinct seasons; precipitation year round; warm summers and cold winters

Mountain Colder than low-level locations found at the same latitude

Subpolar Very cold throughout the year; no true summer; tundra vegetation

Polar Extremely cold and dry throughout the year; permanent ice cover

Mariana Islands

Guam

Palau

Micronesia

Marshall Islands

Hawaii

Line Islands

Papua New Guinea

Solomon Islands

Vanuatu

Samoa

Marquesas

Darwin

Fiji

Cook Islands

New Caledonia

Tonga

Tahiti

Polynesia

Australia

Brisbane

Sydney

Canberra

Adelaide

New Zealand

Perth

Melbourne

Wellington

Hobart

Ronne Ice Shelf

Antarctica

South Pole

Pitcairn Islands

Ross Ice Shelf

CLIMATIC INFLUENCES

The Pacific Ocean exerts a strong influence on temperature and moisture. Because the ocean is slow to warm up and cool down, the small islands, the coastal regions of the continents, and inland areas that receive ocean winds have fairly uniform year-round temperatures, as determined by their latitude. Far from the ocean, the Australian and Antarctic interiors experience extremely hot and extremely cold temperatures respectively.

Variation in precipitation often defines the seasons, especially in the tropics. Those Pacific islands that are constantly affected by the Intertropical Convergence Zone (ITCZ) have high rainfall in every season and are considered to have a tropical wet climate. As the ITCZ shifts north and south with the sun, some islands receive mainly summer rainfall, giving rise to a

tropical seasonal climate. Over northern Australia, the summer monsoon winds combine with the ITCZ to bring high rainfall in summer; when the monsoon retreats and the ITCZ shifts north, the region is much drier in winter. Tropical cyclones also contribute significant precipitation in summer; they develop above the warm Pacific Ocean and move westward over the Pacific islands and into northeastern Australia.

North and south of the tropics, descending air in the subtropical high-pressure system prevents cloud formation, so that very little rain falls and there is intense solar radiation. Australia's arid regions occur where the subtropical high is always present; its semi-arid regions

lie along the southern margins of the desert, where the westerlies bring winter rainfall, or along the northern margins, which receive summer rain from the southern fringes of the monsoon system.

The ocean–atmosphere phenomenon known as El Niño–Southern Oscillation (ENSO), occurring every two to seven years, has a significant impact on the variability of precipitation from year to year. El Niño is associated with persistent droughts in Australia, especially in the east, while La Niña

brings heavier than normal rainfall.

At middle latitudes, westerly winds are carriers of year-round precipitation, which falls either as snow or rain, depending on the season, altitude, and latitude. This combines with cooler temperatures to create the temperate climates of Tasmania and New Zealand. At higher latitudes, air sinking over the South Pole in the polar circulation cell inhibits precipitation, making Antarctica very dry. As the cold air returns north, it produces Antarctica's frigid katabatic winds.

Lamington National Park (left)
In northeastern Australia, dense rain forest thrives in the ample year-round rainfall, mild winters, and very warm summers of Queensland's subtropical climate.

Antarctica (below) Here, temperatures are permanently below freezing. Relatively little snow falls and, with little melt, it remains to become ice, making this southern continent permanently icebound.

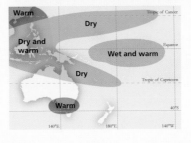

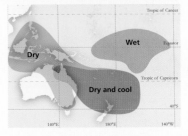

ENSO (above) Summers are normally wet in the tropical Pacific islands and northern Australia. During El Niño events, summers become very dry, while southeastern Australia is very hot (**left**). In winter (**right**), cooler temperatures return but drought conditions persist.

Pinnacles Desert, Western Australia (above) In the Australian interior, hot, dry summers and cooler, dry winters produce desert areas with vast expanses of sand dunes with little or no vegetation.

Fijian Islands (below) In these and other islands of the Pacific, temperatures and precipitation vary little all year. Summers are slightly warmer and wetter, while winters are slightly cooler and drier.

Central Otago, South Island, New Zealand (above) Most of New Zealand is close to the coast, resulting in the moderate summers and cool winters of a temperate climate. Abundant year-round precipitation produces lush pastures that support livestock.

Tropical Australia

In far northern Australia, high sun angles maintain hot temperatures all year, but rainfall varies seasonally. The heaviest rain falls from December to March, when monsoonal circulation moves into the region in summer and tropical cyclones often form over the warm Pacific Ocean. In winter, winds from the continental interior dominate, bringing dry weather.

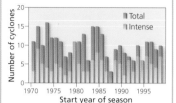
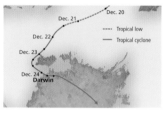

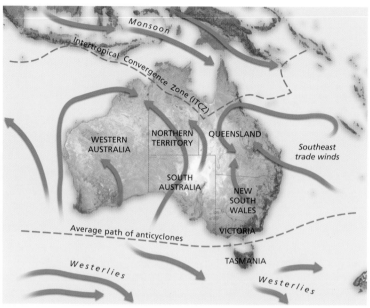

January chart (above) In summer, strong westerly monsoonal winds converge with the ITCZ. These, combined with tropical cyclones, bring very heavy rainfall and strong winds to northern Australia.

Cape Leveque (right) On Western Australia's northwestern coast, the marine influence keeps temperatures consistently warm all year. It is wet from November to April, and dry the rest of the year.

Kimberley Plateau (above) This rugged sandstone plateau in the northwest of Western Australia receives ample rain closer to the coast in the north, but becomes progressively drier inland toward the south, where precipitation is limited by the subtropical high.

Kakadu (above) In these humid tropical wetlands of Australia's Northern Territory, the onset of the monsoon rains is marked by cloudy conditions and thunderstorms that start around early November.

Grasslands (below) Large termite mounds are common in the grasslands that flourish in the wet–dry seasonal cycle of inland northern Australia. The dry season lasts from about May to October.

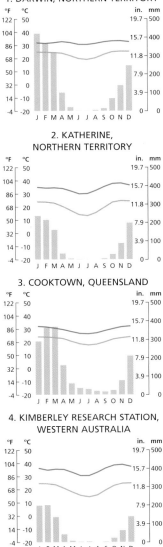

FACT FILE

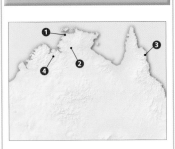

Climographs These graphs show the distinct wet and dry seasons that are typical of tropical seasonal climates. Temperatures are uniformly hot to warm all year. Darwin is wettest and the Kimberley region is the driest.

- Average precipitation
- Max. temperature
- Min. temperature

1. DARWIN, NORTHERN TERRITORY

2. KATHERINE, NORTHERN TERRITORY

3. COOKTOWN, QUEENSLAND

4. KIMBERLEY RESEARCH STATION, WESTERN AUSTRALIA

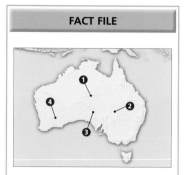
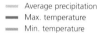
Climographs Summers are very hot, and winters are warm to mild. Rainfall is low, with mostly drier winters; Kalgoorlie tends to be drier in spring and summer. Semi-arid Alice Springs has the most rain, and arid Coober Pedy the least.

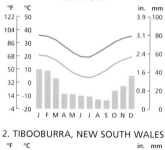

- Average precipitation
- Max. temperature
- Min. temperature

1. ALICE SPRINGS, NORTHERN TERRITORY

°F	°C		in.	mm
122	50		3.9	100
104	40		3.1	80
86	30		2.4	60
68	20		1.6	40
50	10		0.8	20
32	0		0	0
14	-10			
-4	-20	J F M A M J J A S O N D		

2. TIBOOBURRA, NEW SOUTH WALES

°F	°C		in.	mm
122	50		3.9	100
104	40		3.1	80
86	30		2.4	60
68	20		1.6	40
50	10		0.8	20
32	0		0	0
14	-10			
-4	-20	J F M A M J J A S O N D		

3. COOBER PEDY, SOUTH AUSTRALIA

°F	°C		in.	mm
122	50		3.9	100
104	40		3.1	80
86	30		2.4	60
68	20		1.6	40
50	10		0.8	20
32	0		0	0
14	-10			
-4	-20	J F M A M J J A S O N D		

4. KALGOORLIE, WESTERN AUSTRALIA

°F	°C		in.	mm
122	50		3.9	100
104	40		3.1	80
86	30		2.4	60
68	20		1.6	40
50	10		0.8	20
32	0		0	0
14	-10			
-4	-20	J F M A M J J A S O N D		

Australian Interior

Arid and semi-arid climates prevail over much of the continent, from its heart to the west coast, making Australia Earth's driest inhabited continent. These climates are largely the result of the subtropical high-pressure belt, which prevents the formation of rain-bearing clouds and exposes the region to intense solar radiation, so that it is both very hot and very dry.

Simpson Desert (below) This hot, arid, largely uninhabited region stretches some 55,000 square miles (143,000 km²) over the Northern Territory's southeast and parts of Queensland and South Australia. In places its dunes rise to 120 feet (37 m).

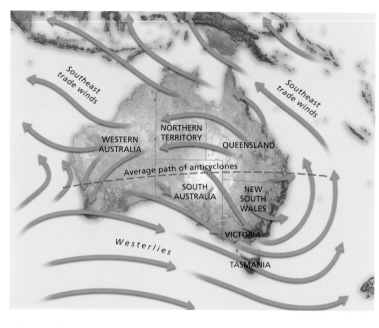

Subtropical high (above) This belt of high pressure separates the trade winds of the tropics from the midlatitude westerlies. Anticyclones moving along this belt bring clear skies and light winds.

Uluru (below) Scrub and grasses grow around this imposing sandstone formation in semi-arid central Australia. The region receives about 12 inches (305 mm) of rainfall annually, mostly in summer.

MacDonnell Ranges (above) These mountains near Alice Springs in central Australia modify the arid climate of the surrounding region to create a slightly wetter, semi-arid microclimate that supports spinifex and shrubs.

Lake Eyre (right) Covering some 460,000 square miles (1.2 million km²) of South Australia, this internal drainage basin is a dry salt pan most of the time, but can fill with water after rare rains to its north.

Subtropical and Temperate Australia

In the east, southern Queensland and northern New South Wales experience a moist, subtropical climate. Much of the rainfall in these regions is produced by afternoon and evening thunderstorms in spring and by tropical cyclones in summer. Farther south, the climate is temperate, with mild summers, cool winters, and ample precipitation all year.

Blue Mountains, New South Wales (right) Although in the subtropical climate zone, this sandstone plateau, rising to 3,870 feet (1,180 m), experiences cooler temperatures and generally more precipitation than neighboring Sydney.

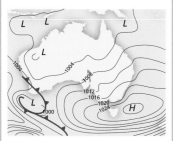

Approach Ahead of the front, high pressure pulls in very warm air from the interior, causing high temperatures over southeastern Australia.

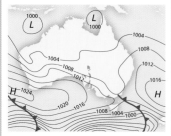

Passing front As the cold front moves across the region, temperatures can drop by more than 18°F (10°C), providing relief from the summer heat.

Melbourne dust storm Strong winds and updrafts may accompany cold fronts. These can lift dry soil, pushing a dust storm ahead of the front.

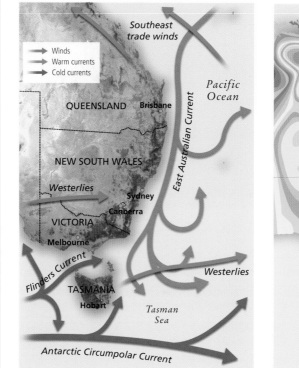

Influences (above) The warm East Australian Current keeps coastal areas warm. In the south, the cold Antarctic Circumpolar Current has a cooling effect. Westerly winds bring storms in winter.

Maryborough sugar cane (below) With hot, moist summers, moderate winters, and ample rainfall, southern Queensland's subtropical climate is well suited to the large-scale cultivation of sugar cane.

Rainfall (above) Between 1970 and 2008, rainfall in parts of eastern Australia declined by up to 2 inches (50 mm) every 10 years. The trend toward decline is consistent with climate change predictions.

Trend in annual total rainfall, 1970–2008

in.	mm
2	50
1.6	40
1.2	30
0.8	20
0.6	15
0.4	10
0.2	5
0	0
-0.2	-5
-0.4	-10
-0.6	-15
-0.8	-20
-1.2	-30
-1.6	-40
-2	-50

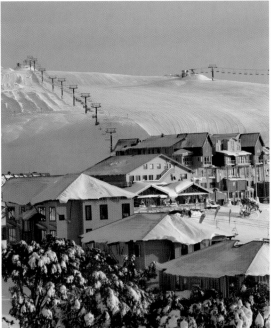

Mount Hotham, Victoria (above)
In Australia's temperate climate zone, this ski resort high in the Australian Alps receives its precipitation as snow in winter, when temperatures remain below freezing.

Tasmanian rain forest (left) Tasmania has a temperate climate, with mild summers and cool, wet winters. It is wettest in the southwest, where enough rain falls to support temperate rain forest.

Bushfires (below) The subtropical and temperate regions of Australia are prone to bushfires during summer heat waves. High evaporation rates dessicate the vegetation, providing ample fuel for the fires.

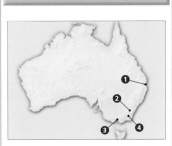
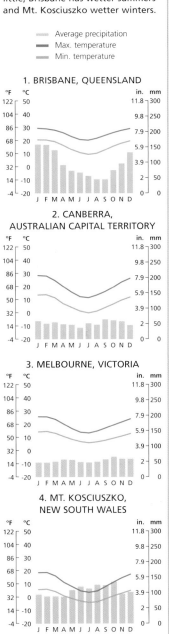

Mediterranean Australia

Narrow coastal fringes in Australia's southwest and south experience conditions typical of a Mediterranean climate. The expansion of the subtropical high-pressure system produces long, hot, dry summers. As the subtropical high contracts to the north in winter, westerly winds with embedded frontal systems dominate and bring cooler, wet weather.

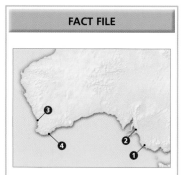

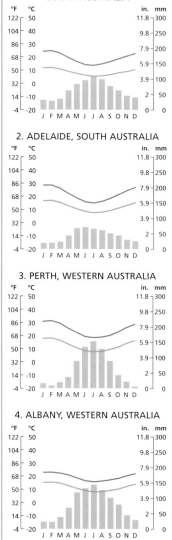

1. MOUNT GAMBIER, SOUTH AUSTRALIA

°F	°C		in.	mm
122	50		11.8	300
104	40		9.8	250
86	30		7.9	200
68	20		5.9	150
50	10		3.9	100
32	0		2	50
14	-10		0	0
-4	-20			
		J F M A M J J A S O N D		

2. ADELAIDE, SOUTH AUSTRALIA

°F	°C		in.	mm
122	50		11.8	300
104	40		9.8	250
86	30		7.9	200
68	20		5.9	150
50	10		3.9	100
32	0		2	50
14	-10		0	0
-4	-20			
		J F M A M J J A S O N D		

3. PERTH, WESTERN AUSTRALIA

°F	°C		in.	mm
122	50		11.8	300
104	40		9.8	250
86	30		7.9	200
68	20		5.9	150
50	10		3.9	100
32	0		2	50
14	-10		0	0
-4	-20			
		J F M A M J J A S O N D		

4. ALBANY, WESTERN AUSTRALIA

°F	°C		in.	mm
122	50		11.8	300
104	40		9.8	250
86	30		7.9	200
68	20		5.9	150
50	10		3.9	100
32	0		2	50
14	-10		0	0
-4	-20			
		J F M A M J J A S O N D		

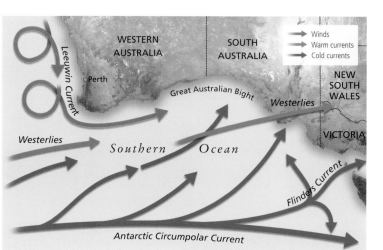

Viticulture (right) Vineyards such as this one in South Australia, shown in fall, flourish in this region. The cool, moist spring promotes growth, and grapes mature over the long, hot summers that characterize the Mediterranean climate.

Cooling effect (left) Winds that blow over the cool currents flowing along the southern and southwestern coasts of Australia lower coastal temperatures. In winter, the cool, moist air penetrates farther inland as the subtropical high retreats north.

Perth, Western Australia (below) This city experiences very high temperatures, blue skies, and dry weather in summer. A sea breeze, known as the Fremantle Doctor, brings cooler air onto the land in the afternoons, providing welcome relief from the heat.

Winter storms (below) Westerly winds frequently carry storms over southern and western Australia in winter. Originating over the Southern Ocean, the storms bring lower temperatures and rain.

Heat waves (right) In summer, hot, dry air masses from the interior can dominate for days at a time, and temperatures can soar to dangerously high levels, as at this Adelaide music festival in 2009.

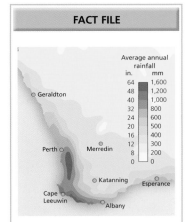

Average annual rainfall

in.	mm
64	1,600
48	1,200
40	1,000
32	800
24	600
20	500
16	400
12	300
8	200
0	0

Geraldton

Perth   Merredin

Katanning   Esperance

Cape Leeuwin   Albany

Higher rainfall The southwestern tip of Western Australia receives much more rain than the rest of Australia's Mediterranean region. This area is farthest from the drying influence of the subtropical high and closest to the rain-bearing winter storms.

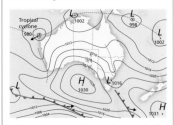

Heat wave conditions These typically begin with a high-pressure system to the south and a trough of low pressure in the west. Winds usually blow from the east and northeast into the trough, bringing with them hot, dry air over several days.

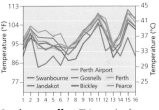

Sea-breeze effect This graph of temperatures at various Perth sites during a 16-day heat wave shows that the sea breeze at a coastal site (green) lowered temperature on some days by as much as 16°F (9°C), compared to the hottest site (blue).

Viticulture decline This region is expected to become warmer and drier, but not as dry as eastern Australia. An increased number of days with very high summer temperatures will have a negative impact on agriculture, especially viticulture, since it will be too hot for the grapes to mature properly.

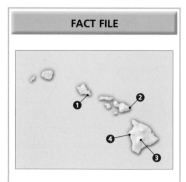
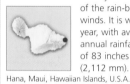

Islands of the Pacific

The thousands of small islands in this region—from the Hawaiian Islands in the north to Micronesia, Melanesia, Polynesia and others in the southwest—all have climates strongly influenced by the ocean. Close to the Equator, they are hot and wet, with no seasonal variation. Farther away, they are hot and wet in summer, and cooler and drier in winter. Sea breezes moderate temperatures along their coasts.

Atmospheric circulation, June–August (right) The ITCZ and its southern arm, the SPCZ, bring heavy rainfall. In islands influenced directly by the expansion of the subtropical high, rainfall is reduced, and the cooling trade winds moderate the high tropical temperatures. Winter storms are carried along a southern midlatitude storm track by the westerlies.

Viti Levu, Fiji (below) This island has year-round high temperatures and heavy rainfall. Its tropical wet climate supports dense forest, as well as crops like sugar cane (foreground) that require heat and plenty of water.

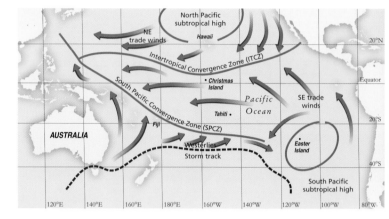

New Caledonia (left) This island has a tropical seasonal climate. Summers are hot and wet, influenced by the SPCZ. In winter the expansion of the subtropical high-pressure system brings drier and cooler conditions.

Mount Yasur, Tanna (right) Rainfall in the subtropical archipelago of Vanuatu varies. The southernmost island of Tanna, famous for its active volcano, is drier than the islands to the north because it lies in the lee of the mountain chain that runs north–south through the islands.

Tropical cyclones (right) In December 2002, Tropical Cyclone Zoe made landfall on the Solomon Islands in Melanesia, laying waste trees and huts, and disrupting communications. Such low-lying tropical islands are vulnerable to the destructive force of these storms, which form over the warm waters of the Pacific in summer.

Christmas Island (below) Just north of the Equator, this tiny island receives very heavy rainfall lasting several days at a time between December and April. The rains are brought by the same monsoon winds that affect northern Australia.

Rangiroa (above) One of the Pacific's many atolls, this Polynesian island near the Equator is wet and warm all year. It is cooler from May to October, with most rain falling from November to April.

FACT FILE

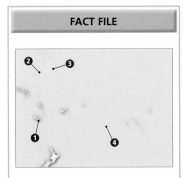

Climographs These cities have hot temperatures all year. Only Noumea is cooler in winter. Summer tends to be wet, and winter and spring dry, except for Ujelang, which is dry in summer. Ujelang has the most rainfall, and Noumea is the driest.

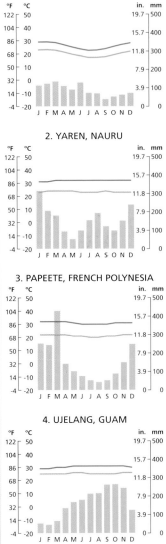

Average precipitation
Max. temperature
Min. temperature

1. NOUMEA, NEW CALEDONIA

2. YAREN, NAURU

3. PAPEETE, FRENCH POLYNESIA

4. UJELANG, GUAM

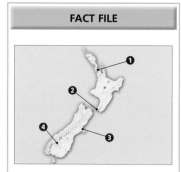
Climographs Precipitation is relatively low and varies little; winter tends to be wetter. Summers are cool to moderate; winters are cool. Auckland has the warmest winters, while Wellington receives the most rain. Christchurch is driest.

— Average precipitation
— Max. temperature
— Min. temperature

1. AUCKLAND, NORTH ISLAND
°F °C   in. mm
122 50   11.8 300
104 40   9.8 250
86   30   7.9 200
68   20   5.9 150
50   10   3.9 100
32   0   2   50
14 -10   0   0
-4 -20
     J F M A M J J A S O N D

2. WELLINGTON, NORTH ISLAND
°F °C   in. mm
122 50   11.8 300
104 40   9.8 250
86   30   7.9 200
68   20   5.9 150
50   10   3.9 100
32   0   2   50
14 -10   0   0
-4 -20
     J F M A M J J A S O N D

3. CHRISTCHURCH, SOUTH ISLAND
°F °C   in. mm
122 50   11.8 300
104 40   9.8 250
86   30   7.9 200
68   20   5.9 150
50   10   3.9 100
32   0   2   50
14 -10   0   0
-4 -20
     J F M A M J J A S O N D

4. QUEENSTOWN, SOUTH ISLAND
°F °C   in. mm
122 50   11.8 300
104 40   9.8 250
86   30   7.9 200
68   20   5.9 150
50   10   3.9 100
32   0   2   50
14 -10   0   0
-4 -20
     J F M A M J J A S O N D

New Zealand

This island nation experiences climates that are strongly influenced by its latitude south of 34°S, the surrounding ocean, and the north–south Southern Alps, which rise to 10,000 feet (3,050 m) in the South Island. While the northern tip of North Island is subtropical, the rest of the country—lying in the path of the rain-bearing westerlies—is temperate or alpine.

Waipoua Forest, Northland (above) Conifer species, including massive kauri trees, flourish in North Island's warm, subtropical climate, where average annual rainfall exceeds 39 inches (1,000 mm).

Precipitation (right) The high country and mountains running the length of New Zealand create a barrier to the prevailing westerly winds. Most precipitation occurs on the western coasts and slopes, particularly in the South Island, influenced by the high Southern Alps. Eastern parts of the country, in the sheltered lee of the mountains, are drier.

Mean annual precipitation 1971–2000

in.	mm
158	4,000
118	3,000
79	2,000
59	1,500
49	1,250
39	1,000
30	750
20	500

Canterbury Plains (below) This low-lying region in east-central South Island is sheltered by the mountain ranges from the rain-bearing westerly winds. Its drier, warmer climate supports livestock, as well as mechanized agriculture.

Russell, Bay of Islands (above) The subtropical climate of the far north makes this region a popular holiday destination. It has warm, humid summers and mild, wet winters, with an annual temperature range of 60°F to 64°F (16°C to 18°C).

Wellington (right) Located in the south of North Island, New Zealand's capital has a temperate climate, but can be very windy. Cook Strait funnels strong winds around the southern coast between North Island and South Island, and in winter the city is also subject to low-pressure systems moving eastward from the Tasman Sea.

Fox Glacier (below) Precipitation falls as snow in South Island's Southern Alps, especially in the west. Mountain peaks are permanently snow-covered, and accessible glaciers are year-round tourist attractions.

Antarctica

Lying almost completely south of the Antarctic Circle, Antarctica is the largest land area covered by a permanent ice cap. It has extreme polar and subpolar climates and temperatures are always below freezing. Winters are long, dark, and frigidly cold, with hurricane-force winds and blizzards. Summers are short and cold. Minimal precipitation falls as snow.

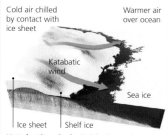

Subpolar climate (below left) Crabeater seals inhabit the shores of the Antarctic Peninsula, where the marine influence moderates temperatures. Although still very cold, it is milder here than farther inland.

Polar climate (below) In the interior, where emperor penguins gather to breed, blizzards can occur at any time of year and temperatures can drop to –40°F (–40°C).

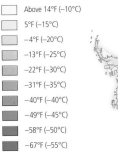
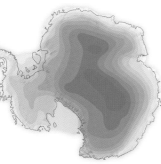

tons/sq. yd.		kg/m²
0		0
		250
0.3		
		500
0.6		750
0.9		1,000

Above 14°F (−10°C)
5°F (−15°C)
−4°F (−20°C)
−13°F (−25°C)
−22°F (−30°C)
−31°F (−35°C)
−40°F (−40°C)
−49°F (−45°C)
−58°F (−50°C)
−67°F (−55°C)

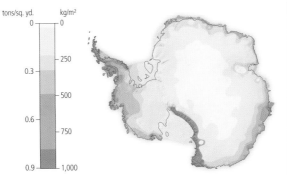

Annual mean temperature (above) The eastern interior of Antarctica, far from the ocean and higher than the rest of the continent, experiences the lowest temperatures.

Annual mean snowfall (above) Antarctica is very dry because the air is too cold to hold much water vapor. Snowfall is higher near the coast and decreases toward the continental interior.

Ross Sea pack ice (left) In winter, a solid ice surface extends from the edge of the Antarctic Continent over the sea far to the north. As temperatures warm in summer, it breaks up to form pack ice.

Dry valleys (above) For a few weeks in summer, the glacial ice in these valleys near McMurdo Sound melts, feeding lakes that remain unfrozen all year beneath a layer of surface ice.

FACT FILE

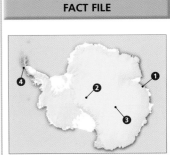

Climographs These stations have long, extremely cold winters and short, cold summers. Very little precipitation occurs, mostly in fall and winter. Stonington receives the most, while Davis is driest. Vostok is the coldest and Casey the warmest.

Average precipitation
Max. temperature
Min. temperature

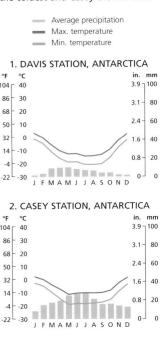

1. DAVIS STATION, ANTARCTICA

2. CASEY STATION, ANTARCTICA

3. VOSTOK STATION, ANTARCTICA

Average temperature

4. STONINGTON ISLAND, ANTARCTICA

CHANGE

CHANGE

HISTORY OF CLIMATE CHANGE

Earth's climate is constantly changing, with periods of glaciation interspersed with warm interglacial periods, triggered by changes in Earth's orbit around the Sun. Our contemporary climate is in one of these warm interglacial periods, with the last ice age ending some 10,000 years ago. It was this Last Glacial Maximum that enabled human settlement of many continents, as glaciations—and the accompanying drop in sea level—exposed land bridges between continents.

Glacial retreat (right) This aerial view of the Greenland Ice Sheet demonstrates the processes driving ice-sheet retreats. As surface snow melts, it forms a darker surface that absorbs more energy from the Sun. This accelerates warming and the movement of ice into the sea. Glacial retreats occur at the end of each ice age.

THROUGH THE AGES

Life began on Earth during a warm period approximately 3.7 billion years ago, when Earth's temperature was about 18°F (10°C) warmer than today. Since then, Earth has experienced four major periods of glaciation of varying degrees, each characterized by alternating ice ages, interspersed with warm interglacial periods, over tens of millions of years. Most plant and animal diversity evolved during these warm periods, and some subsets of species survived each ice age. Our present-day climate is in a relatively warm interglacial phase of the Quaternary Ice Age, the fourth of these periods of intermittent glaciations. Human evolution has taken place entirely during this latest phase.

The Sun is the only external source of energy for Earth. The cyclical nature of Earth's climate over its geological history is driven by changes in its orbit around the Sun. Changes to the shape of Earth's orbit and the angle of Earth's axis relative to its orbit drive overall cyclical patterns in Earth's paleoclimate, or ancient climate. Each cycle spans tens or hundreds of thousands of years. The low point in the amount of energy received from the Sun is hypothesized to have coincided with so-called Snowball Earth, around 700 million years ago, when the planet is thought to have been covered by glacial ice from pole to pole.

This solar-driven climate cycle has been occasionally affected by additional external influences. A notable, but still somewhat controversial, example is the meteor impact approximately 65 million years ago in the Yucatan Peninsula of Mexico. This coincided with the mass extinction of the dinosaurs, the K-T, or Cretaceous-Tertiary, extinction event. Evidence of a meteor impact at that time has been found, giving weight to a mass extinction event theory. The meteor impact ejected a large amount of dust and debris into the atmosphere,

Fossil record (left) Fossils such as this ammonite—an extinct marine animal—help scientists identify the geological period in which a sediment, or rock layer, was deposited. A paleoclimatic record of atmospheric and oceanic temperatures over the history of life on Earth slowly develops.

blocking sunlight. It took months or years for the debris to settle. It is also possible that the meteor impact was sufficiently violent to agitate Earth's subsurface magma and spur a period of enhanced volcanic activity, which would also inject large amounts of solar-reflecting matter into the atmosphere. Both of these occurrences would result in an effective shutdown of photosynthesis and a loss of plant life at the bottom of the food chain, leading to a mass extinction event among larger animals.

In 2000, a new geological epoch—the Anthropocene—was proposed by Nobel laureate Paul Crutzen to recognize the clear dominance of humans in driving Earth's climate and ecosystems. The beginning of this new epoch coincides with the start of the industrial revolution in the late 18th century. Humanity's use of fossil fuels during this period has caused a dramatic and rapid new climate forcing, a change to Earth's energy balance. The marked increase in atmospheric carbon dioxide has no precedent in approximately the last 1 million years, according to currently available ice-core records. It is uncertain what long-lasting impact this new epoch will have on Earth's geological history, but it may rival or even exceed the meteor impact K-T extinction event, which brought the age of the dinosaurs to an end.

Silent witness (below) The Bungle Bungle Range in Western Australia has witnessed 250 million years of geological history. Sedimentary rock, deposited as sand, was uplifted and gradually eroded to leave the distinctive beehive shapes. Recent local annual rainfall increases of up to 10 inches (250 mm) are attributed to a combination of global warming and pollution from Asia.

Ice cores Air bubbles become trapped in vertical layers of glacial ice. They record the atmospheric composition above the glacier at the time each layer was deposited. This includes information on air temperature and trace gas concentrations.

Extracting cores When the face of an ice sheet is exposed—as in this ice mine below the South Pole—a horizontal core can take a large sample from one time.

Dusty year A dark stripe is visible in this vertical core from the Quelccaya Ice Cap in Peru. The stripe is from dust deposited during a windy El Niño year.

Glacial pace An engineer uses GPS equipment to measure the changing positions of poles—and thus the speed of glacial flow—at an Icelandic glacier.

Natural Clues

Nature provides clues to past climates, or paleoclimatology, in many forms. A record of past temperatures and atmospheric carbon dioxide concentrations is preserved in layers of glacier ice, and craters and erosion can expose layers of rock that provide chemical indications of past climate. These natural records help to reconstruct a timeline of ancient-to-modern climate.

Meteor crater (above) The Barringer Crater (also known as Meteor Crater) in the Arizona desert was formed by a meteor impact during the Pleistocene epoch, about 50,000 years ago, when the area was grassland or woodland.

GRAND CANYON LAYERS

265 million years ago
Kaibab Limestone—contains the remains of sea creatures

270 million years ago
Toroweap Sandstone—formed from sand deposited by a sea

275 million years ago
Coconino Sandstone—contains the remains of a vast desert

280 million years ago
Hermit Shale—formed from silt deposited by a river system

300 million years ago
Supai Group—formed from sandstone deposited by rivers and oceans

340 million years ago
Redwall Limestone—contains the remains of later marine creatures

375 million years ago
Temple Butte Limestone—holds remains of creatures who lived in warm seas

540 million years ago
Bright Angel Shale—formed from silt deposited in the sea as it flooded

560 million years ago
Tapeats Sandstone—formed from beaches created by a rising sea

More than 2 billion years ago
Vishnu Schist—a metamorphic rock that was part of a mountain range

265 mya
270 mya
275 mya
280 mya
300 mya
340 mya
375 mya
540 mya
560 mya
More than two billion years ago

Grand Canyon (above) The Colorado River eroded layers of rock over the past few millennia to carve the 1-mile (1.6 km) deep canyon, exposing over 2 billion years of geological and climatic history. Fossils of primarily marine creatures are contained in the rock layers.

Vostok ice core This 10,800-foot (3,300 m) deep vertical ice core from East Antarctica has been analyzed for temperature and the greenhouse gases carbon dioxide and methane. The core spans 400,000 years and four major glacial periods.

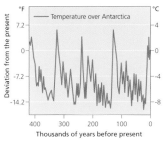

Temperature Glacial conditions existed for most of that time, with temperatures 7–14°F (4–8°C) cooler than at present, punctuated by brief warm periods.

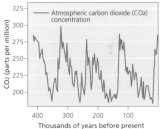

Carbon dioxide Fluctuating CO_2 concentrations have been tracked for the past four glacial cycles. High CO_2 levels correlate with high temperature.

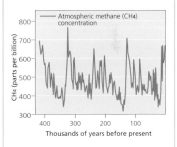

Methane Concentrations of methane have also followed temperature. Current concentrations, at 1,700 ppb, are more than double that of preindustrial times.

Layering Layers deposited over long geological timescales are exposed at the edges of glaciers, such as the Hubbard Glacier in Alaska. Fresh snow accumulates each year and is compressed by subsequent years' snowfalls. Distinct layers are visible: occurrences such as volcanic eruptions are recorded as a dark layer of ash in the ice.

Natural Clues continued

Clues to past climates can be measured in glacial ice or shown in naturally exposed rock. We can also obtain further paleoclimatic information from fossilized plants and animals. More recent climate histories can be gained from boreholes dug into Earth's crust, mineral formations such as stalactites and corals, and even living plants.

Borehole sites (right) Boreholes provide a direct measurement of Earth's recent surface temperature history, by measuring temperature change with depth. The rate of this change will be different if the surface has consistently warmed or cooled than if it has remained constant.

Bristlecone pine This 4,000-year-old tree in California's White Mountain Range records four millennia's worth of California's sunlight, rainfall, and soil moisture in its growth rings.

Stalactites These mineral formations are found in caves where groundwater has trickled through cracks in the rock. Residual calcium bicarbonate forms solid icicle-shaped structures.

Coral reefs Australia's Great Barrier Reef is composed of carbonate minerals similar to that of stalactites, which store information about past climates.

Giant sequoia These sequoias in Redwood National Park, California, contain a record of past climate in the width of their annual growth rings. These widths depend on diverse factors: temperature, rainfall, soil conditions, wind, and tree age.

• Borehole sites

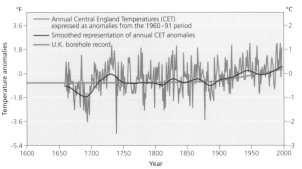

Borehole validation (left) The longest available instrumental record of temperature in the world is the Central England Temperature (CET) thermometer record, which commenced in 1659. The CET annual measurements are consistent with a borehole-based temperature reconstruction for the U.K., demonstrating the accuracy of this method.

Stromatolites (below) These stromatolites, from Shark Bay in Western Australia, are layered carbonate structures formed by cyanobacteria in shallow water. The earliest fossil stromatolites are nearly 3 billion years old. The chemical composition of the carbonate material provides climatic information, particularly about humidity and rainfall.

FACT FILE

Fossilized evidence Fossils indicate the structure of ancient plants and animals and the locations in which they thrived. Fossils can be dated by locating them in layers of sediment and identifying the "index species" with which they occur.

Ferns This fossilized Carboniferous fern lived 300 million years ago, when the world was covered in fern forests.

Shells Fossil snail shells incorporate information about moisture and temperature at the time they formed.

Tree rings Climatic conditions are encoded in tree growth rings; wider rings indicate favorable conditions.

Trilobites The range and evolutionary differences of the 600 species of trilobite correlates to environmental conditions.

Early Ice Ages

Over the first 4.6 billion years of Earth's history, there have been four major periods of glaciation, each with extended ice cover and cycles of ice ages lasting tens of millions of years. The second glaciation period, occurring 850–635 million years ago (mya), began very rapidly and was probably the most severe, resulting in an ice-covered Snowball Earth. Simple organisms survived these glaciations, but the dinosaurs emerged in the warm interglacial period between 250 mya and the present Quaternary Ice Age.

Father of glaciology (right) Swiss-born scientist Louis Agassiz (1807–1873) spent his career studying paleontology and glaciology. His observations in Europe and North America led him to formulate a theory that Earth had experienced an ice age, with widespread glacial cover.

The first 4.6 billion years (below) This timeline reconstructs average surface temperatures since Earth was formed. The horizontal line (blue) shows the present-day average temperature. The four major periods of extended glaciation are seen as dips below the horizon. The globes illustrate changes in continental placement and other major events.

Snowball Earth: 790–635 mya Some scientists believe Earth was completely covered by ice during this ice age.

Cambrian explosion: 540–490 mya The rapid diversification of life-forms included the first mineralized organisms, which later became fossilized.

Gorner Glacier (above) One of the glaciers studied by Agassiz is the Gorner Glacier, in the Swiss Alps. The visible dark seams are examples of moraines. This glacial debris accumulates at the margins or outlets of glaciers, leaving behind evidence of their existence long after they recede.

Temperature

Present-day average temperature

Widespread ice sheets

Brief ice age

ERA	PRECAMBRIAN				PALEOZOIC E
Period		Cambrian	Ordovician	Silurian	De

4,600   ← 1,000   550   500   450   400

Million years ago

Pangea: 270 mya During this long period of glaciation, the supercontinent Pangea was formed. Supercontinent formation and breakup is cyclical.

Jurassic: 200–146 mya In the Jurassic period, much of what is now Europe was submerged. The warm jungle climate was about 5.4°F (3°C) warmer than the present.

K-T extinction event: 65 mya The Cretaceous-Tertiary, or K-T, extinction event marking the end of the Mesozoic era led to the extinction of more than half of all species, approximately 30 percent of genera, including the dinosaurs.

Climate warms, dinosaurs appear

K-T extinction event, dinosaurs disappear

Extended ice age

Quaternary

	MESOZOIC ERA				CENOZOIC	
niferous	Permian	Triassic	Jurassic	Cretaceous	Tertiary	

| 300 | 250 | 200 | 150 | 100 | 50 | Present |

Fossilization Dinosaur skeletons located today in museums were preserved through a process of fossilization. The bones of the animals became hard mineral rocks, and the encasing sedimentary rock locked the skeleton in place.

Underwater A dead dinosaur beneath the surface of the water would not be scavenged by other large animals, leaving an intact skeleton.

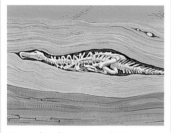

Covered and compressed Subsequent layers of sand and silt are deposited on top of the dinosaur skeleton, preventing it from being disturbed by currents or washed away.

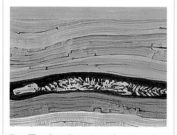

Fossilization Over time, the minerals and organic material in the bones exchange with the much harder minerals in the sand around them.

Exposed Millions of years later, the fossilized skeleton is revealed when the encasing rock is shifted to the surface and eroded away, or exposed by mining.

Extinctions

Life on Earth has been punctuated by extinction events, during which a large percentage of species disappeared in a short time. Over the past 540 million years—the period for which fossil records exist—extinction events have marked the ends of the Cambrian, Ordovician, Permian, Triassic, and Cretaceous periods. The most recent event marked the extinction of the dinosaurs.

Extinction rates (below) Over the past 500 million years, five major events occurred in which over 25 percent of existing genera perished. The Great Dying was the largest extinction, ending the age of mammal-like reptiles and allowing dinosaurs to dominate.

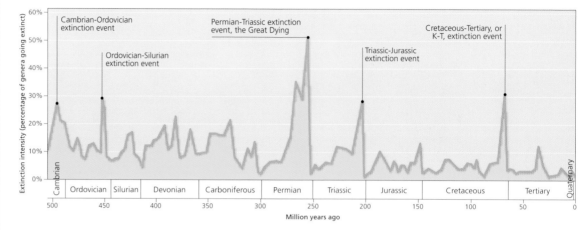

Dinosaur fossils A technician works on a cliff face at the Dinosaur National Monument, Utah, U.S.A. The longest dinosaur skeleton on record, a *Diplodocus*, was discovered in this prehistoric riverbed.

Meteor impact A leading theory to explain mass extinction events is that the impact of large meteors caused global clouds of debris, blocking sunlight from reaching Earth, effectively shutting down photosynthesis and killing plant life.

Volcanic activity A massive meteor impact could also spur increased volcanic activity, emitting large amounts of sulfur dioxide into the atmosphere. Combined with water vapor, sulfur dioxide would form a thick haze, also blocking sunlight.

Acid rain The sulfur dioxide expelled by the volcano turns to sulfuric acid in the haze of the upper atmosphere and descends through the wetter lower atmosphere. This corrosive precipitation would further threatened any surviving plant life.

FACT FILE

Size matters Computer models are used to simulate the impact and effect of mass extinction theories. Extinction events did not kill off all life on Earth. Many smaller animal species survived, while the largest animals perished.

Dimetrodon The Permian period saw the spectacular evolutionary development of many species of reptile, including this mammal-like synapsid, *Dimetrodon*. This top predator of its age died out before the arrival of dinosaurs.

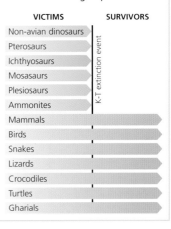

Shunosaurus This sauropod dinosaur, a class that includes the largest animals ever to live on Earth, was an early plant eater from the middle Jurassic period.

K-T EXTINCTION EVENT

During the K-T extinction event (65 mya), all species of dinosaur became extinct, but some mammals, birds, and smaller reptiles were left to evolve into larger species.

VICTIMS	SURVIVORS
Non-avian dinosaurs	
Pterosaurs	
Ichthyosaurs	
Mosasaurs	
Plesiosaurs	
Ammonites	
Mammals	
Birds	
Snakes	
Lizards	
Crocodiles	
Turtles	
Gharials	

K-T extinction event

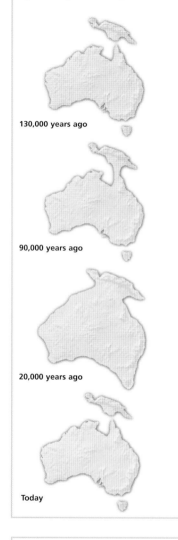

130,000 years ago

90,000 years ago

20,000 years ago

Today

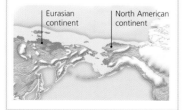

Eurasian continent | North American continent

After the Last Ice Age

The climate of the past 2 million years has seen significant variability, affecting the evolution of large animals and the migration of animal and human populations. Earth is currently in a period of episodic glaciation with the most recent ice age peaking at the Last Glacial Maximum about 20,000 years ago and ending about 10,000 years ago. Our current climate is interglacial, with glaciers in Greenland and Antarctica in gradual retreat.

Vegetation zones (below) During the Last Glacial Maximum, around 20,000 years ago, much of North America and northern Eurasia was covered by ice sheets and tundra. At lower latitudes, multiple climate regimes existed—from temperate grasslands to tropical rain forest.

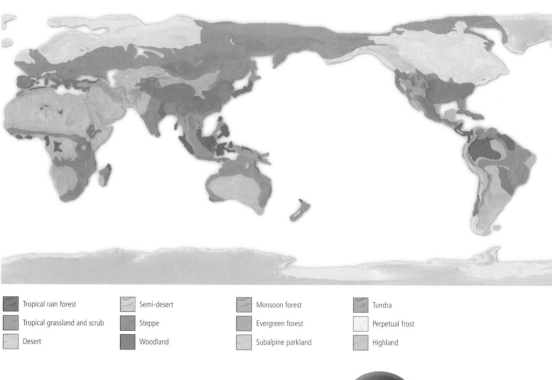

■ Tropical rain forest	■ Semi-desert	■ Monsoon forest	■ Tundra
■ Tropical grassland and scrub	■ Steppe	■ Evergreen forest	■ Perpetual frost
■ Desert	■ Woodland	■ Subalpine parkland	■ Highland

The past 2 million years This timeline reconstructs average surface temperatures over the past 2 million years. The horizontal line (blue) represents the present-day average temperature. There have been relatively rapid oscillations between warm interglacial and cold glacial periods, when the polar ice sheets extended toward the Equator.

Last Glacial Maximum: 20,000 years ago During this last major ice age, ice covered all of Canada and most of northern Europe.

Appearance of *Homo sapiens*

Holocene climate optimum warm period, development of farming

Present-day average temperature

Temperature

Ice ages occur roughly every 100,00 years

Last glacial maximum

Period		Quaternary
Epoch	Pleistocene	

← 1,600   800   600   400   200   ← 20   ← 11   8

Thousand years ago

Little Ice Age (above) From the 16th to the mid-19th centuries, the Northern Hemisphere experienced a slight cooling of about 1.8°F (1°C). Glaciers advanced in the European Alps and the River Thames in England froze over. Cold conditions were depicted in European art, such as this 1683–84 engraving of the Thames.

Mammoths in America (below) The Columbian mammoth, a member of the elephant family, lived in North America during the late Pleistocene epoch. The mammoth was one of the final examples of megafauna from the last ice age to go extinct, around 10,000 years ago.

Today The polar ice cap extent is limited to the Arctic Ocean and the Antarctic continent.

Medieval climate optimum warm period

Little Ice Age in Europe

Holocene

4   2   1   Present

Evidence of change Animal and plant remains, ruins of previous settlements, and even depictions in art provide evidence of climate change. Some of these changes have been quite recent, within the past 1,000 years.

Nile crocodile Today, crocodiles are found along the Nile River, but this animal inhabited the entire Sahara region during the warm middle Holocene epoch, when the area had more lakes.

British wine During the Medieval warm period, Britain's climate was mild enough to grow grapes to make wine. British viticulture largely disappeared as the climate again cooled.

Greenland's optimum Currently, Greenland is almost entirely covered by glaciers, but in the 10th century it had a sufficiently hospitable climate to support a large human settlement.

Sunspots The Little Ice Age in Europe corresponded with a period of very low sunspot activity. Because high sunspot activity is associated with increased energy output from the sun, it has been speculated that this period of low activity may have caused the Little Ice Age.

NEW YORK CITY

Megacities such as New York City, with their built-up infrastructure and high population density, are among the places most vulnerable to climate change. Rapid warming has been observed in urban areas of New York. The city's coastal location also makes it vulnerable to changes in sea level. Sea level projections suggest that some of Manhattan could be submerged by this century's end.

WITNESS TO CHANGE

The last ice age saw the Laurentide Ice Sheet reaching as far south as the American Midwest and New England. As it retreated northward, the ice sheet carved out Niagara Falls and the Great Lakes. In New York City's Central Park, grooves left by these glaciers can still be seen. The Last Glacial Maximum also saw the formation of the land bridge across the Bering Strait that enabled human settlement of the Americas.

Modern climate in New York City, as elsewhere, has been variable. Major blizzards struck in 1888, 1914, 1960, 1978, and 2006, with an all-time record snowfall of 27 inches (69 cm) in 2006. The same year also saw a record heat wave in July and August, worsened by power outages. Heat waves have occurred frequently in New York's history, with an average 14 days per year with temperatures over 90°F (32°C)

since 1971. Seven of the 10 years with the highest number of hot days in the past century have occurred since 1980. Precipitation records in New York City show a slight trend to more extreme precipitation events in recent years.

A decadal warming trend is also clear. Urban areas, with their industry, intense energy use, air pollution, and loss of vegetation to built infrastructure, experience this warming especially acutely. Increasingly frequent summer heat waves overtax the electricity grid and medical services.

New York City authorities are preparing municipal services for the changes projected by global climate models. These changes include an increase in the number and duration of heat waves, increased coastal flooding as sea levels rise, and more intense precipitation events but also more frequent droughts.

Frigid weather (above) The Blizzard of 1888 produced temperatures in March well below 0°F (–18°C), with high wind gusts and deep snowdrifts stranding several cities on the eastern seaboard. New York City suffered the most damage, particularly to its harbor. Here, trains cross the snow-covered Brooklyn Bridge after the blizzard.

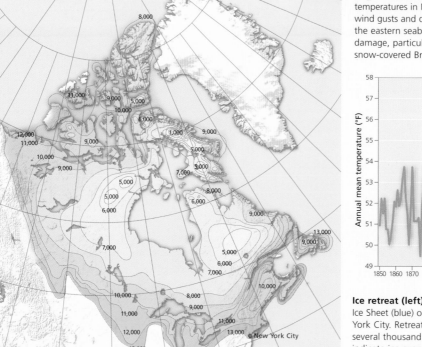

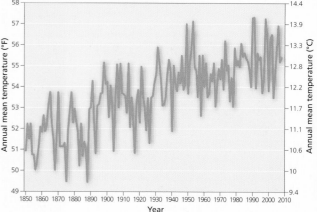

Ice retreat (left) The Laurentide Ice Sheet (blue) once reached New York City. Retreat occurred over several thousand years. Contours indicate ice exent across North America in years before the present.

Warming trend (above) The annual mean temperature of Central Park, recorded between 1850 and 2008, shows warming of around 3.6°F (2°C). Warming was most rapid prior to 1950, reflecting the pattern of urbanization.

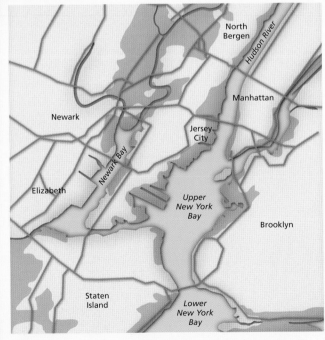

Hot 1961 (above) Heat waves are felt most acutely in urban areas, where concrete and asphalt retain heat and residents have fewer places to escape the scorching temperatures. These women kept cool during the September 1961 heat wave by moving a Central Park bench into the water.

Manhattan pollution (below) Local climate in cities is further complicated by air pollution. This haze layer can build up and prevent sunlight from reaching the surface. However, any cooling effect from this smog is negated by the impact on human health of these fine particles.

North Bergen

Hudson River

Manhattan

Newark

Jersey City

Newark Bay

Elizabeth

Upper New York Bay

Brooklyn

Staten Island

Lower New York Bay

Threat from the sea (above) Some models predict a rise in sea level of 3.3 feet (1 m) this century. The areas shaded in green show the ocean inundation of the New York City region that would accompany such a rise.

NATURAL PATTERNS

Over the geologically brief period of recorded human history, significant climate variability has been observed. Some natural climate cycles have impacted on human civilization and we can learn about these from historical documents or archaeological remains. Other natural climate cycles, such as the major glaciation–interglacial oscillations that occur approximately every 100,000 years, are not on a timescale that our species has experienced, but leave geological clues to past climates. Understanding natural climate patterns and timescales is critical to the study of human impact on this background variability.

Wildfire (above) A warming climate brings hotter temperatures and drier conditions. When densely forested regions have hotter summer temperatures and earlier spring snow melts, wildfire frequency increases. This fire, north of Fairbanks, Alaska, occurred during the wildfires of 2004 that destroyed 10 percent of Alaska's boreal forest area.

Water shortage (right) A hotter and drier climate brings water shortages. This dry dam bed in Montpellier, France, was photographed in May 2001, when a heat wave swept through Europe. Several years of above-average temperatures and below-average rainfall led to increased demand for water.

CYCLES OF CHANGE

Earth's climate varies naturally, over long and short timescales. Essentially all natural climate variability arises from the Sun, our single external source of energy. Earth's orbit around the Sun alters in several predictable cyclical patterns that affect how much sunlight reaches different latitudes and, hence, drives climate changes. The shape of Earth's orbit changes in cycles of 400,000 and 100,000 years, driving the major glaciations. Earth's tilt and gyroscopic wobble also vary, in 23,000- and 41,000-year cycles respectively. The variations in solar radiation on Earth's surface drive the cyclical expansion and retreat of the polar ice caps. These changes have led alternately to ice ages, with their accompanying mass extinction events, and the balmy tropical climate under which the dinosaurs evolved and thrived.

Another impact on Earth's climate comes from continental drift. The tectonic activity that moves plates of Earth's crust and alters the location and size of the oceans naturally influences climate systems. The last supercontinent—Pangea—existed about 270 million years ago; since then the continents have been moving apart, leaving behind separated oceans with distinctive circulation patterns. Earth's climate and weather is governed by massive atmospheric heat cycling that is influenced by ocean circulation, so moving continents necessarily affect climate. Furthermore, the changing position of the continents impacts the formation and stability of the polar ice caps.

Natural changes also occur on much shorter timescales. Sunspot activity, which affects solar output, occurs in a somewhat irregular cycle of approximately 11 years. The El Niño Southern Oscillation dramatically affects temperature and precipitation across the equatorial Pacific and around the globe. Much current climate research endeavors to determine the potential impacts of human activity on this and other natural cycles.

Tropical cyclone (below) The ocean–atmosphere heat engine can generate tropical cyclones when a low-pressure area forms above warm surface waters. This twin cyclone system developed between Iceland and Scotland in November 2006, a rare high-latitude cyclone that can form in regions of high temperature contrasts.

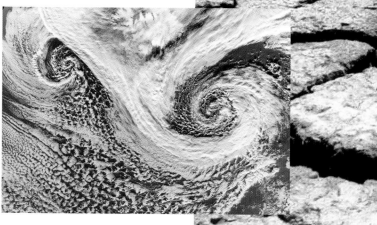

Natural Changes

The primary driver of Earth's climate variability is the Sun. Sunlight provides our external energy and the distribution of this energy on Earth's surface varies with our planet's orbit. The output of solar energy also varies with sunspot activity. This changing solar energy output and distribution has caused the expansion and retreat of the polar ice sheets. Sporadic volcanic activity is another natural determinant of Earth's variable climate.

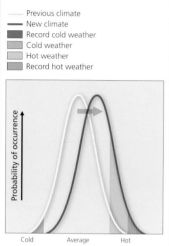

Increase in average temperature
This shifts the entire curve to a higher temperature, with more hot weather and hotter extremes, and less cold weather.

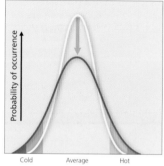

Increase in climate variability This flattens the curve without changing the average, resulting in more hot and cold days, and more hot and cold extremes.

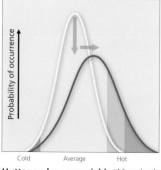

Hotter and more variable When both changes occur simultaneously, the curve flattens and shifts to a higher average temperature, with more hot weather.

Tilt (below) The tilt of Earth's axis, which gives us our seasons, varies between 22 and 24.5 degrees during a 41,000-year cycle. This causes up to 15 percent change in solar radiation at the Poles, altering the energy balance at the polar ice caps, and giving rise to glaciation cycles.

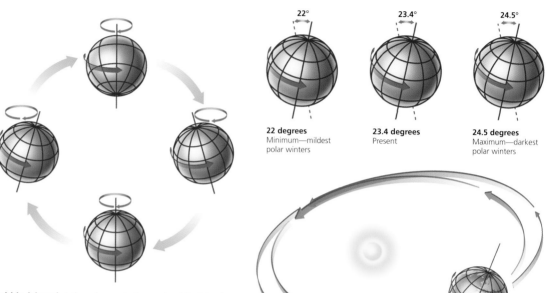

22° 23.4° 24.5°

22 degrees
Minimum—mildest polar winters

23.4 degrees
Present

24.5 degrees
Maximum—darkest polar winters

Wobble (above) A slow change in the angle of Earth's axis causes Earth to wobble like a gyroscope. This precessional cycle takes approximately 23,000 years to complete, and varies the amount of sunlight received by the North and South Poles.

Abandoned (below) These ancient cliff dwellings of the Anasazi people, in Colorado, were abandoned in the late 13th century, possibly due to drought or other climactic disturbances. This was during the time of the Medieval Warm Period in Europe.

Orbit (above) Earth's annual orbit changes shape gradually—over 400,000- and 100,000-year cycles—from an oval path (blue) to a rounder path (red). This results in Earth being further or closer to the Sun, hence receiving varying amounts of solar energy.

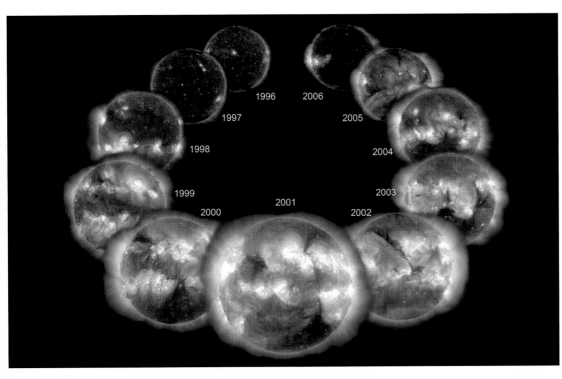

1996 2006

1997 2005

1998 2004

1999 2003

2000 2001 2002

Solar output (above) Solar activity was observed by satellite during a full solar cycle of 11 years, from 1996–2006. During this cycle, the Sun's magnetic field changes. This causes variations in sunspot activity and flares, which alters solar energy output.

Ice carving (right) One symptom of the changing energy balance of Earth is the formation and retreat of glaciers. Geological features such as Watkins Glen, in western New York, are evidence of the last North American glacial retreat.

1. Patagonia The Patagonian glacier spans a mountainous region in Chile and Argentina. Long parallel valleys have been cut into the mountains as the glacier recedes, and the ice that is left is riddled with deep crevasses. The outlets of this massive glacier system are receding at varying rates.
Patagonia, South America

2. Great Lakes The Great Lakes in northeastern U.S.A. were carved by the Laurentide Ice Sheet, which retreated at the end of the Last Glacial Maximum. Much of the topography of North America's lake and river systems was carved by this massive glacier.
Great Lakes, U.S.A.

3. Bitter Springs The Bitter Springs formation, in northern Australia, is a rich rock bed where many Proterozoic fossils have been found, including cyanobacteria and marine microfossils up to 850 million years old. These finds help reveal the complexity and evolutionary timescale of early marine life.
Bitter Springs, Australia

4. The Sahel A semi-arid tropical savanna south of the Sahara desert, the Sahel has a transitional climate and seasonal wetlands. In the past 6,000 years, its climate has dried, causing shrinking lakes, desertification, and a decline in animal populations.
The Sahel, Africa

Natural cooling Forces other than the Sun can cause natural variability in Earth's climate. Volcanic eruptions emit large quantities of ash and aerosols into the upper atmosphere. These diminish the amount of solar radiation reaching Earth, causing a rapid but temporary cooling effect.

El Niño and La Niña

El Niño and La Niña designate weather conditions experienced at either extreme of the Southern Oscillation, a massive coupling of oceanic and atmospheric heat transfer across the Pacific Ocean near the Equator. These conditions cycle irregularly: every two to seven years for El Niño, which rarely lasts longer than a year and alternates with opposing La Niña events.

Water temperature Strong El Niño weather conditions occurred from 1997 to 1998. These images show the evolution of the depth of ocean temperature during the event, with temperatures ranging from 46.4°F (8°C), in blue, to 86°F (30°C), in red.

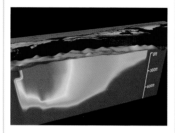

January 1997 During the normal conditions that precede El Niño, the ocean is warmest throughout its depth in the western Pacific Ocean.

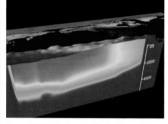

June 1997 Trade winds slacken with the onset of El Niño, reducing cold upwelling in the eastern Pacific and allowing warm water to spread east.

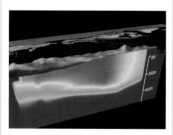

November 1997 Warm water sinks deeper and reaches South America as El Niño develops fully. The water near Australia's east coast begins to cool.

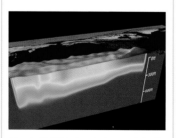

March 1998 Trade winds are minimal when El Niño peaks. Warm water off the coast of South America causes localized heavy rainfall and flooding.

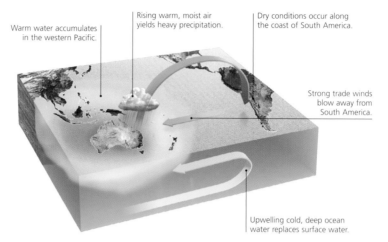

Warm water accumulates in the western Pacific.

Rising warm, moist air yields heavy precipitation.

Dry conditions occur along the coast of South America.

Strong trade winds blow away from South America.

Upwelling cold, deep ocean water replaces surface water.

Normal (above) Trade winds push warm surface water westward, resulting in the upwelling of cold water along the eastern boundary of the Pacific Ocean. Evaporation causes heavy rainfall over Australia.

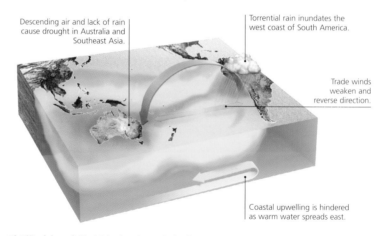

Descending air and lack of rain cause drought in Australia and Southeast Asia.

Torrential rain inundates the west coast of South America.

Trade winds weaken and reverse direction.

Coastal upwelling is hindered as warm water spreads east.

El Niño (above) Diminished surface winds allow warm water to spread across the Pacific, preventing the normal cold, deep ocean water upwelling. Heavy precipitation falls in South America.

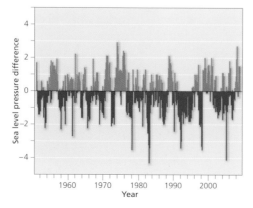

Southern Oscillation Index Readings taken in Tahiti and Darwin, Australia, measure the strength of trade winds, which flow from high to low pressure. El Niño events are characterized by negative (blue) values and La Niña by positive (red). In 2008, La Niña conditions prevailed.

FACT FILE

1. Drought in Australia Parts of Australia experience extreme drought during El Niño events. Strong El Niño events in 2003–04 and 2006–07 led to severe and extended droughts, with 2004 the worst year on record for

 eastern Australia. New policies, such as gray water recycling, have been introduced to lessen the demand for water.

Australia

2. Drought in Southeast Asia Dry conditions in Southeast Asia also accompany El Niño events. The most severe drought on record occurred in late 2004 and early 2005, after the 2004 wet season ended early. Thailand

 suffered the greatest crop losses, primarily sugar, but agriculture declined throughout the region, leading to food shortages.

Southeast Asia

Floodwater Villagers attempt to slow a mudslide with stones outside Lima, Peru. Strong El Niño conditions result in severe flooding in Peru and Ecuador. Scientists speculate that a warming climate will intensify El Niño events, increasing the threat to these coastal regions.

LA NIÑA

La Niña often follows El Niño, especially a strong El Niño event, and has basically the opposite effect. It is characterized by unusually cold ocean temperatures, shown in purple, extending further west across the Pacific than usual.

El Niño and La Niña continued

The influence of El Niño and La Niña events are felt globally. As the conditions reach their peak around January, widespread warming (El Niño) or cooling (La Niña) is observed, along with disturbances to rainfall patterns. The more destructive warm phase, El Niño, produces extreme flooding in the eastern Pacific and severe drought in the western Pacific.

1. Rain in the Americas The effects of El Niño vary across the world, with the U.S. eastern seaboard experiencing more intense rains. This is an example of teleconnections: related climatic anomalies can be observed across large geographical distances.

Eastern U.S.A.

2. Drought in southeast Africa During the strong 2004 El Niño event, many drought-stricken countries in southeast Africa faced food shortages 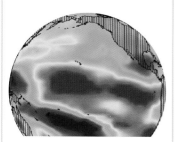 in January and February. Soil dried out after several countries received less than 50 percent of average rainfall.

Southeast Africa

Water vapor Satellite analysis of atmospheric water vapor provide striking evidence of peak El Niño conditions. High concentrations of water vapor (red) over the Pacific Ocean are due to warmer than normal ocean surface temperatures.

February 3, 1998

February 9, 1998

Philippine drought (right) El Niño brings severe drought to the Philippines. At the end of a prolonged El Niño event in 1998, a farmer salvages what remains of his crop of kangkong, or water spinach, from a dried pond outside General Santos City.

Weather extremes (left) The weather patterns resulting from El Niño and La Niña events generate regions of wetter and drier than normal conditions. At the peak of an El Niño event, in December–February, the normal seasonal warming in the equatorial Pacific is strongly pronounced, and unusually warm conditions are observed in several regions at higher latitudes. Peak La Niña conditions during this time of year are characterized by colder than normal temperatures in the same regions.

Dry and warm
Warm
Dry
Wet and warm
Wet
Wet and cool
Cool

PEAK EL NIÑO CONDITIONS, DECEMBER–FEBRUARY

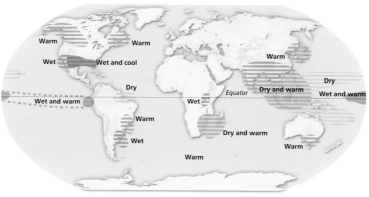

PEAK LA NIÑA CONDITIONS, DECEMBER–FEBRUARY

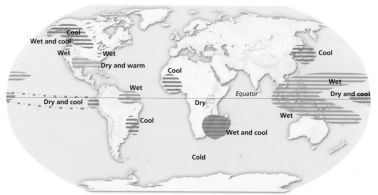

Ecuadoran flood (below) The wet conditions that characterize El Niño events along the west coast of South America can manifest as severe flooding. This house in Daule, Ecuador, was isolated by floods during a severe El Niño event in March 2002. Floods cause many deaths and damage local agricultural economies.

Marine life The El Niño phenomenon was first noted by Peruvian and Ecuadoran fishermen. They noticed that the occurrence of an occasional warm ocean current along the coast in December and January correlated with a diminished abundance of fish.

Anchovy Anchovies thrive in the cool, nutrient-rich upwelling water off the coast of South America. Peruvian fishermen notice that anchovies disappear during El Niño events.

Pelagic red crab Thousands of these crabs washed ashore in San Diego, California, in May 2002 when an El Niño event was building. They ride the warm El Niño ocean currents.

Tropical fish During the very strong 1997–98 El Niño event, several species of tropical fish appeared in the Bahia de San Quintin lagoon in Mexico. This is normally a cold-water lagoon.

Californian landslide Increased rainfall also occurs on the west coast of the U.S.A. during El Niño events. In March 1998, the effects were felt in a Los Angeles, California, suburb, where luxury homes slid down a hillside eroded by heavy rain.

1. Melting One alarming trend is the accelerated melting of the Greenland Ice Sheet, the second-largest body of ice in the world after the Antarctic Ice Sheet. It is losing mass more rapidly than Antarctica.

A complete melting of the Greenland Ice Sheet would increase global sea levels by about 23 feet (7 m).
Greenland

2. Desertification Hotter and drier weather in the Mediterranean region is leading to desertification, turning arable land into infertile desert. This

process is primarily driven by overgrazing and poor land-use practices, but is also accelerated by drought.
Mediterranean coast, northern Africa

3. Snow retreat Retreat of the iconic "snows of Kilimanjaro," in Tanzania, provides another indication of climate change. The year-round snow cover on 15,100-foot (4,600 m) high Mount

Kilimanjaro, the highest peak on the African continent, is receding rapidly, and may be gone by as early as 2015.
Mount Kilimanjaro, Tanzania

4. Biodiversity stress Despite its small size, Costa Rica boasts five percent of the world's species. Loss of biodiversity in this rich tropical ecosystem is at least

partly due to climate change. In 1989, one "marker species," the golden toad of Monteverde Cloud Forest, disappeared.
Costa Rica, Central America

5. Ice shelf collapse The Larsen Ice Shelf, in Antarctica, is a series of three shelves extending from a land anchor out over the Weddell Sea. They are very vulnerable to the warming ocean

currents underneath. Larsen A collapsed in 1995 and Larsen B in 2002. Larsen C is also threatened by warming oceans.
Larsen Ice Shelf, Antarctica

2000 and Beyond

Earth's climate around the turn of the 21st century shows clear evidence of rapid change occurring. Certain climate hot spots have already witnessed pronounced climate change, while other parts of the world have not yet been noticeably disrupted. The relatively gradual nature of climate change in these locations can lead to a dangerous false sense of security.

Lakefront property (above) Human impacts on local climate are manifold. Shallow Lake Chad, in central Africa, has been shrinking in recent decades due to increasing demand for water by the 20 million people living near it. Overgrazing around the lake has also led to decreased vegetation and desertification, causing the lake to recede faster.

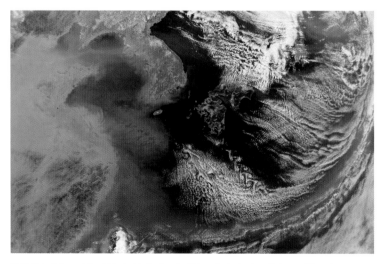

Pollution (above) Rapid industrialization in China, fueled primarily by coal-burning power plants, results in a massive plume of haze and pollution, captured here in a February 2004 satellite image. In addition to polluting the air, the haze diminishes the amount of sunlight reaching Earth's surface, cooling local climate and reducing plant productivity.

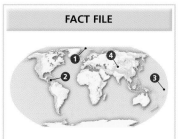

1. Ice cap retreat Arctic temperatures have increased at a more rapid rate than the global average. As a result, the size of the permanent polar ice cap has decreased by about 20 percent

since 1979. It is projected that the Arctic region may be ice-free in summer by the end of the 21st century.

The Arctic

2. Coral bleaching Another sign of a warming, carbon-dioxide-enriched climate is coral bleaching. The more acidic and warmer oceans are also

less hospitable to reef ecosystems. Fish densities have declined significantly in Caribbean reefs since about 1995.

Caribbean

3. Sea-level rise A warming climate causes a rise in sea levels, due to melting land ice and the thermal expansion of warming oceans. Low-lying countries such as Tuvalu, occupying four reef

islands and five atolls with a peak elevation of 15 feet (4.5 m), already experience the effects of this rise acutely.

Tuvalu, Pacific

4. Glacial retreat Central Asia's Tian Shan mountains show another sign of a rapidly warming climate. The glacier atop this 24,000-foot (7,300 m) high

mountain range has been retreating since the 1970s, which correlates with increasing summer temperatures.

Tian Shan mountains, China

Breaking the ice Photographed on January 11, 2008, this enormous piece of ice broke off the Knox Coast, in the Australian Antarctic Territory, to become an iceberg in the Southern Ocean. Increasingly, rapid ice shelf collapse accompanies a warming of the water surrounding Antarctica.

CHANGING CLIMATE

Human civilization, characterized by the rise of agriculture and animal husbandry, arrived on the scene approximately 10,000 years ago, and began to have a significant impact on global climate and ecosystems about 200 years ago. The current epoch has been dubbed the Anthropocene, in recognition of the dominant effect of human activities on the planet's ecosystems. This human footprint extends from the atmosphere, oceans, and land, to the ice at the Poles.

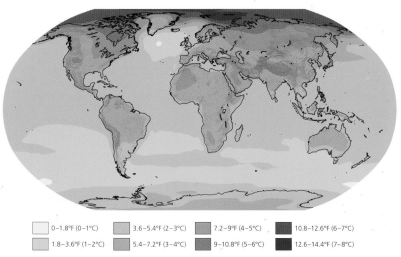

PROJECTED INCREASE IN SURFACE TEMPERATURE BY 2099

0–1.8°F (0–1°C)	3.6–5.4°F (2–3°C)	7.2–9°F (4–5°C)	10.8–12.6°F (6–7°C)
1.8–3.6°F (1–2°C)	5.4–7.2°F (3–4°C)	9–10.8°F (5–6°C)	12.6–14.4°F (7–8°C)

Climate model (left) This model produced by the Intergovernmental Panel on Climate Change (IPCC) predicts warming from now until 2099, based on a future emissions scenario where population peaks in 2050 and new, efficient technologies spread.

Ecosystem shift (right) A red tide off the South Island, New Zealand, is an ecosystem shift that accompanies climate change. Higher ocean temperatures and stimulated photosynthesis due to elevated CO_2 cause shifting location and timing of algal blooms.

Unusual snowfall (below) Significant seasonal and interannual variability is expected on top of the overall global warming trend. A snowstorm, in February 2008, at the Olympian Zeus archaeological site in Greece exemplifies climate variability.

HUMAN DISTURBANCE

Since the advent of agriculture, humans have been disturbing Earth's climate system. In recent years, however, our growing population—6.8 billion and counting—and industrialization have caused unprecedented and accelerating changes. The discovery of the energy source in fossil fuels enabled the industrial revolution, dramatically improving quality of life but ushering in an era of environmental destruction.

Chief among the industrial pollutants is carbon dioxide (CO_2), the prime perpetrator of the greenhouse effect that causes global warming. Thus far, our emissions have increased atmospheric CO_2 to 384 parts per million (ppm), far exceeding the natural range of the past 650,000 years (180–300 ppm). This has resulted in an increase in global average temperature of 1.3°F (0.7°C) since industrialization. The Arctic has warmed at a more rapid rate and is the site of some of the most dramatic evidence of climate change: the area of the polar ice cap has diminished and the Greenland Ice Sheet is melting rapidly, contributing to global sea-level rise.

The fouling of the atmosphere extends to other industrial pollutants.

These cause public health problems, acidic precipitation that damages ecosystems, and atmospheric particles that change overall precipitation patterns. Shifting weather patterns over a warming ocean render coastal communities more vulnerable to storms. Rising sea levels, caused by melting ice sheets and thermal expansion of the oceans, submerge fertile coastal land or inundate it with salt water.

A warmer planet also experiences more frequent drought and the desertification of formerly fertile lands, stressing food resources. As glaciers melt, a critical freshwater supply is diminished. Wildfires become more regular occurrences, further polluting the atmosphere and removing trees, a valuable sink for atmospheric CO_2. These are the range of challenges that face us in the Anthropocene epoch.

Mount Kilimanjaro Retreating glacial cover on Mount Kilimanjaro provides highly visible evidence of a warming planet. Permanent snow and ice on the summit have almost completely disappeared. The ice cap formed over 11,000 years ago.

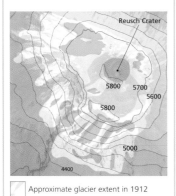

Reusch Crater

5800   5700
       5600
5800

5000

4400

░ Approximate glacier extent in 1912

▨ Glacier extent in 2003

— Rim of summit plateau

Retreat Satellite photographs of Mount Kilimanjaro's crater capture 10 years of glacial shrinkage. Over 80 percent of ice has been lost over the past century, and the summit may be ice-free by 2015.

February 17, 1993

February 21, 2000

June 2, 2003

Global Changes

Evidence of global warming can be seen today around the world in temperature trends, retreating ice caps, shrinking mountain glaciers, warming oceans, rising sea levels, biodiversity loss, and failing human health. Erratic weather patterns have also been observed, such as heat waves, droughts, floods, and coastal storms of escalating frequency and intensity. These events are harbingers of things to come as global temperatures rise.

Severe weather (below) A common feature in the predictions of all global climate models is an increase in severe weather events as the climate warms. There will be droughts in some regions and floods in others. Heat waves and melting glaciers also attest to the warming trend.

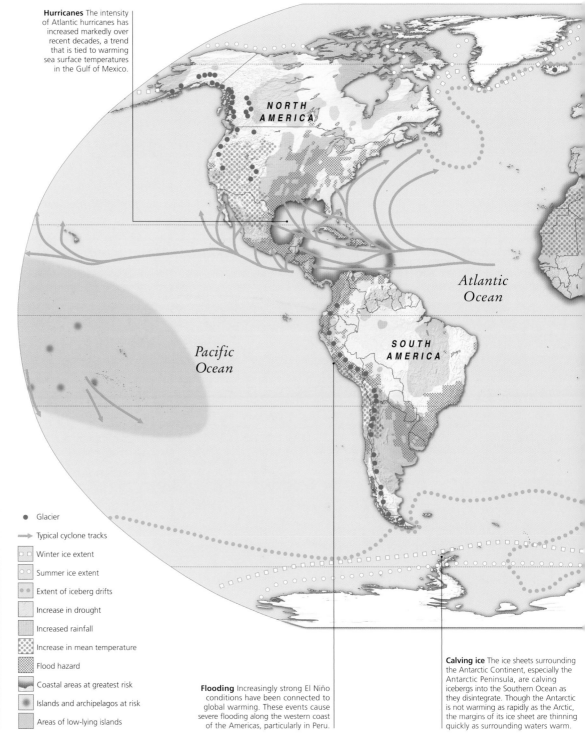

Hurricanes The intensity of Atlantic hurricanes has increased markedly over recent decades, a trend that is tied to warming sea surface temperatures in the Gulf of Mexico.

NORTH AMERICA

Atlantic Ocean

Pacific Ocean

SOUTH AMERICA

● Glacier

⟶ Typical cyclone tracks

▫ Winter ice extent

◦ Summer ice extent

◦ Extent of iceberg drifts

░ Increase in drought

▨ Increased rainfall

▨ Increase in mean temperature

▨ Flood hazard

▨ Coastal areas at greatest risk

◉ Islands and archipelagos at risk

░ Areas of low-lying islands

Flooding Increasingly strong El Niño conditions have been connected to global warming. These events cause severe flooding along the western coast of the Americas, particularly in Peru.

Calving ice The ice sheets surrounding the Antarctic Continent, especially the Antarctic Peninsula, are calving icebergs into the Southern Ocean as they disintegrate. Though the Antarctic is not warming as rapidly as the Arctic, the margins of its ice sheet are thinning quickly as surrounding waters warm.

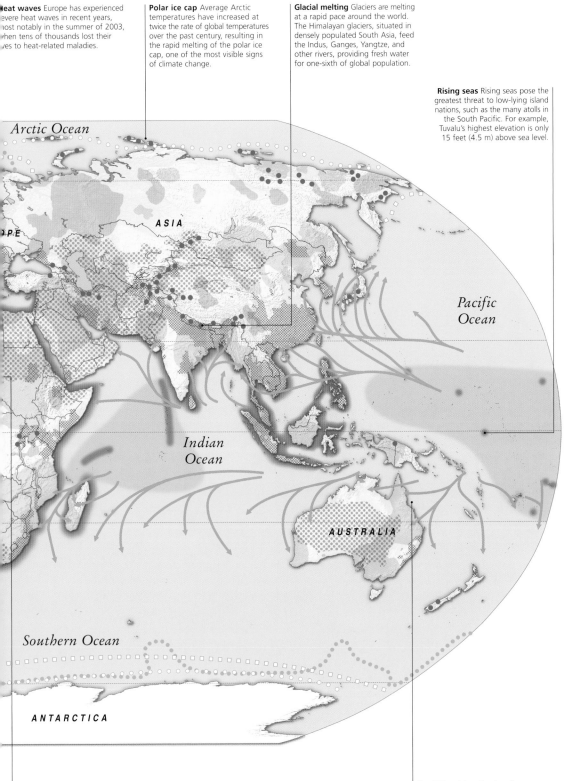

Heat waves Europe has experienced severe heat waves in recent years, most notably in the summer of 2003, when tens of thousands lost their lives to heat-related maladies.

Polar ice cap Average Arctic temperatures have increased at twice the rate of global temperatures over the past century, resulting in the rapid melting of the polar ice cap, one of the most visible signs of climate change.

Glacial melting Glaciers are melting at a rapid pace around the world. The Himalayan glaciers, situated in densely populated South Asia, feed the Indus, Ganges, Yangtze, and other rivers, providing fresh water for one-sixth of global population.

Rising seas Rising seas pose the greatest threat to low-lying island nations, such as the many atolls in the South Pacific. For example, Tuvalu's highest elevation is only 15 feet (4.5 m) above sea level.

Arctic Ocean

ASIA

EUROPE

Pacific Ocean

Indian Ocean

AUSTRALIA

Southern Ocean

ANTARCTICA

Drought The Sahel region of sub-Saharan Africa experiences frequent droughts, caused by overgrazing and poor land management. Drought frequency has increased in recent years due to climate change, human-produced atmospheric aerosols, and warming seas shifting regional precipitation patterns.

Coral bleaching Coral reefs are highly sensitive to changes in ocean temperature and acidity. Widespread bleaching death of reef ecosystems, such as the Great Barrier Reef, is an indication of destructive changes happening beneath the surface.

Warming The Intergovernmental Panel on Climate Change predicts a global average warming trend of 3.6–7.2°F (2–4°C) over the coming century. This will cause sea levels to rise, more frequent extreme weather, and the spread of disease.

Rising seas Sea levels have risen at a rate of 0.12 inches (0.3 cm) per year since 1993. Sea levels are projected to rise 7–23 inches (18–58 cm) over the next century, submerging low-lying islands.

Extreme weather Tropical cyclones are common in the Caribbean, South Pacific, and Bay of Bengal, but in June 2007, Cyclone Gonu formed in an unusual location: the Arabian Sea.

Health risks A warming climate and more frequent flooding contribute to the spread of disease, especially in areas with poor sanitation and difficult living conditions, such as Bangladesh.

Population Density

The world's population is becoming increasingly concentrated in major urban areas. Today, there are about 25 megacities, each inhabited by more than 10 million people. In 1950, there was only one: New York City. As of 2008, over half of the world's population resides in urban areas. This concentration of people has efficiency benefits for industry and transport, but brings with it problems of housing and maintaining adequate air and water quality.

New Delhi (right) The metropolitan area of Delhi, which includes New Delhi, has the second-highest population in India after Mumbai. Around 12 million people live in an area of 570 square miles (1,500 km²). It has the highest number of vehicles of any city in the world—5.5 million in 2008.

FACT FILE

1. Mumbai India's largest city, Mumbai, has a current population of around 14 million people. The city's location on the west coast and its natural harbor led to its establishment as a maritime hub. It developed into a financial and commercial center, growing rapidly as people came in search of work.

Mumbai, India

2. Shanghai With 19 million people, Shanghai is the most populous city in the world's most populous country, China. The city prospered initially as a trade hub. Since economic reforms were introduced in 1990, the population has increased by about 25 percent.

Shanghai, China

3. Istanbul Bridging Europe and Asia, Istanbul served as an imperial capital for nearly 2,000 years. As long ago as A.D. 500 it had a population of 500,000. Its growth has accelerated in recent years, doubling since 1990. It now ranks as one of the world's most populous cities, with more than 12 million inhabitants.

Istanbul, Turkey

4. São Paulo The largest city in the Southern Hemisphere and in South America is São Paulo. Its 11 million inhabitants come from diverse ethnic backgrounds, with people of Italian, Portuguese, African, and Arabian ancestry making up four of the largest communities.

São Paulo, Brazil

5. Mexico City Nine million people inhabit densely populated Mexico City. Its metropolitan area population of 22 million is the largest in the Western Hemisphere. High population density, the burning of wood fuel, and its location in a high valley make Mexico City's air quality among the world's worst.

Mexico City, Mexico

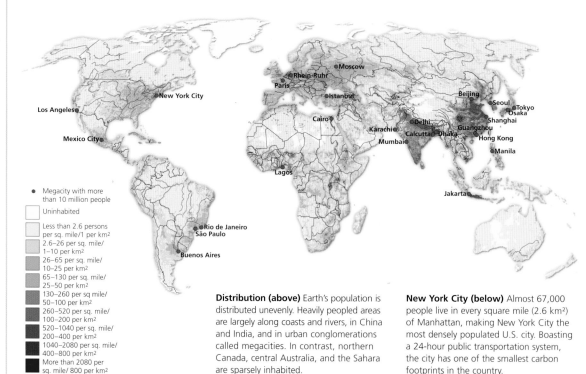

- Megacity with more than 10 million people
- Uninhabited
- Less than 2.6 persons per sq. mile/1 per km²
- 2.6–26 per sq. mile/1–10 per km²
- 26–65 per sq. mile/10–25 per km²
- 65–130 per sq. mile/25–50 per km²
- 130–260 per sq mile/50–100 per km²
- 260–520 per sq. mile/100–200 per km²
- 520–1040 per sq. mile/200–400 per km²
- 1040–2080 per sq. mile/400–800 per km²
- More than 2080 per sq. mile/800 per km²

Distribution (above) Earth's population is distributed unevenly. Heavily peopled areas are largely along coasts and rivers, in China and India, and in urban conglomerations called megacities. In contrast, northern Canada, central Australia, and the Sahara are sparsely inhabited.

New York City (below) Almost 67,000 people live in every square mile (2.6 km²) of Manhattan, making New York City the most densely populated U.S. city. Boasting a 24-hour public transportation system, the city has one of the smallest carbon footprints in the country.

Population projections The United Nations has documented historical trends and prepared projections for various aspects of the world's population. These include rates of fertility, continental differences, and rural and urban anomalies.

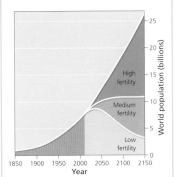

Fertility rates Projections vary with fertility assumptions. High fertility assumes 2.8 children per woman; medium, a replacement level of 2.1 children; and low, 1.6 children.

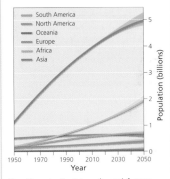

Continents Past trends and future projections vary between continents. Europe's birth rate is currently below the replacement rate, while Africa has the largest rate of population increase.

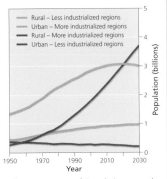

Urban versus rural Population growth over the coming decades will be more pronounced in urban areas in the less industrialized regions. Populations in all rural areas are predicted to decrease.

Detrimental practices Farming can have a major impact on agricultural greenhouse gas emissions. Because much carbon is stored in the ground, low-till or no-till practices are more climate friendly, as are any methods that reduce burning fossil fuels.

Nitrous oxide Fertilizer makes land more productive. Yet over-fertilization releases excess nitrogen as the harmful greenhouse gas nitrous oxide.

Methane Rice paddies are a major source of methane, a greenhouse gas. The soggy soils are ideal growth media for microbes that produce methane.

ARTIFICIAL IRRIGATION

Irrigation enables crops to grow in areas where they would not naturally flourish. Satellite images taken in 1993 (below) and 2002 (bottom) of the Harran Plains, Turkey, show the effect of an irrigation tunnel, which delivers water to local cotton crops.

Agricultural Impacts

Agriculture represents one major way that humans modify Earth's environment. It accounts for about one-third of greenhouse gas emissions, including carbon dioxide, methane, and nitrous oxide. These emissions can be mitigated by sustainable farming practices, but must be weighed against the need to feed a growing world population. Increasing crop cover also affects local climates.

Land use (below) This false-color satellite image shows dramatic differences in land-use patterns across the U.S.–Mexico border. Red indicates the lush agricultural fields in southern California, contrasted with barren fields in Mexico (tan and blue).

GREENHOUSE GAS EMISSIONS FROM AGRICULTURE

CO₂ emissions, billion tons (billion tonnes)

> 150 (> 136)	25–49 (23–44)
100–149 (91–135)	< 25 (< 23)
50–99 (45–90)	No data

Greenhouse gases (above) Emissions from agriculture depend on farming practices, human diets, and population. People in industrialized countries eat four times as much beef as those in developing countries, and cattle are a major source of methane. Rice grown in flooded paddies also releases methane, and fertilizers produce nitrous oxide.

Feeding the world (below) Creating new farmland requires supplanting the native ecosystem. This change, in many cases from tropical forest to crops, has two major effects: the reflectivity of Earth's surface increases, shifting the energy balance of incoming solar radiation; and the hydrological cycle is disrupted.

Rainmakers Crops have different water respiration rates than native forests, shifting the cycling of water between plants and the atmosphere.

A brighter planet Most crops have an overall brighter color than dense green forests, and generally increase the reflectivity, or albedo, of Earth's surface.

Greenhouse gas Fertilizers used to improve crop productivity contain excess nitrates, which results in the crops releasing nitrous oxide, a potent greenhouse gas.

Drought When large areas of cropland displace forests, water evaporation into the atmosphere can be significantly diminished, causing drought.

Sedimentation Deforestation destabilizes soils and removes root systems. The resulting runoff causes river sedimentation, further changing the landscape.

Eutrophication When excess fertilizer runs off into marine ecosystems, the nitrogen facilitates rapid, massive algal blooms and die-offs, depleting the ocean of the oxygen upon which marine species rely.

Causing problems The services performed by forests are myriad, as are the detrimental effects of deforestation. Recognition of the unintended consequences of logging has helped to develop a sustainable forestry movement.

Clear-cutting The practice of cutting down all trees in a harvest area reduces the forest's ability to regenerate. A more sustainable practice is to leave younger trees to grow and replace those logged.

Soil erosion Tree root systems play a critical role in holding together soils that are exposed to water or wind. Erosion can progress when root systems are damaged, silting nearby waterways.

Flooding Without the water-absorbing benefit of trees, flooding is another consequence of deforestation. More rainwater reaches the soil surface and runs off into rivers, causing inundation.

Burning Deforestation has a twofold negative impact on climate: it removes trees that store carbon as biomass and instantly releases the carbon to the atmosphere as carbon dioxide when the forests are burned.

Deforestation

Forests provide many valuable services: trees replenish the atmosphere with water, regulate rainfall, cool the climate, prevent erosion and landslides, and enrich soil with nutrients. Forests contain much of Earth's biodiversity in plant and animal life, and improve air quality by absorbing oxidants and trapping pollutants. Most importantly, they consume carbon dioxide and release oxygen. Yet the need for wood and cropland is leading to rapid deforestation.

Global clearing (below) Over the past 2,000 years, massive deforestation has occurred to create additional farmland and for logging. Nearly half of the world's forested area disappeared during this period and more is under threat as demand for wood products and crops increases.

DISAPPEARING FORESTS

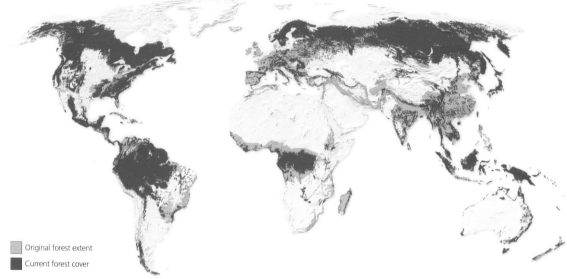

- ☐ Original forest extent
- ■ Current forest cover

Healthy forest (above) A healthy forest demonstrates the natural water cycle in a rain forest. The small crop clearing in the center is not large enough to affect the cycle. Trees absorb water from the soil, emit it to the atmosphere, clouds form, and it rains.

Deforestation (below) The water cycle is altered by deforestation. Less humidity rises from the deforested area. This diminishes rainfall, leading to more tree deaths. The removal of trees and their root systems results in soil runoff from the land into a nearby river.

Cutting and planting Deforestation eliminates an efficient sink for atmospheric CO_2. Forests convert CO_2 to O_2, slowing the buildup of atmospheric carbon dioxide. Reforestation is an effective carbon offset that mitigates climate change.

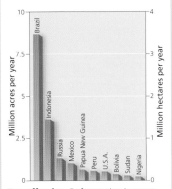

Top offenders Deforestation is worst in rain forest regions. Brazil and Indonesia clear more area each year than the rest of the world combined.

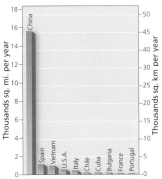

Reforestation Many countries are now planting trees to provide environmental benefits. Trying to combat severe erosion and desertification, China leads the way.

Wetting down This pile of logs at a sawmill in Idaho, U.S.A., needs to be watered to reduce heat buildup as the cut trees await processing. Amid concerns about climate change, the logging industry has increased its emphasis on sustainable forestry.

Biofuel crops Efforts to produce greener transportation fuels, such as biodiesel and ethanol, has led to the clearing of massive tracts of Brazilian rain forest to grow soybean and sugarcane crops. An analysis of environmental impact would need to balance reduced fossil fuel use against the perils of deforestation.

SP_OT ▶ Amazon Rain Forest

▶**REGION:** Amazon Rain Forest, South America. Forested areas extend into Brazil, Bolivia, Peru, Ecuador, Colombia, Venezuela, Guyana, Suriname, and French Guiana

▶**THREATS:** Deforestation, land-use changes, soil erosion, increased sedimentation, water pollution, wildfires, animal trade, poaching

▶**CLIMATE IMPACTS:** Increased CO_2 release from forest burning, drought, increased atmospheric temperature, more mosquito-borne diseases

▶**ENDANGERED SPECIES:** Giant otter, hyacinth macaw, blue-headed macaw, black spider monkey, pink river dolphin; one-fifth of tree species threatened

Sixty percent of the world's existing tropical rain forest is found in the basin of the Amazon River, a rich jungle ecosystem in South America. In addition to hosting more than 40,000 plant species and the greatest diversity of birds, freshwater fish, and butterflies on Earth, the Amazon Rain Forest provides an important climate change mitigation service: approximately one-tenth of the world's terrestrial conversion of carbon dioxide occurs there.

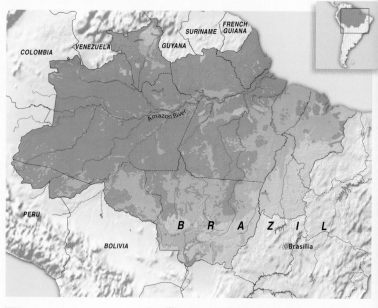

☐ Areas of deforestation
☐ Tropical forest
☐ Other vegetation

Brazilian Amazon (above) The total area of Amazon Rain Forest is 2.3 million square miles (6 million km²), 60 percent of which lies within Brazil. Since 1970, more than one-tenth of Brazil's rain forest has been deforested.

FOREST LOSS

The Amazon processes a significant portion of anthropogenically emitted carbon dioxide through photosynthesis when trees and plants absorb atmospheric carbon dioxide and convert it to oxygen and organic matter. Unfortunately, land-use changes in the region are counteracting this beneficial effect.

Deforestation to accommodate the development of farming and grazing land and human settlement have dramatically reduced the extent of the Amazon Rain Forest. Nearly 20 percent of the Amazon has been deforested to date, with a current rate of loss of approximately 9,300 square miles (24,000 km²) per year. The deforestation problem has been exacerbated in recent years by the rise in the global price of soybeans, which provides economic incentive to clear forest for soybean

farming to produce biofuel. Brazil is now the world's second largest soybean producer, after the U.S.

Widely practiced slash-and-burn methods for forest removal emit massive amounts of smoke and ash, while actively releasing carbon dioxide into the atmosphere. Around 20 percent of global greenhouse gas emissions arise from deforestation, with a large fraction occurring in the Amazon. The Intergovernmental Panel on Climate Change states that reducing or preventing deforestation will have a significant impact on reducing carbon dioxide levels.

The region is also threatened by a warming climate. A temperature increase of 3.6°F (2°C) could kill 20–40 percent of the rain forest in a century. Projected drought conditions would further contribute to the loss of forest cover, accelerating global warming and leading to species loss.

Smoke plume (right) This satellite image shows high concentrations (red) of carbon monoxide emitted by agricultural burning in the Amazon Basin in September 2005. The upwardly-convected pollution cloud is transported across the Atlantic Ocean. Fires in sub-Saharan Africa are also visible.

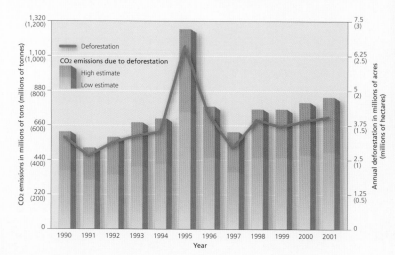

Emissions (left) A steady increase in annual Amazon deforestation from 1990–2001 correlates with a rise in carbon dioxide emissions. Peak deforestation in 1995 followed economic reforms that stabilized the Brazilian currency.

Endangered October 2005 saw the worst drought conditions recorded in the Brazilian Amazon for 50 years. Lack of rain decreased river flow, beaching a pink river dolphin along the shore of Furo do Lago Cristo Reis, near Manaus, Brazil.

Graph (bottom left):

Legend:
- Deforestation
- CO_2 emissions due to deforestation
  - High estimate
  - Low estimate

Left axis: CO_2 emissions in millions of tons (millions of tonnes): 0, 220 (200), 440 (400), 660 (600), 880 (800), 1,100 (1,000), 1,320 (1,200)

Right axis: Annual deforestation in millions of acres (millions of hectares): 0, 1.25 (0.5), 2.5 (1), 3.75 (1.5), 5 (2), 6.25 (2.5), 7.5 (3)

X-axis (Year): 1990 1991 1992 1993 1994 1995 1996 1997 1998 1999 2000 2001

Amazon River Originating as a glacial stream in the Andes, the Amazon flows 4,000 miles (6,400 km) to the Atlantic Ocean, primarily through sparsely populated jungle. It has the largest flow volume of any river in the world. In the rainy season, parts of the river swell to 130 miles (210 km) wide.

Slash and burn (below) The process of clearing land by burning immediately releases into the atmosphere carbon that was contained within the vegetation as CO_2, contributing significantly to global greenhouse gas emissions.

Timber (above) After clear-cutting, trees litter the ground. The environmental and climatic ramifications are significant even if this biomass is never burned. As the cut trees decay, they emit stored organic carbon as carbon dioxide.

2005 drought (below) Boats were stranded when water receded in Alter do Chao, Brazil. Climate models predict significant warming and drying of the region by 2100, which could result in the conversion of much of the rain forest to dry savanna.

Fossil Fuels

The primary human influence on Earth's climate is through the burning of fossil fuels to power industry, produce electricity, and for transportation. Fossil fuels are composed of carbon that has been compressed into the concentrated forms of petroleum (oil) or coal. Burning fossil fuels releases carbon into the atmosphere in the form of carbon dioxide, a powerful greenhouse gas.

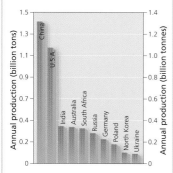
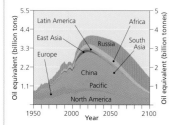
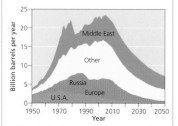

Oil production (above) The production, distribution, refining, and retailing of petroleum products is the world's largest industry. This Norwegian State Oil Company rig, under construction in the North Sea, is part of the ever-expanding infrastructure to exploit newly discovered reserves of oil in increasingly difficult-to-access locations.

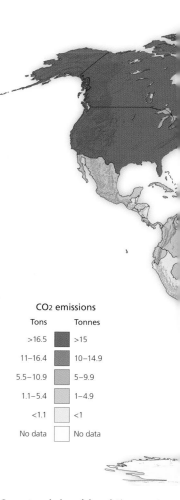

CO₂ emissions

Tons	Tonnes
>16.5	>15
11–16.4	10–14.9
5.5–10.9	5–9.9
1.1–5.4	1–4.9
<1.1	<1
No data	No data

Current emissions (above) The current CO_2 emissions per person show that the U.S.A., Canada, and Australia are countries with the most energy-intensive lifestyles. Countries in the developing world have relatively low per capita CO_2 emissions, but also the fastest increase in emissions.

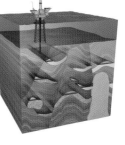

Oil and gas formation (below) Oil and gas are both extremely energy-rich forms of carbon. Formed over millions of years from the compressed fossilized remains of prehistoric organisms—usually marine creatures—this valuable fuel source is known as a "fossil fuel."

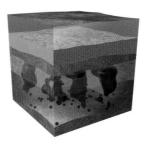

1 Early carbon-based marine creatures fall to the ocean floor and are covered by layers of silt.

2 Over time, the layers become sedimentary rock, compressing the fossils into oil and gas.

3 The oil and gas percolate through porous rocks and collect in reservoirs.

4 The reservoirs are trapped by impermeable rock layers until their extraction by drilling.

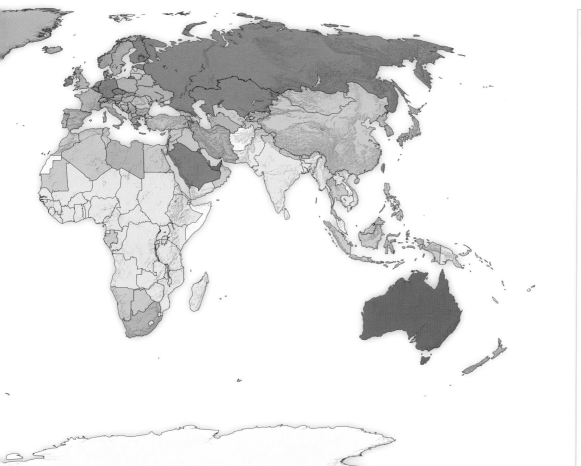

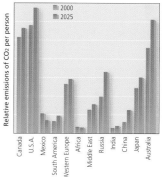

Relative emissions of CO_2 per person

■ 2000
■ 2025

Canada | U.S.A. | Mexico | South America | Western Europe | Africa | Middle East | Russia | India | China | Japan | Australia

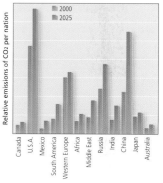

Relative emissions of CO_2 per nation

■ 2000
■ 2025

Canada | U.S.A. | Mexico | South America | Western Europe | Africa | Middle East | Russia | India | China | Japan | Australia

Century of emissions (right) Although emissions are stabilizing in the industrialized world, they continue to grow rapidly in the developing world. Because fossil fuels take a long time to form, they cannot be regenerated at the current rate of usage.

Coal formation (below) In contrast to oil and gas, coal is formed by the compression of vegetation. The compaction process drives off moisture and organic gases, leaving solid carbon containing little water.

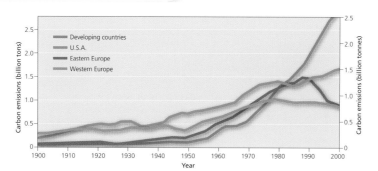

Developing countries
U.S.A.
Eastern Europe
Western Europe

Carbon emissions (billion tons) | Carbon emissions (billion tonnes)

1900 1910 1920 1930 1940 1950 1960 1970 1980 1990 2000
Year

1 Dead plant material collects underwater, where it is protected from decay, and forms peat.

2 Compaction first forms lignite, or brown coal, which contains about 45 percent water.

3 Further compression forms the more energy-dense bituminous, or black, coal.

4 The highest grade of coal is anthracite coal, laboriously extracted from deep mines.

FACT FILE

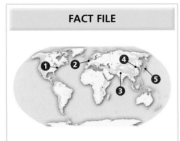

1. Washington, D.C. The U.S. capital is not a major industrial or manufacturing center. However, many industry advocacy firms are headquartered in Washington to influence government. Recently, the
 U.S. government decided to regulate CO_2, signaling a change in climate policy that will affect many industries.
Washington, D.C., U.S.A.

2. London England's capital saw rapid growth during the industrial revolution in the late 18th and early 19th centuries as the steam engine enabled dramatic
 modernization. From about 1830 to 1920, London was the world's largest city and is still a major business center.
London, England

3. New Delhi The capital of India is one of northern India's largest commercial and financial centers, with an expanding service sector in banking, information technology, media, telecommunications,
 and tourism. With an English-speaking workforce, India has attracted many multinational corporations.
New Delhi, India

4. Beijing China's capital is a megacity of over 10 million inhabitants. Since the economic reforms of the 1990s, the country has seen rapid expansion
 and urbanization, creating air pollution problems arising from the coal-powered manufacturing of electricity.
Beijing, China

5. Tokyo The world's most populous metropolitan area is Japan's capital, with 35 million residents. Tokyo is a major international financial center and the headquarters of several large
 companies. Japan's economic boom in the 1980s and 1990s came mainly from high-technology industries.
Tokyo, Japan

Industry

Global industry accounts for a large part of the human influence on climate change as many of the processes rely on fossil fuel combustion. Energy- and water-intensive industries, such as textiles and paper, are a major drain on resources. The climate impact of most industries can be mitigated by moving away from fossil fuels and toward renewable and carbon-free energy sources.

Steel production (below) Being very energy intensive, blast furnaces and steel mills release significant industrial carbon emissions in steel-producing countries and approximately four percent of global man-made greenhouse gas emissions.

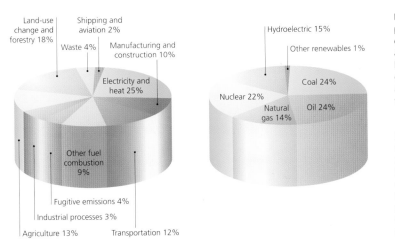

Land-use change and forestry 18%
Shipping and aviation 2%
Waste 4%
Manufacturing and construction 10%
Electricity and heat 25%
Other fuel combustion 9%
Fugitive emissions 4%
Industrial processes 3%
Agriculture 13%
Transportation 12%

Hydroelectric 15%
Other renewables 1%
Coal 24%
Nuclear 22%
Natural gas 14%
Oil 24%

Emissions by sector (far left) Industrial processes directly account for three percent of global carbon dioxide emissions. However, a large portion of the emissions produced by other sectors—manufacturing and construction, transportation, electricity and heat, shipping and aviation, and waste—are also attributable to industry.

Electricity production (left) This chart shows the proportion of energy sources used in global electricity production, which has a major impact on greenhouse gas emissions. Coal-sourced electricity has the highest carbon emissions per energy produced, oil is second, natural gas is third. Hydroelectric, nuclear, and renewable energy sources have lower CO_2 emissions.

Auto industry (above) Transportation produces 12 percent of global greenhouse gas emissions. Therefore, innovations in this industry are important to reducing emissions, and the race is on to create more fuel-efficient vehicles. A car is retrieved from an automatic palette at the Volkswagen factory in Wolfsburg, Germany.

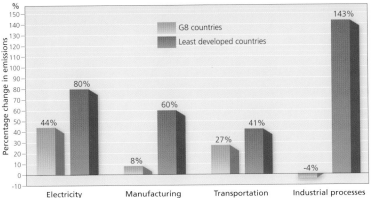

%
150
140
130
120
110
100
90
80
70
60
50
40
30
20
10
0
-10

Percentage change in emissions

G8 countries
Least developed countries

44% 80% — Electricity and heating
8% 60% — Manufacturing and construction
27% 41% — Transportation
-4% 143% — Industrial processes

Emissions growth (left) Since 1990, the emissions of greenhouse gases from industrial processes have decreased in the industrialized G8 countries—Canada, Italy, France, Germany, Japan, Russia, the U.K., and the U.S.—but dramatically increased in developing countries.

Enhanced Greenhouse Effect

Greenhouse gas levels have increased significantly since the industrial revolution and they will affect Earth's climate long after emissions cease. The primary anthropogenic greenhouse gases are carbon dioxide, methane, nitrous oxide, ozone, and halocarbons, such as carbon tetrachloride. The combustion of relatively inexpensive fossil fuels generates carbon dioxide, a long-lived greenhouse gas that warms the planet.

FACT FILE

Greenhouse gases The global warming potential of gas molecules such as carbon dioxide and methane depends on their atmospheric concentrations, but also on how effectively they absorb infrared radiation, and on their lifetime.

PROPORTION OF GREENHOUSE GASES

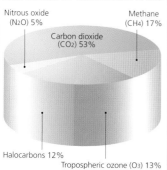

Nitrous oxide (N2O) 5%
Methane (CH4) 17%
Carbon dioxide (CO2) 53%
Halocarbons 12%
Tropospheric ozone (O3) 13%

METHANE EMISSIONS BY COUNTRY

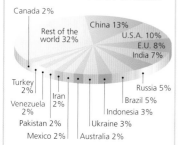

Canada 2%
Rest of the world 32%
China 13%
U.S.A. 10%
E.U. 8%
India 7%
Turkey 2%
Iran 2%
Venezuela 2%
Russia 5%
Brazil 5%
Pakistan 2%
Indonesia 3%
Mexico 2%
Ukraine 3%
Australia 2%

ATMOSPHERIC LIFETIMES

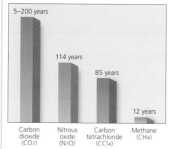

5–200 years
114 years
85 years
12 years

Carbon dioxide (CO2) | Nitrous oxide (N2O) | Carbon tetrachloride (CCl4) | Methane (CH4)

INCREASE SINCE INDUSTRIALIZATION

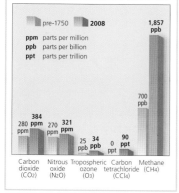

pre-1750 | 2008

ppm parts per million
ppb parts per billion
ppt parts per trillion

1,857 ppb
700 ppb
384 ppm
280 ppm
270 ppm
321 ppm
25 ppb
34 ppb
0 ppt
90 ppt

Carbon dioxide (CO2) | Nitrous oxide (N2O) | Tropospheric ozone (O3) | Carbon tetrachloride (CCl4) | Methane (CH4)

1 Natural greenhouse effect (below)
Without any greenhouse gases in Earth's atmosphere, its surface temperature would be −0.4°F (−18°C). The natural greenhouse effect is essential to the habitability of Earth.

Carbon cycle Animals exhale CO2 gas and plants convert atmospheric CO2 into oxygen (O2). The decay of dying plants converts organic matter into CO2.

3 Human influences (right)
Humans have modified the natural radiation and carbon balance by altering Earth's surface, and by burning fossil fuels to power industries, transportation, and electricity.

Balance Water vapor equilibrates between the atmosphere and the ocean, which can dissolve some CO2 as carbonic acid. Water vapor is a natural greenhouse gas that responds to temperature as a climate feedback.

Warming Water vapor, CO2, and other greenhouse gases in the atmosphere absorb infrared energy emitted by Earth and re-radiate some portion back down, warming the surface.

2 Natural radiation balance (right)
Water vapor, clouds, and CO2 present in Earth's atmosphere create a so-called greenhouse effect by trapping some solar radiation, rather than allowing it to radiate out into space. Greenhouse gas molecules absorb some of the infrared (heat) energy emitted from Earth's surface.

Incoming and outgoing Solar radiation reaches Earth as visible light, of which approximately 30 percent is directly reflected back to space, either by clouds or reflective parts of Earth's surface.

Outgoing After solar radiation heats the surface, Earth emits infrared energy, at a lower energy level than the Sun's radiation.

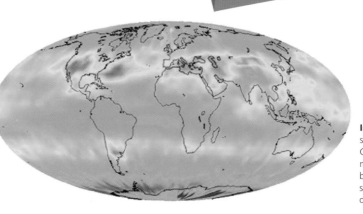

In the air This July 2003 NASA satellite image shows an uneven spatial distribution of global CO2 at an altitude of 5 miles (8 km), in the mid troposphere. Levels vary from 365 (dark blue) to 385 (red) parts per million. Emissions sources are centered around populated areas or places that use slash-and-burn deforestation.

Emissions Human-induced additions to CO₂ build up in the atmosphere as the biosphere and oceans cannot absorb sufficient quantities to match the accelerating emissions.

Deforestation Clearing trees to make room for farmland or urban development removes large tracts of carbon sinks. When trees are burned, their organic matter is converted into atmospheric CO₂.

Livestock Cattle are a major source of the potent greenhouse gas methane. Other methane sources include landfills, natural gas, and rice cultivation.

Urbanization Cities generally have few carbon-consuming trees and high concentrations of CO₂ emissions from vehicles, factories, and power plants.

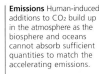

Drilling Procuring oil is an energy-intensive process that produces CO₂ emissions. It extracts more carbon than will eventually be burned.

Tankers Shipping is a major source of anthropogenic CO₂ and marine fuels are often bottom-of-the-barrel petroleum products.

Transportation Motor vehicles produce about 12 percent of fossil-fuel-based emissions. In the U.S.A., road emissions constitute one-third of greenhouse emissions.

Power plants Burning fossil fuels, especially coal, to make electricity releases vast amounts of carbon. Regrettably, coal use is widespread, as it is the cheapest energy source.

4 Radiation balance changed (right)
Human activity has increased greenhouse gas levels, resulting in a larger fraction of outgoing infrared radiation being trapped in Earth's atmosphere and cast back down toward the planet's surface.

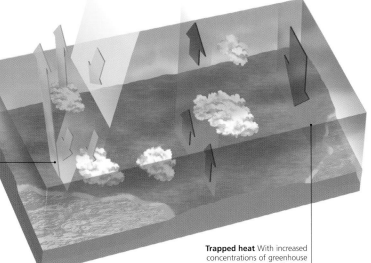

Incoming and outgoing Solar radiation is essentially constant, but land-use changes, such as turning forests into farmland, alters surface reflectivity. This modifies how much the surface is warmed by sunlight.

Trapped heat With increased concentrations of greenhouse gases, more infrared radiation is trapped and re-radiated, causing greater surface warming.

FACT FILE

Industrial processes Metal-ore mining is an environmentally destructive, energy-intensive process. The direct burning of oil and coal, to power automobiles or generate electricity, is the main contributor of man-made atmospheric CO₂.

Mining About 90 percent of copper ore is extracted by open-pit mining. The Rio Tinto mines in Spain produce rich copper and iron supplies, but nearly all local vegetation has been destroyed.

Burn off Black carbon particles ejected into the atmosphere by burning oil wells, such as these in Kuwait, are light-absorbing. They heat up the atmosphere and decrease air quality.

Electrical power Coal-fired power plants are the most carbon-intensive way to produce electricity. Generating electricity accounts for about one-quarter of all greenhouse gas emissions.

Acid Rain and Aerosols

Atmospheric aerosol results from volcanic activity or from human-induced emissions of soot, sulfur oxides and nitrogen oxides. Small acidic particles become hydrated by water vapor in the atmosphere and are deposited downwind as damaging acid rain. While airborne, aerosols can affect climate by decreasing the amount of sunlight reaching Earth. This exerts a local cooling effect and diminishes the photosynthetic activity of plant life.

Acid rain (below) Sulfur oxides (SOx) emitted by volcanoes and the combustion of some fossil fuels react with water vapor in the atmosphere to form sulfuric acid aerosol. Nitrogen oxides (NOx), emitted from all combustion engines, become nitric acid aerosol. These combine to form acid rain.

AUTOMOBILE FUMES

Vehicle exhaust contains nitrogen oxides and unburned hydrocarbons that can lead to aerosol formation. These aerosols produce the haze that regularly diminishes visibility in urban areas.

SOx and NOx emitted by coal-burning power plant

Acidified clouds contain sulfuric and nitric acids

Acid rain and snow

SOx emitted by volcanoes

SOx emitted by tankers burning low-grade fuel

NOx emitted by vehicles

Trees die off from acidic precipitation

River acidified

Marine life dies off

Acid rain runoff penetrates soil and groundwater

Tree die-off (below) Spruce trees in Krkonose National Park, in the Czech Republic, were killed by acid rain. During the Soviet era, booming communist industrial economies were powered by burning massive quantities of high-sulfur coal, and produced acidified rain downwind.

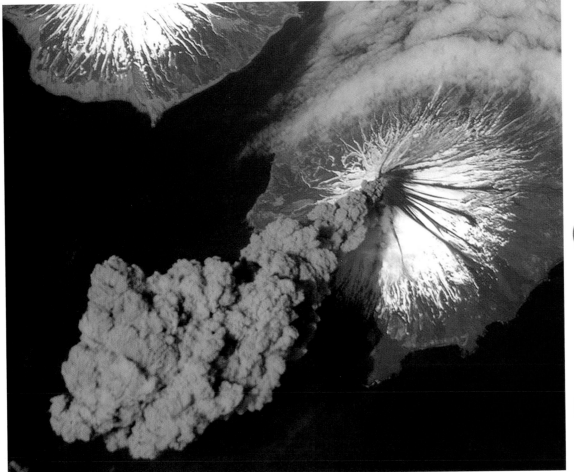

Volcanic eruption (above) Cleveland Volcano, in Alaska, erupted on May 23, 2006. Volcanic activity is the predominant natural source of sulfur oxides in the atmosphere. Sulfur oxides—which include sulfur dioxide—are some of the main producers of acid rain.

Mount St. Helens This volcano in the state of Washington, U.S.A., erupted on May 18, 1980. Images show the mountain before (above, in 1973) and after (above right, in 1983) the eruption. Vegetation had still not recovered three years after the blast.

Toxic emissions (right) Annual sulfur dioxide (SO₂) emissions from Mount St. Helens exceeded 2,200 tons (2,000 tonnes) per day in the year it erupted, and continued to be high for several years. This created sustained acid rain in Washington State for many years.

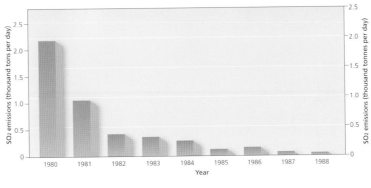

FACT FILE

Climate impact Mount Pinatubo, in the Philippines, erupted in June 1991. A satellite monitored the dispersion of Pinatubo aerosols over several years, demonstrating the long-term climate cooling potential of volcanic eruptions.

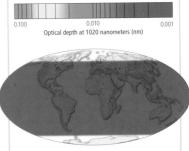

0.100   0.010   0.001
Optical depth at 1020 nanometers (nm)

Pre-eruption Baseline aerosol levels, measured as optical depth, were low around the world. Optical depth is a measurement of the transparency of the atmosphere due to aerosol particles.

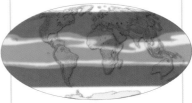

Eruption In the month following the eruption, aerosol optical depth showed nearly a 100-fold increase in the tropics. Earth's rotation distributed aerosols longitudinally around the globe.

Two months later Aerosol optical depth in the stratosphere remained high and had spread to higher latitudes. The eruption affected global temperatures, with late 1991 being unexpectedly cool.

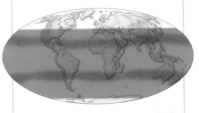

Four years later Enough rain-out had occurred by 1994 to diminish the aerosol optical depth and reduce its influence on global temperatures. Yet the effects of the eruption had not fully disappeared.

FACT FILE

Air pollution A major side effect of urbanization, air pollution is forecast to worsen and negatively impact health. Warmer air promotes the formation of pollutants such as ozone, and is also likely to spur more energy usage to cope with the heat.

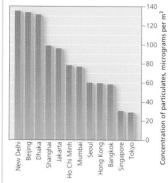

Asia The particulate air pollution of 12 Asian cities was measured in 2004. These airborne particles are capable of penetrating deep into the lungs and causing illness.

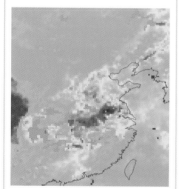

China Beijing is one of the world's most polluted cities. This satellite image shows the spatial extent of carbon monoxide (red), a gas released by the burning of fossil fuels, over eastern China.

India This smoggy sunrise over Delhi, in 1997, shows the effect of air pollution on visibility. Air pollution in India arises from vehicle emissions and the domestic burning of biomass, or organic matter.

Pollution

Increasing air, water, and soil pollution is a by-product of our modern lifestyle that renders the natural environment less resilient to the impacts of climate change. Air pollution, in particular, is predicted to worsen in a warming climate, while some air pollutants themselves affect local climate by either reflecting or absorbing the sunlight reaching the surface.

Oil spills (below) Environmental damage due to fossil fuel usage includes oil spills during transportation. This map locates global oil disasters since 1967. Most spills occur in the ocean, polluting coastal water and harming birds and marine animals.

GLOBAL OIL DISASTERS

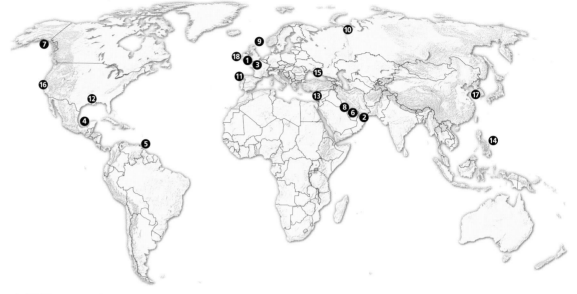

1. **1967** *Torrey Canyon* oil tanker spill, off the southwest coast of England
2. **1972** *Sea Star* oil tanker spill, Gulf of Oman
3. **1978** *Amoco Cadiz* oil tanker spill, off the coast of Brittany, France
4. **1979** Ixtoc oil well disaster, Gulf of Mexico
5. **1979** Collision of *Atlantic Empress* and *Aegean Captain* oil tankers, off Trinidad and Tobago
6. **1983** Nowruz oil field disaster, Persian Gulf

7. **1989** *Exxon Valdez* oil tanker spill, Prince William Sound, Alaska, U.S.A.
8. **1991** Burning oil fields during Gulf War, Kuwait
9. **1993** *Braer* oil tanker spill, Shetland Islands, U.K.
10. **1994** Ruptured oil pipeline, Usinsk, Russia
11. **2002** *Prestige* oil tanker spill, off northwest coast of Spain
12. **2006** Citgo Refinery oil spill, Lake Charles, Louisiana, U.S.A.

13. **2006** Jiyeh Power Station oil spill, Lebanon
14. **2006** Guimaras oil spill, Philippines
15. **2007** Kerch Strait oil spill, Strait of Kerch, Ukraine and Russia
16. **2007** San Francisco Bay oil spill, California, U.S.A.
17. **2007** *Hebei Spirit* oil spill, Yellow Sea, South Korea
18. **2009** West Cork oil spill, off southern Ireland

Mopping up The South Korean Army works to clean up the country's worst oil spill, in December 2007. Around 12,000 tons (10,900 tonnes) of oil leaked when a barge collided with an oil tanker, the *Hebei Spirit*, off South Korea's west coast.

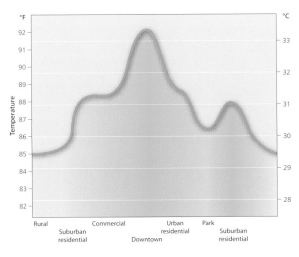

Heat island effect (above) Concrete buildings and sidewalks trap heat, making cities several degrees warmer than surrounding rural areas. With little open space and vegetation, and dense populations with heavy power usage, urban areas are major sources of heat.

Plastic waste (above) Another pollution problem is waste disposal. Plastic, as a petroleum by-product, is a limited resource. It is also effectively permanent, with no possibility of biodegrading. Unless recycled, it can remain in the environment indefinitely.

Waste management The Danktopa market, in Cotonou, Benin, West Africa, is polluted with dumped garbage. Contaminants in this garbage stream can leach into the town's water supply and promote the spread of disease.

FACT FILE

1. Smog London's great smog event of 1952, believed to have caused the death of 12,000 inhabitants, drew international attention to the hazards of urban air pollution. Postwar London

had only low-grade coal to burn during the December cold snap, producing dangerous levels of sulfur dioxide.

London, England

2. Pesticides Improvements to agriculture through the development of fertilizers and pesticides have a dangerous downside, tragically

demonstrated in 1984. A leak at a pesticide plant in Bhopal released toxic gas that killed 16,000 people.

Bhopal, India

3. Mining The Ok Tedi gold and copper mine in Papua New Guinea is one of the world's largest copper producers. Since the 1990s, mine tailings and runoff have polluted the

Fly River, killing fish and leaving a layer of contaminated sludge along the shores of 120 downstream villages.

Papua New Guinea

4. Burning oil During the 1991 Gulf War, Kuwaiti oil wells were ignited. Some of the fires burned for eight months, spewing black carbon into

the atmosphere and effectively controlling the weather of the Persian Gulf for part of that year.

Kuwait

5. Oil spill The 1989 *Exxon Valdez* oil spill in Prince William Sound, Alaska, was one of the most devastating, due to its remote location and the site's status as a habitat for salmon, sea otters, seals, and seabirds.

Crude oil covered 11,000 square miles (28,000 km²) of ocean and coast before clean-up operations began.

Alaska, U.S.A.

Ozone Depletion

The ozone layer protects Earth from the Sun's harmful ultraviolet radiation. It is vulnerable to certain industrially emitted chemicals containing chlorine and fluorine. The gases responsible for the growing ozone holes were banned by the Montreal Protocol, which came into force in 1989. The extent of the holes is stabilizing and the ozone layer is beginning to show signs of recovery.

Natural balance Ozone in Earth's stratosphere exists in a delicate natural balance of production and loss, with peak ozone at an altitude of around 12 miles (20 km). This balance is disturbed by man-made emissions of ozone-depleting gases.

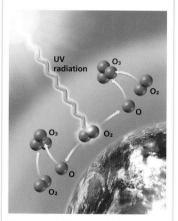

Formation High-energy UV radiation breaks apart molecules of oxygen (O_2). These single oxygen atoms bond with other oxygen molecules to form ozone (O_3), composed of three oxygen atoms.

OZONE DESTROYERS

Substances	Sources/uses
CFCs	Refrigerants, solvents, aerosol spray cans
Methyl bromide	Pesticide
Polar stratospheric clouds	Atmospheric nitrogen oxides (NOx)
Carbon tetrachloride (CCl_4)	Chemical solvent
Methyl chloroform (CH_3CCl_3)	Chemical solvent

ARCTIC HOLE

The ozone hole in the Northern Hemisphere is much smaller than in the Southern Hemisphere. There is a seasonal cycle, with minimum ozone observed over each Pole in spring. In March 2000, Arctic ozone was at its lowest level in eight years.

Ozone layer (below) Chlorofluorocarbons (CFCs), used as refrigerants, solvents, and propellants, destroy ozone molecules (O_3). Holes in the ozone layer allow a larger amount of damaging UV radiation (purple waves) to penetrate the lower atmosphere and reach Earth's surface.

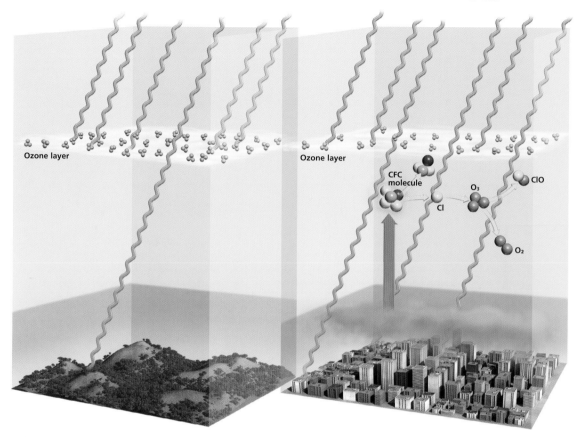

1 The ozone layer is located at an altitude of 12–25 miles (20–40 km). It absorbs most high-energy ultraviolet (UV) radiation, protecting plants and animals from DNA damage.

2 CFCs are long-lived in the atmosphere, rising to the ozone layer before they are broken apart by high-energy radiation. This liberates a chlorine atom (Cl), which then breaks apart an ozone molecule (O_3).

Polar clouds Ozone depletion is most pronounced over Earth's Poles. When sunlight returns in spring, CFCs that built up over the winter rapidly deplete the ozone layer. This is further accelerated by polar stratospheric clouds, which support additional chemical reactions that cause ozone destruction.

Smallest extent 1981

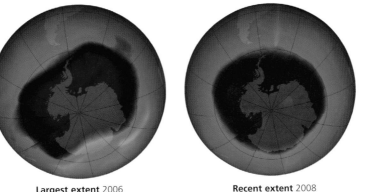

Largest extent 2006

Recent extent 2008

Ozone hole (above) The Antarctic ozone hole peaks in September. In 1981, CFCs had not yet built up to high levels. By 2006, CFC emissions were in global decline, but due to their long atmospheric lifetime the ozone hole reached its maximum size that year.

Hole size (right) The appearance of the ozone hole over Antarctica is a dramatic example of human activity impacting the atmosphere. The size of the hole steadily increased through the 1980s and 1990s, with extremes occurring in 1981 and 2006.

Ozone distribution (below) This satellite image from October 2008—in the Southern Hemisphere's spring—shows the greatest destruction of ozone is over the South Pole, but regions of low ozone (blues and purple) also extend to the lower latitudes.

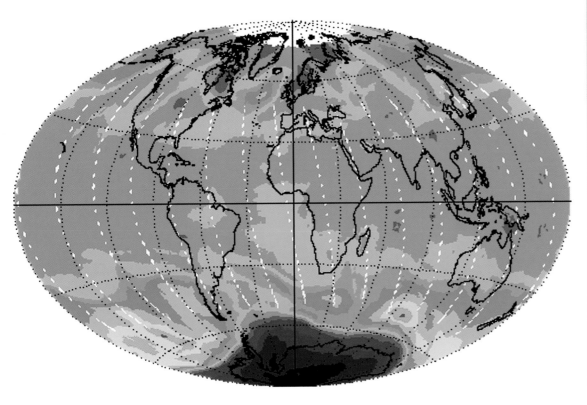

Man-made emissions Destruction of ozone is accelerated by the emission of gases containing chlorine, such as CFCs, used in refrigeration and air conditioning. These substances came into use about 60 years ago and began to build up in the atmosphere.

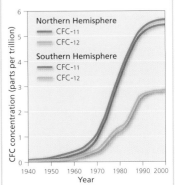

CFCs Two major chlorofluorocarbon gases emitted into the atmosphere by industrial processes increased rapidly after the 1950s. Their production was banned in the late 1980s.

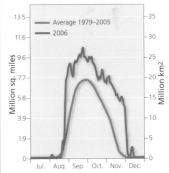

Annual peak The Antarctic ozone hole peaks in size in September, Southern Hemisphere springtime. In 2006, the hole reached its largest size: 10.6 million square miles (27.5 million km²).

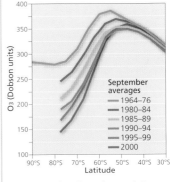

High latitudes The Dobson unit is a basic measure of ozone concentration. Maximum ozone depletion is shown to occur at high latitudes: closer to the South Pole in the Southern Hemisphere.

Changing Precipitation

Increasing average global temperatures will lead to changes in precipitation, a consequence of shifting atmospheric circulation and increasing evaporation. Regional patterns of change are variable, but some global trends are clear: there will be an overall increase in average precipitation and an increase in the intensity of severe precipitation events.

Less snow (right) Decreasing snowfall in the Sierra Nevada mountains of California, U.S.A., has been worrying local authorities as the spring snowpack, upon which much of northern California relies for its water, has been decreasing in mass.

1. Urbanization Fine particles emitted by human activity suppress precipitation. In Phoenix, Arizona, winter rainfall levels have been found to correlate with weekly vehicle usage and emissions.

Also, the built-up city environment and urban heat island effect have changed local storm patterns.
Phoenix, Arizona, U.S.A.

2. Ski industry Scotland is researching the effects of climate change on skiing and other winter sports. Snowfall is predicted to drop dramatically in

the next 40 years. If ski slopes are moved to higher altitudes, the risks of avalanches and accidents increase.
Scotland

3. Winter tourism The glacier-covered Alps, home to much of Europe's winter sports industry, have keenly felt the effects of diminished snow cover. Average snow levels throughout the

region are half what they were 40 years ago, and the Alps' substantial winter tourism industry is struggling to adjust.
The European Alps

4. Polar warming Siberia, along with all polar regions, has experienced notably rapid warming. The naturally dry climate worsens the situation. Low

annual snowfall means that the melting snow cover may not be fully replenished by fresh falls.
Siberia, Russia

5. Extreme weather Myanmar has a tropical monsoon climate, with a cool dry season and hot summers featuring heavy rainfall. Cyclone Nargis, which hit in May 2008, was the country's most

devastating natural disaster, highlighting the costs of increased extreme weather events predicted in a warming climate.
Yangon, Myanmar

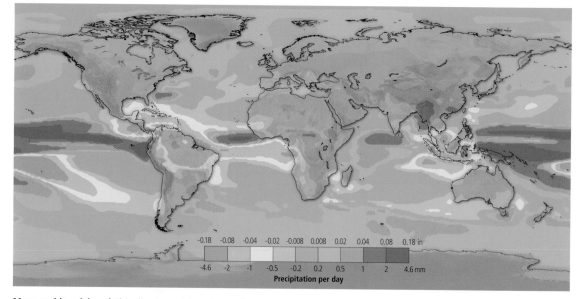

-0.18 -0.08 -0.04 -0.02 -0.008 0.008 0.02 0.04 0.08 0.18 in

-4.6 -2 -1 -0.5 -0.2 0.2 0.5 1 2 4.6 mm
Precipitation per day

More and less (above) This climate model predicts a change in annual average precipitation from 1960–1990 to 2070–2100. Some land areas will be drier, including parts of South America and Southeast Asia, and some wetter, such as East Asia and Alaska.

More rain (below) One region predicted to experience increasing annual rainfall in a warming climate is East Asia. On June 7, 2008, Hong Kong shop owners assessed the damage after a storm that dumped more than 7.5 inches (190 mm) of rain overnight.

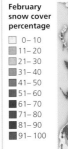

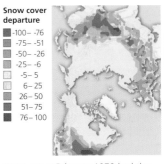
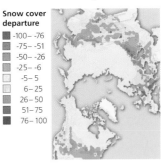
Surprise snow An expected feature of a changing climate is increased frequency of extreme precipitation. In February 2006, residents of Amman, Jordan, experienced a freak snowstorm that closed schools and stranded people in high-altitude areas.

1. Grinnell Glacier The Grinnell Glacier has retreated significantly since it was first measured in 1850, when the area was 1.11 square miles (2.87 km²). Today, the glacier measures only 0.34 square

miles (0.88 km²). Glaciologists estimate that the glaciers in Glacier National Park, Montana, might disappear by 2040.

Grinnell Glacier, Montana, U.S.A.

2. Quelccaya Ice Cap Peru is home to the largest system of tropical glaciers in the world. The Quelccaya Ice Cap has experienced a ten-fold increase in the

rate of retreat since formal measurements began in the 1960s, causing concern that it could vanish within five years.

Quelccaya Ice Cap, Andes, Peru

3. Langjökull Glacier Iceland's second largest glacier, the Langjökull Glacier has an area of 370 square miles (960 km²).

Icelandic glaciers, covering 11 percent of the country, are retreating at a rate of 8–12 square miles (20–30 km²) per year.

Langjökull Glacier, Iceland

4. Himalayas Melting of the Himalayan glaciers, which contain the largest store of fresh water outside the polar ice sheets, will have dire consequences

for South Asia. With accelerated melting, the region will suffer devastating floods, then diminished water availability.

Himalayas

DISTRIBUTION OF FROZEN FRESH WATER

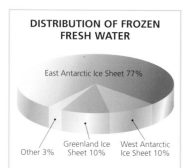

East Antarctic Ice Sheet 77%

Other 3%

Greenland Ice Sheet 10%

West Antarctic Ice Sheet 10%

Melting Glaciers

The observed acceleration of glacial melting is of great concern to climatologists. Particularly nearer the Equator, glaciers have a fragile existence that is threatened by even slight warming or decreased precipitation. Glacial retreat has high year-to-year and inter-decadal variability that depends on local precipitation, weather patterns, and temperatures.

Receding (right) Greenland's Jakobshavn Isbrae Glacier is the world's fastest retreating glacier. Red lines show its ice front from 1851 to 2006. The white area shows calved icebergs piling up in the fjord. The glacier's retreat threatens to accelerate sea-level rise.

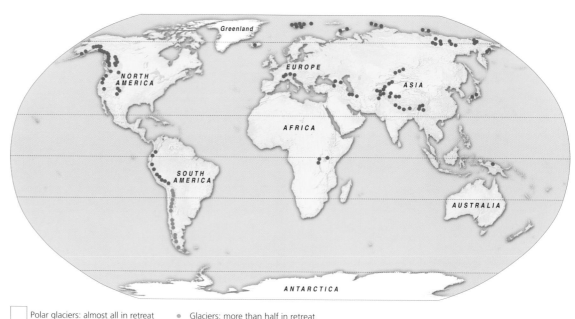

Greenland

NORTH AMERICA

EUROPE

ASIA

AFRICA

SOUTH AMERICA

AUSTRALIA

ANTARCTICA

☐ Polar glaciers: almost all in retreat
● Glaciers: almost all in retreat
● Glaciers: more than half in retreat
● Glaciers: some in retreat

Locations (above) Temperate glaciers essentially trace mountains with sufficiently cold peak temperatures to support year-round snow. Closer to the Poles, glaciers are possible at lower altitudes.

Vanishing snow (below) In coming decades, warmer temperatures and decreased precipitation may cause the glacier on Africa's highest peak, Tanzania's Mount Kilimanjaro, to disappear completely.

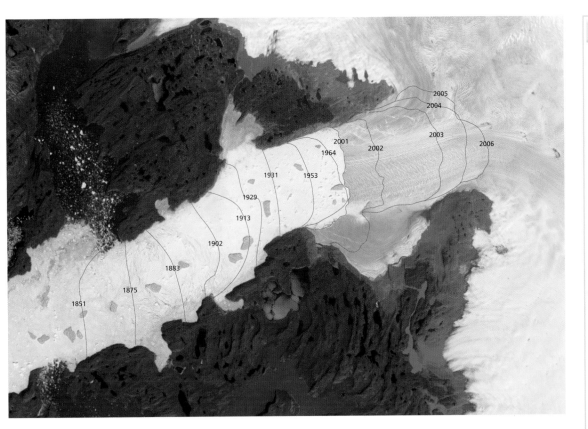

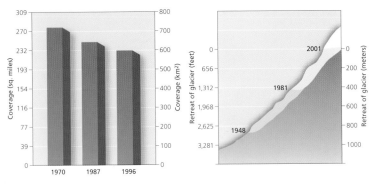

FACT FILE

Diminished size As glaciers retreat, their shape and velocity change in response to local climate variations and where meltwater is channeled. Crevasses and meltwater lakes are among the topological features that can affect a glacier's movement.

Alaska	
Canada and U.S.A. (excluding Alaska)	
Norway	
European Alps	
Russia	
Tien Shan and Pamirs	
Tibet	
Tropical glaciers	
South America	
New Zealand	

1700 1750 1800 1850 1900 1950 2000
Year

Maximum   Decrease ⟶   Minimum

Retreat After the end of the Little Ice Age, around 1850, retreat has been seen in the all world's glacier systems, with variability due to local climates.

Cracks Crevasses form when ice masses collide or travel at different speeds. When crevasses fill with water, it can lubricate the glacier's base, speeding its travel.

Meltwater At the base of mountain glaciers, meltwater rivers can carve the landscape. If in contact with the glacier, meltwater may alter the glacier's velocity.

Glacial lake (above) The Cordillera Blanca glacier system in Peru has lost more than one-quarter of its permanent ice pack since the 1970s. Glacial melt provides water for irrigation and hydroelectricity.

Coverage (far left) This bar graph shows the retreat of area covered by the Cordillera Blanca glacier system, with a 17 percent loss between 1970 and 1996.

Moving up (left) Warming has led to rainfall replacing snowfall in the region. Glacial extent is retreating rapidly to higher elevations.

Ocean flow The circulation of ocean currents depends on the balance of water temperature and density in various parts of the ocean. This massive oceanic conveyor belt could be disrupted by salinity changes caused by a melting Arctic Ice Cap.

Current pattern The Atlantic Ocean flow moves tropical surface water northward, warming Western Europe. The water cools and sinks in the Arctic, before traveling back toward the Equator.

Disruption The melting Greenland Ice Sheet releases buoyant fresh water into the Arctic, halting the northward transfer of warm water and cooling Europe.

SUBGLACIAL LAKE

Lake Vostok is a subglacial lake in Antarctica. These are formed when the temperature of the continental surface maintains a reservoir of liquid water beneath the frozen ice sheet. The glacier continues to flow over the top of the lake.

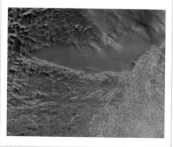

Ice Shelves and Floating Ice

Most of the fresh water in the world comes from snowfall and is stored primarily as ice sheets and ice shelves on Antarctica and Greenland. Floating ice in the Arctic, formed mainly from compacted snow, is another freshwater reserve. Earth's ice cover is shrinking especially rapidly at the Poles, impacting sea levels and ocean salinity and circulation.

Northwest Passage (below) A navigable passage between the Atlantic and Pacific oceans formed in the Arctic Archipelago in August 2007 and lasted for several weeks. A permanently ice-free passage is expected to become a reality in coming decades.

B-15A In March 2000, Iceberg B-15 **(above)** calved from the Ross Ice Shelf, on the coast of Antarctica. With an initial area of over 4,200 square miles (11,000 km²)—larger than Jamaica—it was the biggest iceberg on record. It broke up in warmer ocean waters, with the biggest piece—B-15A—drifting on the circumpolar currents **(right).** Free-floating ice, such as these icebergs, contributes to sea-level rise and affects local ocean salinity as its fresh water melts.

Balleny Islands

March 15, 2006

February 13, 2006

December 29, 2005
November 4, 2005
October 27, 2005

B-15A
fragmentation

October 11, 2005

September 16, 2005

Victoria Land

July 25, 2005

May 15, 2005

Ross Sea

April 15, 2005

Impact with
Drygalski Ice Tongue

February 15, 2004
February 2, 2004
July 18, 2002

Iceberg B-15

Ross Ice Shelf

Glacial flux (below) Glaciers are in a dynamic state of ice accumulation (blue arrows) and loss (red arrows). As snow falls, the glacier gains mass in the center, which compacts and presses outward, forcing into the sea. Overall loss of ice mass occurs when the rate of outflow exceeds snowfall accumulation.

Increasing mass Snowfall is the source of polar ice. Officially a desert, Antarctica receives on average 2 inches (50 mm) of snow per year, adding mass to the center of the continent.

Surface water Meltwater ponds, formed by localized warming on the surface of the ice sheet, accelerate melting. Liquid water has a lower albedo than ice, thus more solar radiation is absorbed.

Calving Ice mass is lost at the edges of the ice sheet as icebergs calve into the ocean.

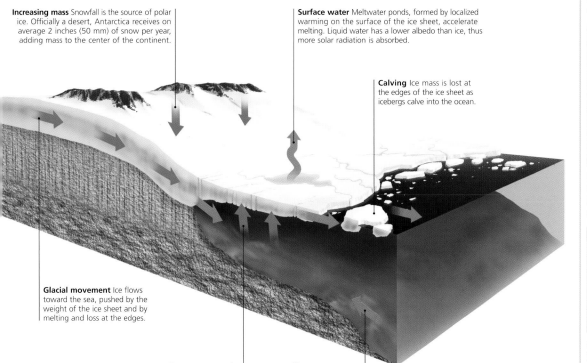

Glacial movement Ice flows toward the sea, pushed by the weight of the ice sheet and by melting and loss at the edges.

Freezing Ice mass can be added when cold surface seawater freezes to the underside of an ice shelf.

Upwelling Warm currents upwelling underneath the ice speed up the melting of floating ice shelves.

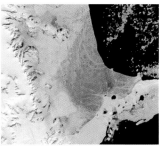
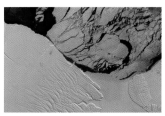

The Arctic

SPOT HOT

> ▶**REGION:** The area contained within the Arctic Circle (66°30'N), including the Arctic Ocean and portions of the U.S.A., Canada, Greenland, Iceland, Norway, Sweden, Finland, and Russia
>
> ▶**THREATS:** Rapidly increasing land and ocean temperatures, stratospheric ozone depletion, oil drilling, conflict over navigation rights
>
> ▶**CLIMATE IMPACTS:** Thinning polar ice pack, melting glaciers, rising sea levels, decreased water salinity, increased methane release
>
> ▶**ENDANGERED SPECIES:** Caspian seal, fin whale, blue whale; two-thirds of the global polar bear population could disappear by 2050

The Arctic region has some of the most observable consequences of climate change, notably the rapidly melting polar ice cap. Climate warming is quickest in the Arctic, with temperatures increasing at double the average global rate. This rapid warming is due to "a climate feedback," where highly reflective snow and ice is being replaced by dark open water, increasing the absorption of the Sun's heat.

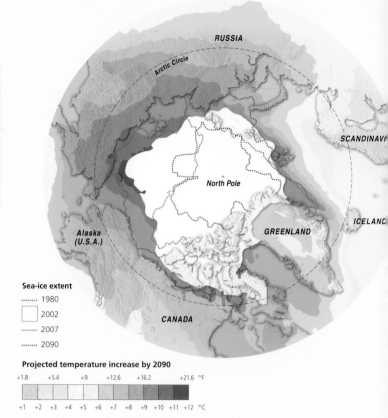

Sea-ice extent

- ┈┈┈ 1980
- ☐ 2002
- ┈┈┈ 2007
- ┈┈┈ 2090

Projected temperature increase by 2090

+1.8	+5.4	+9	+12.6	+16.2	+21.6 °F

| +1 | +2 | +3 | +4 | +5 | +6 | +7 | +8 | +9 | +10 | +11 | +12 °C |

Temperature and ice (above) This map shows the actual and projected changes to the extent of September Arctic sea ice, as well as predicted temperature increases by 2090. However, some scientists forecast that, due to accelerating polar warming, the Arctic will experience ice-free summers long before 2090.

NORTHERN EXPOSURE

The search for a Northwest Passage may be resolved with help from a warming climate. The Arctic has an annual cycle of snow and ice extent that reaches its minimum at the end of the Northern Hemisphere summer. Superimposed on this cycle is a self-reinforcing warming trend. Melting ice yields warmer Arctic waters, which in turn causes thinning of the polar ice cap, in addition to its already decreasing extent. This observed shrinking of the annual ice extent has led some scientists to speculate that the Arctic may soon experience ice-free summers.

Accelerated Arctic warming presents great challenges to human and animal populations. Inuit hunters and polar bears alike will struggle to find food as fish, seal, and caribou shift their range. The Arctic is also threatened by commercial interests, most notably by an oil industry eager to exploit newly accessible fossil fuel reserves as permafrost melts and sea ice recedes.

The effects of a melting Arctic will by no means remain localized. Arctic warming is destabilizing Greenland's glacial cover. As a land-based ice mass, a total melting of Greenland's glaciers would cause a sea-level rise of 23 feet (7 m) around the world. The addition of this fresh water to the North Atlantic could disrupt global ocean currents and cause far-reaching climate shifts.

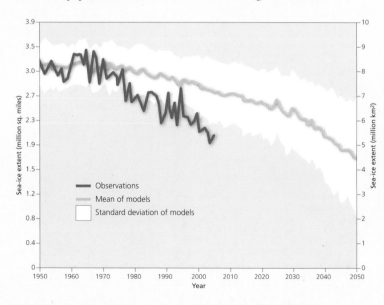

Ice in decline (left) Global climate models project a gradual decrease in September Arctic sea-ice extent, accelerating with global warming. Recently, observed sea-ice decrease has exceeded all predictions.

Vulnerable species (above) Native to the Arctic, polar bears spend much of their lives on frozen sea ice, an increasingly threatened habitat. Formerly most imperiled by hunting, the greatest danger to the world's polar bears is now global warming.

Soaking up sunlight (above left) A Russian ice pack shows the effects of glacial melting on reflectivity. While bright snow reflects most sunlight back to space, darker pools of water absorb sunlight, expanding and further accelerating local warming.

Ice cave (above) An ice cave can be formed by a meltwater river flowing under the glacier's surface or through a fissure. Meltwater can also lubricate the base of a glacier, hastening its flow toward the sea, where it calves into icebergs and is lost.

Threatened communities (left) Indigenous Arctic communities, such as this Inuit hamlet in Canada, are vulnerable to the effects of Arctic climate change. Traditional food sources diminish and melting permafrost speeds the erosion of coastal villages.

Oil industry (below) The construction of the Trans-Alaska oil pipeline disrupted local animal migrations and disturbed the delicate tundra ecosystem. Ironically, its stability is now threatened by melting of the permafrost into which its supports are drilled.

SP⟨H O T⟩ ▶ The Antarctic

▶**REGION:** Antarctica, Earth's southernmost continent, is almost entirely contained within the Antarctic Circle (66°30'S)

▶**THREATS:** Increased land and ocean temperatures, stratospheric ozone depletion, increased greenhouse gases (CO_2, CFCs), commercial fishing

▶**CLIMATE IMPACTS:** Loss of ice shelves, rising sea levels, decreased water salinity, increased UV radiation, release of pollution locked in ice

▶**ENDANGERED SPECIES:** Blue whale, four species of albatross; sperm whale and macaroni penguin vulnerable; one-celled Antarctic marine plants harmed by stratospheric ozone depletion

The Antarctic is not warming as rapidly as the Arctic because its continental mass is surrounded by circumpolar ocean currents and winds that largely isolate it from the rest of the world. However, recent warming on the Antarctic Peninsula and collapsing ice shelves have resulted in localized warming that is changing the ice cover and climate of the Southern Hemisphere polar region.

ICY CONTINENT

Until recently, it was believed that Antarctica was not warming significantly, but current satellite data shows that the continent has been warming an average of 0.2°F (0.1°C) per decade, comparable to average global warming. This is in marked contrast to the Arctic, where warming rates are double global averages. Temperatures over the Antarctic Peninsula have been rising more rapidly, increasing 5.4°F (3°C) over the past 50 years.

The Antarctic interior is expected to retain its snow and ice cover longer than the Arctic, due to the continental support below the glaciers. Further, while ice-sheet collapses are observed around Antarctica and the continent is losing ice on its coastlines, the interior may be gaining ice mass as warming surrounding waters provide fuel for increased precipitation.

Ice-shelf collapse is accelerated by the warming seas around the continent, part of a self-reinforcing cycle where dark water absorbs more of the Sun's energy and warms further. The loss of the buttressing effect of these ice shelves speeds glacial flow into the ocean, which will eventually overtake increased precipitation to lead to a net loss of ice in the Antarctic. If the most fragile portion, the West Antarctic Ice Sheet, were to be sufficiently destabilized by these losses to completely melt, it would contribute 16 feet (5 m) to global sea-level rise.

Projected temperature increase by 2090

+1.8 +5.4 +9 +12.6 °F

☐ Current ice shelf

+1 +2 +3 +4 +5 +6 +7 °C

A warmer Antarctica (above) Overall, Antarctic warming over the next 80 years is projected to be slower than Arctic warming. However, where ice shelves collapse, such as is occurring with alarming frequency along the Antarctic Peninsula, warming will be accelerated, as reflective ice gives way to absorbing dark water.

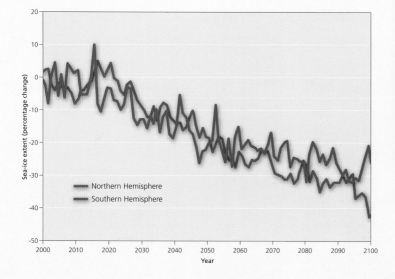

Sea-ice extent (left) This climate model projects a steady decrease in the extent of sea ice in both Arctic and Antarctic regions. Because the Southern Hemisphere has much more continental ice relative to floating ice, it will retain ice cover longer.

Population at risk (above) King penguins live in large groups on Antarctic offshore islands, feeding on fish and squid. Rising sea temperatures around the disintegrating ice sheets diminishes their available marine prey, reducing penguin survival rates.

Maritime hazard (left) When ice shelves collapse, calving at the margin of the ice sheet is accelerated. This produces icebergs that float on the circumpolar currents, making the waters around Antarctica particularly treacherous to navigate.

Volcanic ash (above) Mount Belinda, one of Antarctica's volcanic mountains, erupted intermittently from 2001 to 2007, staining surrounding icebergs and glaciers with dark ash. This increases solar absorption, causing more melting of snow and ice.

Ice shelves under stress (below) Several large Antarctic ice shelves, including Wilkins and Larsen A and B, have collapsed dramatically in recent decades. Ice-shelf disintegration can be hastened by the formation of crevasses and stress fractures in glaciers as they flow toward the sea. These cracks allow water to seep into the glacier, eroding it further.

Rising Sea Levels

Most sea-level rise is due to thermal expansion of the warming oceans. The remaining rise is caused by melting land glaciers, such as the Greenland and Antarctic ice sheets, adding water to the oceans. Melting floating sea ice at the North Pole does not contribute to sea-level rise but does speed the warming of the oceans due to the albedo effect.

Unequal rise (below) Satellite data for the period 1993–2008 shows the pattern of annual sea-level change. As much of the sea-level rise is attributed to the thermal expansion of water, this map also identifies where the oceans are heating most rapidly.

Past and future Global sea-level rise is difficult to measure and even more difficult to project. Scientists do, however, agree that the rate of rise has accelerated in recent years as the melting of the ice sheets has quickened.

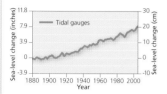

Past century Global sea levels, measured by tidal gauges over a century, have risen an average 0.08 inches (2 mm) per year for a total of about 8 inches (20 cm).

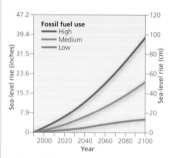

Projected The IPCC projects 21st-century sea levels to rise about 7 inches (18 cm) if fossil fuel use is low and as much as 37 inches (95 cm) with high usage.

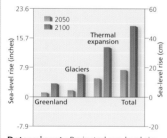

Determinants Projected sea-level rise by 2100 is dominated by the thermal expansion of warming oceans, with glacial melt contributing secondarily.

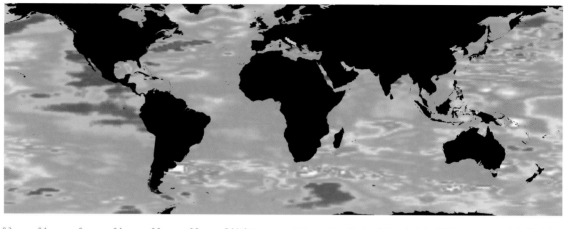

| -0.2 | -0.1 | 0 | 0.1 | 0.2 | 0.3 | 0.4 inches |

-5   0   5   10 mm
Change in sea level per year

Submersion (below) Funafuti Atoll is home to nearly half of the nation of Tuvalu's 12,000 people and could be submerged within the next few decades as sea levels rise. Inhabitants of Tuvalu and other low-lying Pacific Island states may become climate refugees.

PREDICTED CONTRIBUTION TO SEA-LEVEL RISE	
Greenland Ice Sheet melting:	0.008 inches (0.2 mm) per year
Antarctic Ice Sheet melting:	0.008 inches (0.2 mm) per year
Glaciers and ice caps melting:	0.03 inches (0.8 mm) per year
Ocean thermal expansion:	0.06 inches (1.6 mm) per year
Total contribution to sea level rise:	0.1 inches (2.8 mm) per year

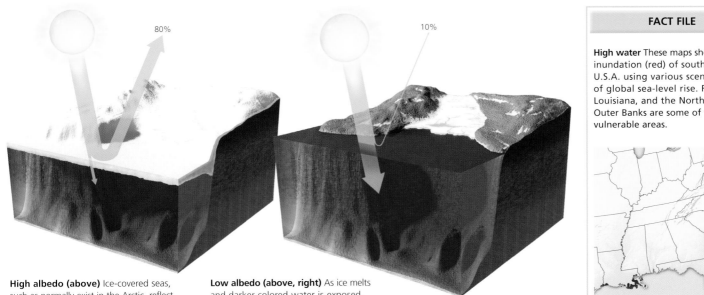

High albedo (above) Ice-covered seas, such as normally exist in the Arctic, reflect almost 80 percent of incoming solar radiation directly back out to space, with little energy absorbed by the surface.

Low albedo (above, right) As ice melts and darker-colored water is exposed, the surface albedo falls below 10 percent, starting a cycle of increased warming as more heat is absorbed by the surface.

Greenland meltwater (above) Around 10 percent of the planet's fresh water is stored in the Greenland Ice Sheet, the edges of which are thinning rapidly, aided by increasing glacial meltwater. Melting Greenland ice contributes to sea-level rise.

Inundation (left) Melting permafrost and erosion in the Alaskan village of Shishmaref—an island community of 600 Inupiat Eskimos—destroyed this home in September 2006. Melting sea ice causes higher than normal storm surges to inundate the island and erode the shore.

FACT FILE

High water These maps show ocean inundation (red) of southeastern U.S.A. using various scenarios of global sea-level rise. Florida, Louisiana, and the North Carolina Outer Banks are some of the most vulnerable areas.

3.3-foot (1 m) rise

6.6-foot (2 m) rise

13.1-foot (4 m) rise

WEATHER WATCH

Unpredictable The melting rate of ice sheets in West Antarctica and Greenland is difficult to predict. This can lead to large discrepancies in projected sea-level rises. Recently, these ice sheets have melted more rapidly, causing a faster sea-level rise of 0.12 inches (3.1 mm) per year.

Coasts under Threat

Approximately 40 percent of the world's population lives within 60 miles (100 km) of a coast. Sea levels have risen an average of 8 inches (20 cm) in the past 100 years, and are projected to rise further this century. Rising seas threaten to submerge low-lying islands and coastal regions, but impacts are felt beyond those obviously threatened areas. Rising oceans inundate freshwater estuaries, increase coastal erosion, and bring storm surges further inland.

Coastal areas (below) Storm surges from rising sea levels will reach further inland and cause widespread flooding. This map shows the number of people in each country that are predicted to be affected by 2100. Populations in low-lying Southeast Asia are particularly vulnerable.

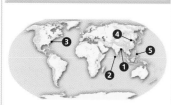
1. Lohachara Island The low-lying Sundarbans delta, at the mouth of the Ganges River, is a large mangrove forest, much of which has been converted to agricultural use. Lohachara Island was completely submerged in 2006, becoming the first inhabited island to be claimed by rising sea levels.

Lohachara Island, India

2. Maldives A group of 250 inhabited islands in the Indian Ocean, the Maldives has an average elevation of only 5 feet (1.5 m) above sea level and a highest point of 7.5 feet (2.3 m). It plans to purchase land from other countries to house its climate refugees.

The Maldives

3. Outer Banks A string of barrier islands, the Outer Banks protect vital ecosystems and coastal infrastructure from the Atlantic Ocean. Sea-level rise projections on this part of the coast are up to 3.3 feet (1 m) by 2100, which would raise salinity of fresh water inland, erode beaches, and intensify coastal flooding.

Outer Banks, North Carolina, U.S.A.

4. Tianjin At the outlet of the Yangtze and Huang He rivers into the Pacific Ocean, Tianjin is part of a commercial port area. The city lies in a low swamp with a monsoonal climate. Water levels in the area have risen due to higher sea levels and surface subsidence.

Tianjin, China

5. Manila Bay Approximately 70 percent of Filipinos live in coastal areas and rely on fish and marine products for much of their diet, making this country among the most vulnerable to sea-level rise. Projected changes in the next 70 years could threaten 0.5–2.5 million people in the Manila Bay area.

Manila Bay, Philippines

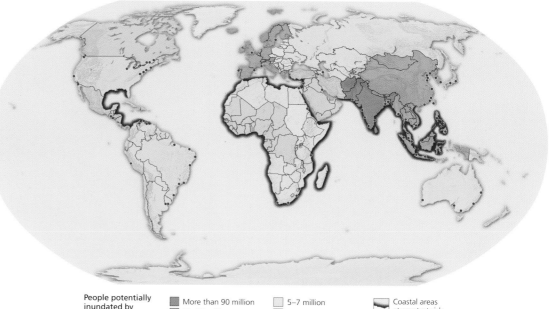

People potentially inundated by sea-level rise and storm surges, 2100

- More than 90 million
- 50–75 million
- 10–20 million
- 5–7 million
- 3–4 million
- Fewer than 1 million
- Coastal areas at greatest risk
- • Major cities at risk

The Netherlands Largely below sea level, the Netherlands uses a system of sophisticated dikes and channels to keep the North Sea at bay. Additional upgrades will be required to counteract rising sea levels.

Venice Piazza San Marco (St. Mark's Square) flooded in February 2009 as a result of the convergence of a high tide and a strong weather system. Such flooding will become more common.

Philippines (above) Due to regular flooding, coastal villagers build houses on stilts, but these simple structures are vulnerable to storm surges that accompany frequent cyclones and tsunamis.

River Thames (below) The Thames Barrier, a flood control structure, closes to protect London from exceptionally high tides or storm surges. Since 1990, the barrier has closed around four times a year.

SPOT ▶ Bangladesh

▶**REGION:** Bangladesh, South Asia

▶**THREATS:** Flooding, deforestation, soil erosion and sedimentation, water pollution, groundwater contaminated by naturally occurring arsenic, overpopulation, poaching

▶**CLIMATE IMPACTS:** Rising sea levels, increased storm activity, more mosquito-borne diseases, increased CO₂ release from forest burning

▶**ENDANGERED SPECIES:** Bengal tiger, Asian elephant, clouded leopard, Ganges River dolphin, mangroves

---

Bangladesh, situated in the Ganges River Delta, is one of the most climate-threatened regions of the world. Much of its land is less than 40 feet (12 m) above sea level. Coastal mangrove forests provide a rich habitat for species such as the endangered Bengal tiger and protect the delta's fertile farmland from ocean inundation. Deforestation and rising seas from a warming climate bring the ocean further inland each year.

GREEN DELTA

With a large and relatively poor population living in a low-lying region, Bangladesh faces many threats in a warming climate. Rising seas bring more intense monsoonal rains and tropical cyclones, which the residents of countries like Bangladesh have scant resources to manage. Severe inundation challenges basic sanitation methods and encourages the spread of disease, while flooding of inland reservoirs with seawater creates a scarcity of fresh water. Agricultural lands are also damaged, generating further socio-economic problems of food supply. Shrimp farming, a significant industry for Bangladesh's economy, is particularly vulnerable to climate impacts.

More than 42 million people live along Bangladesh's coastline, therefore the human cost of storm surges, cyclones, and coastal flooding is great. The Ganges River Delta, also known as the Green Delta, lies on the Bay of Bengal and is the largest in the world. Two-thirds of its 220-mile (354 km) coastline is in Bangladesh, with the other third in neighboring India. Snowmelt from Himalayan glaciers finds its way to the Ganges Delta. This has increased in recent years and has become an additional source of flooding in Bangladesh. In the aftermath of coastal storms and flooding, outbreaks of water-borne diseases are common as sanitation systems break down. Furthermore, as temperatures increase, the range of tropical diseases like malaria is also expected to increase, rendering the Bengali population more vulnerable to disease outbreaks.

The Sundarbans, a protective barrier of mangrove forests along the fertile Ganges River Delta, is threatened by tropical cyclones and erosion. Upstream, fresh water is diverted for farming and drinking water, which means that less fresh water is reaching the mangroves. Rising sea levels and saltwater inundation continue to test the hardiness of this extremely valuable ecosystem. Deforestation to create additional farmland is another constant threat to the mangrove forest. In recognition of its very important biodiversity and ecology, the Sundarbans in Bangladesh was designated a UNESCO World Heritage site in 1997. It adjoins India's Sundarbans National Park, inscribed by UNESCO in 1987.

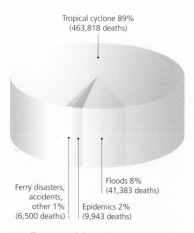

Tropical cyclone 89%
(463,818 deaths)

Floods 8%
(41,383 deaths)

Ferry disasters, accidents, other 1%
(6,500 deaths)

Epidemics 2%
(9,943 deaths)

Deadly surges (above) From 1987–2001, the primary cause of death in Bangladesh was tropical cyclones in the Bay of Bengal. The bay funnels storm surges into the delta, causing destruction inland. Monsoons deliver another deadly phenomenon—flooding.

Population density (above) Much of Bangladesh's population lives near the coast and in the low-lying delta, vulnerable to even moderate sea-level rise. As the population density coincides strongly with areas of low elevation, a sea-level rise of 5 feet (1.5 m) would impact 15 percent of the population of 150 million.

Elevation

<33 ft/ <10 m	>33 ft/ >10 m	**Population density**
		0–260 per sq. mile/ 0–100 per km²
		260–1,300 per sq. mile/ 100–500 per km²
		More than 1,300 per sq. mile/500 per km²

Bengal tiger (above) A famous inhabitant of the Ganges River Delta is the majestic Bengal tiger. The population of wild Bengal tigers in the Indian subcontinent has decreased 90 percent over the past century, primarily due to habitat losses and poaching.

Health hazard (above) Given the simple living conditions and high population density, flooding in Bangladesh creates severe sanitation problems. This increases the spread of infectious diseases such as pneumonia, tuberculosis, and measles, and mosquito-borne diseases such as malaria.

Protective mangroves (below) The Sundarbans (dark green) is the largest remaining tract of mangrove forest in the world. It protects the agricultural farmland (light green) upstream in the Ganges Delta region that supports the large population of Bangladesh.

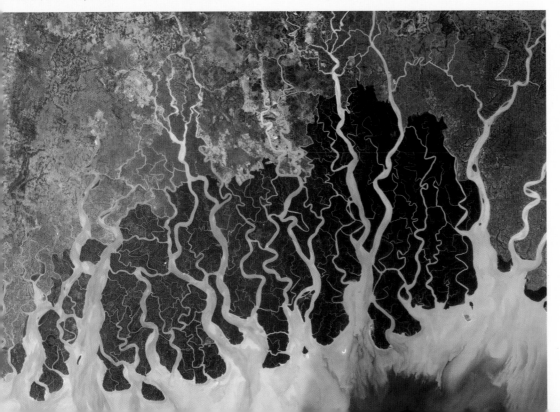

Fish and shrimp (above) Fish is a major source of protein for the four million residents of the Sundarbans. Shrimp aquaculture ponds surround the protected mangroves. This is a potential threat to local water quality and biodiversity.

Ocean Acidification

Oceans mitigate the effects of global warming by absorbing about one-third of anthropogenic emissions of carbon dioxide. While this is good news for the atmosphere, the oceanic sink of CO_2 increases seawater's acidity. Since the industrial revolution, ocean acidification has badly damaged coral reefs and is projected to have further ecological consequences.

Acidity As atmospheric carbon dioxide concentrations rise, oceans directly assimilate some portion of that CO_2. This rapidly converts to carbonic acid (H_2CO_3), increasing the ocean's acidity and decreasing carbonate ion levels.

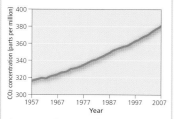

Concentration Fossil fuel emissions and deforestation raise atmospheric CO_2 concentrations, increasing ocean acidity and threatening marine life.

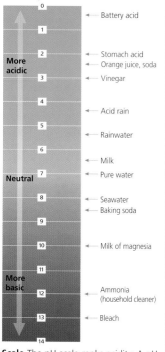

Scale The pH scale ranks acidity. A pH of 7.0 is neutral. Lower numbers are acidic, higher numbers basic, or alkaline. Seawater is naturally slightly basic.

Fertilizing the ocean Plankton convert CO_2 to calcium carbonate, the basis of coral reefs. This process is limited by the amount of iron in the ocean. Recent experiments have attempted to "fertilize" the upper layers of the ocean by dumping large quantities of iron to create a phytoplankton bloom to consume CO_2.

Carbon in the ocean (below) As acidity increases, the ocean's ability to build coral reefs and exoskeletons for marine life, such as some types of plankton, is diminished. As plankton is near the bottom of the oceanic food chain, other species that feed on plankton will be lost.

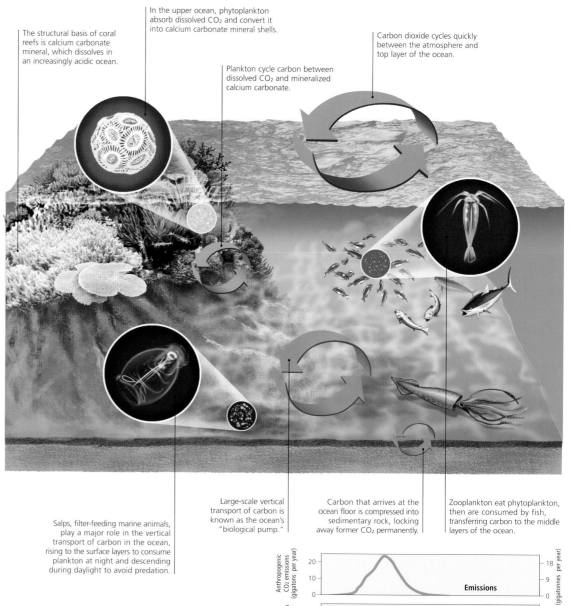

The structural basis of coral reefs is calcium carbonate mineral, which dissolves in an increasingly acidic ocean.

In the upper ocean, phytoplankton absorb dissolved CO_2 and convert it into calcium carbonate mineral shells.

Plankton cycle carbon between dissolved CO_2 and mineralized calcium carbonate.

Carbon dioxide cycles quickly between the atmosphere and top layer of the ocean.

Salps, filter-feeding marine animals, play a major role in the vertical transport of carbon in the ocean, rising to the surface layers to consume plankton at night and descending during daylight to avoid predation.

Large-scale vertical transport of carbon is known as the ocean's "biological pump."

Carbon that arrives at the ocean floor is compressed into sedimentary rock, locking away former CO_2 permanently.

Zooplankton eat phytoplankton, then are consumed by fish, transferring carbon to the middle layers of the ocean.

Projections (right) In this model, CO_2 emissions are predicted to increase until 2200 before decreasing. Atmospheric CO_2 would then be absorbed by the ocean over many centuries, due to the long timescale of ocean mixing. Seawater has a natural pH of 8.16, but recent acidification has lowered it by around 0.1 pH unit. The pH scale is logarithmic, not linear, thus a change of even 0.1 pH unit can have drastic effects on marine life.

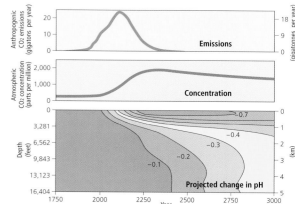

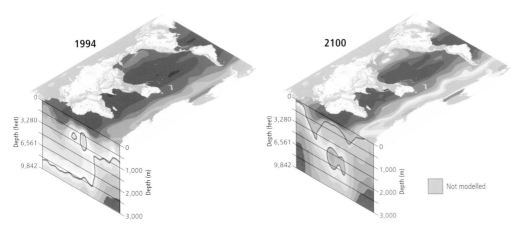

1994   2100

Not modelled

Carbonate saturation (above) Calcium carbonate thrives in water saturated with carbonate ions. The boundary which divides supersaturated (green) from undersaturated water (red) is predicted to shift shallower from 1994 to 2100 as atmospheric and oceanic CO_2 levels rise.

Algal analysis (below) Research at Biosphere 2, an artificial ecosystem in Arizona, U.S.A., verified that increasing atmospheric carbon dioxide concentration reduces coral reef growth. Divers collect algae to determine the amount of carbon absorbed.

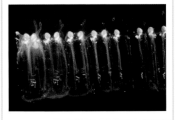

Warming Seas

As atmospheric temperatures rise, oceans absorb a great deal of energy and warm up. Heat content in the upper 2,300 feet (700 m) of seawater steadily increased in the latter half of the 20th century. Warmer ocean temperatures cause coral reef death through bleaching; thermal expansion of water, leading to sea-level rise; and an increase in the intensity of tropical storms and hurricanes.

FACT FILE

Coral bleaching Coral polyps secrete a calcium carbonate skeleton, which is covered by a thin layer of living tissue. Warming oceans disrupt the delicate codependence between the coral and zooxanthellae—algae that convert sunlight into sugars.

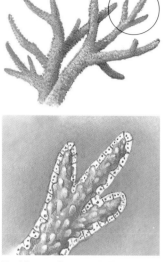

Healthy coral Zooxanthellae live within healthy coral polyps. These are symbiotic algae that provide food and oxygen in exchange for protection.

Bleached coral Warmer temperatures stress the dependent relationship, resulting in the digestion or expulsion of the zooxanthellae and coral death.

Dead coral The ghostly white calcium carbonate skeleton of dead coral provides a foundation for opportunistic filamentous algae.

Ocean heat (below) Changes to ocean temperatures are indicated for the period 1955–2003. Red shading shows increased heat content and blue shading, decreased. The greatest warming has occurred in the Atlantic and off the eastern South American and southern African coasts.

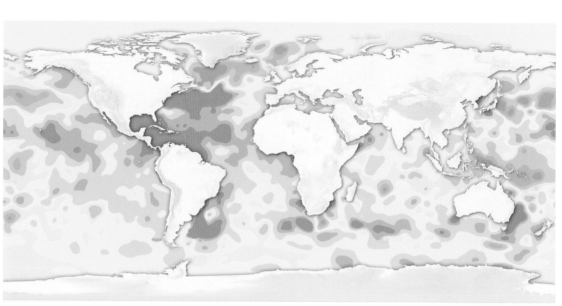

Bleached reef Environmental stressors such as warming water, harmful ultraviolet (UV) radiation, and ocean acidity can kill and whiten reefs. Living coral is colored by photosynthetic pigments in symbiotic organisms.

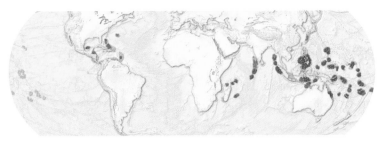

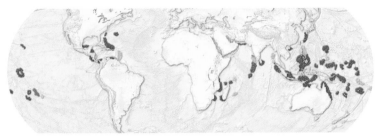

Coral under threat Two climate models, based on a 100-year increase of 3°F (1.7°C) **(left)** and 5.4°F (3°C) **(below left)**, show where coral bleaching is likely to occur by around 2055. Red circles indicate areas of greatest thermal stress. Coral reefs around Oceania are particularly threatened.

Coral reef thermal stress

- Low
- Medium
- High

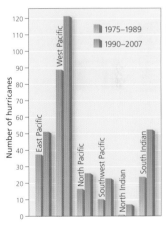

Krill kill (left) At the base of the Antarctic food chain, krill are an important food source for penguins and other birds in the polar marine environment. Warming seas diminish coastal ice-shelf extent, reduce krill habitat, and disrupt the ecosystem, forcing penguins to migrate from the area to find food.

Hurricane intensity Hurricanes derive their energy from the heat released when moist air rises from warm surface ocean waters and condenses into massive cloud systems. Warming oceans provide a breeding ground for more intense hurricanes.

Atlantic track A satellite image from July 2003 shows warm sea surfaces across the equatorial Atlantic after the longest days of sunshine in the region.

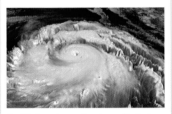

Pacific hurricane Hurricane Linda, in September 1997, was the strongest Pacific hurricane on record, with wind speeds of 185 miles per hour (298 km/h).

On the rise An increase in the number of intense hurricanes (Category 4 and 5) from 1975–1989 to 1990–2007 has been observed in all ocean basins.

Algal blooms Warming oceans support more frequent algal blooms. When a marine ecosystem tips out of balance, algae multiply and deplete nutrients and oxygen in the surface waters. Chinese fishermen clean up the coast southeast of Beijing in July 2008.

Great Barrier Reef

SP**O**T ▶

▶**REGION:** Great Barrier Reef, Australia

▶**THREATS:** Water pollution, increased sedimentation, overfishing, tourism, crown-of-thorns starfish

▶**CLIMATE IMPACTS:** Warming seas, rising seas, ocean acidification, coral bleaching, increased storm activity

▶**ENDANGERED SPECIES:** Coral, blue whale, sei whale, loggerhead turtle, green turtle, hawksbill turtle, leatherback turtle

The Great Barrier Reef is the largest coral reef system and the largest biological entity on Earth. This massive structure was built by tiny coral polyp organisms. The rich biodiversity of the reef is a major reason why UNESCO named it a World Heritage site, in 1981. The Great Barrier Reef Marine Park was established in 1975 to protect the reef from threats associated with pollution, overfishing, and tourism.

DELICATE ECOSYSTEM

This magnificent network of coral reefs and islands, set in tropical waters, supports a wide array of marine and other species, including whales, dolphins, porpoises, turtles, crocodiles, sharks, mollusks, frogs, birds, snakes, and over 1,500 species of fish. Many are endemic to the Great Barrier Reef and, because of environmental threats to the reef system, now have vulnerable or endangered species status.

This massive structure off the northeast coast of Australia is the largest site on the World Heritage List. With an area of 133,000 square miles (344,000 km²), spanning nearly 14 degrees of latitude, it is made up of a series of 760 separate fringing reefs and more than 2,000 other reefs and cays. The shallow tropical waters are an ideal habitat for a wide variety of marine life, but the environmental pressures of tourism, fishing, water pollution, and warming seas leading to coral reef bleaching pose increasing threats to this habitat.

Coral reef bleaching is one of the most severe ecological side effects of climate change observed to date. Mass coral bleaching events—that is, coral reef death—occurred due to rising ocean temperatures in the summers of 1998, 2002, and 2006, when warm El Niño conditions added to already warmer ocean surface temperatures. As seas continue to warm, bleaching events may occur annually, even in the absence of El Niño conditions. If some coral organisms remain, recovery is possible, but more than half of the reef has become bleached, leading to a clear overall decline in healthy reef habitat. Some scientists project a loss of 95 percent of the reef's living coral by 2050.

In addition to warming waters and coral bleaching, the ecosystem of this 8,000-year-old living reef is threatened by other factors. It is vulnerable to chemical and silt runoff from farming; overfishing of key species which interrupts the marine food chain; and destruction of reef habitats by tourist boats.

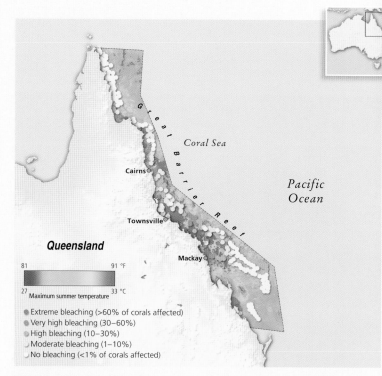

Mass bleaching (above) In 2002, the Great Barrier Reef experienced its worst coral bleaching on record. More than half of the reef suffered some degree of bleaching. From January to March of that year, the entire reef was 3.6°F (2°C) or more above normal water temperatures, the likely cause of the bleaching.

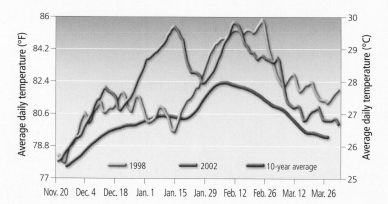

Warm seas (left) Sea temperatures around the reef peak in January and February—the Southern Hemisphere summer. Extensive coral bleaching occurred in 1998 and 2002, both El Niño years featuring warmer seas.

Reef resident (above) Epaulette sharks are commonly found in shallow coral reef waters. Growing to about 3 feet (1 m) long, their slender body allows them to forage through narrow reef crevices.

Symbiosis (top) Healthy coral polyps house zooxanthellae (here, greenish-yellow). Warming waters erode this relationship, inducing corals to expel zooxanthellae. As these algae also host the pigments that give reefs their color, this expulsion causes bleaching.

Record keeping (above) Scientists survey the reef annually to maintain a record of bleaching. A region of bleached coral is shown off the Keppel Islands in Queensland. Bleaching can be reversible if some coral organisms remain to recolonize the reef.

Aerial art (above) The Great Barrier Reef is Earth's only structure made by living organisms that is visible from space. Broad continental shelves support extensive coral reef systems in several areas of the tropics, but the shallow surface waters in which coral reefs thrive are highly susceptible to warming and therefore coral bleaching.

Reef pollution (left) A silted flood plume from Maria Creek disperses into the Coral Sea toward the Great Barrier Reef. This polluted runoff is one environmental threat to the reef, now protected by the Great Barrier Reef Marine Park. This protected status means that fishing and tourism, which also threaten the reef, is restricted.

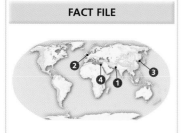

1. Pakistan In 2007 and 2008 unusually intense monsoonal rains pounded the Balochistan region, causing widespread flooding. Local houses are usually mud constructions and many were severely

damaged. Brick schools and mosques were utilized for shelter. The 2007 flood affected up to a million Pakistanis.

Balochistan, Pakistan

2. U.K. Heavy rains caused severe flooding across much of the U.K. in 2007. The deluge flooded homes and businesses, damaged crops,

claimed 13 lives, and displaced thousands more. The disaster highlighted the lack of flood-prevention measures.

U.K.

3. North Korea In 2007, North Korea also experienced severe flooding, after a week of torrential rain in August. Pyongyang residents waded waist-deep along major boulevards, and railways

and roads were inundated around the country. Several hundred people were killed and 300,000 left homeless.

Pyongyang, North Korea

4. Egypt The Nile River experiences regular seasonal flooding that irrigates the delta and deposits soil. This is beneficial to agriculture. However,

rising sea levels will inundate the fertile Nile Delta with salt water, stripping the soil of nutrients and destroying crops.

Nile Delta, Egypt

Severe Flooding

A warming climate will bring drought to some regions, but growing evaporation from warmer oceans means more overall precipitation. Many areas are expected to experience an increased frequency of severe flooding. Coasts become inundated due to rising storm activity over a warmer ocean, and rivers flood due to greater snowmelt and stronger monsoonal rains.

Germany (right) A low-pressure system in the North Sea in November 2007 caused flooding on the Elbe River in Hamburg, after making landfall on the German coast with wind speeds of up to 81 miles per hour (130 km/h).

India Two satellite images show the Kosi River, a tributary of the Ganges River, before **(top right)** and after **(right)** a major flood in 2008. Monsoonal rains regularly flood rivers in northern India, but this particularly severe episode burst the riverbanks and sculpted a new, straightened river channel, displacing more than a million people.

Peru (above) Flooding in January 2007 caused a landslide in Chanchamayo, Peru, 220 miles (350 km) east of Lima. During El Niño events, the west coast of South America is warm and rainy, and Peru regularly experiences extreme flooding.

China (below) In July 2007, the Yangtze River rose over 80 feet (25 m), inundating this park in Wuhan, Hubei Province, in central China. As Himalayan glaciers melt, the rivers they feed are prone to flooding. Monsoon rains also contribute to rising river levels.

FACT FILE

1. Ghana Countries throughout west, central, and east Africa experienced the worst flooding in recorded history in September 2007. The flooding displaced 400,000 people in Ghana alone, submerged much of Africa's productive farmland, and raised concerns about water-borne diseases.

Ghana, West Africa

2. U.S.A. The Mississippi River flooded the states of Wisconsin, Iowa, Missouri, and Louisiana in June 2008, due to heavy rainfall that augmented the spring snowmelt. Floodwaters swamped thousands of homes, cresting at heights second only to the Great Flood of 1993.

Mississippi River, U.S.A.

3. Haiti Severe flooding devastated Haiti in September 2004 in the wake of Tropical Storm Jeanne, causing mudslides and resulting in 2,400 deaths. The nation is chronically threatened by flooding as the widespread practice of deforestation makes mountains and hillsides more vulnerable to erosion.

Haiti

WORLDWIDE TRENDS

The number of flood disasters has risen markedly since 1960. The greater incidence of deadly floods can be attributed to increasing precipitation, coastal storms, and glacial melt, combined with higher population density along rivers.

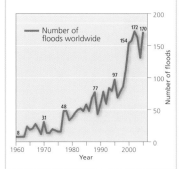

Increased Storm Intensity

A correlation has been observed between storm intensity and increasing sea surface temperatures, suggesting that global warming will spawn more frequent and more severe tropical storms. The Atlantic hurricane season of 2005 was particularly devastating, with a record number of storms (28), 15 of which became hurricanes, including four Category 5 hurricanes.

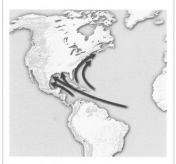

Tracks A subtropical high-pressure ridge steers tropical storms and hurricanes westward and northward into the Gulf of Mexico or up the Atlantic seaboard.

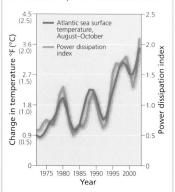

Intensity The Power Dissipation Index measures Atlantic storm intensity and is correlated to the oscillating, overall increasing trend in sea temperatures.

SOUTH ATLANTIC CYCLONE

Due to lower water temperatures and high winds, the South Atlantic does not usually produce the right conditions for cyclone formation. However, in March 2004, a large tropical cyclone—unofficially named Cyclone Catarina—occurred off the coast of Brazil. This Category 2 storm killed up to 10 people.

Storm gallery The 2005 Atlantic hurricane season had more storms than two regular seasons combined. When Hurricane Wilma exhausted the original list of 21 names, the Greek alphabet was used to name storms for the first time.

Cindy On July 5, 2005, Cindy made landfall over Louisiana and Alabama.

Dennis Over a million people in Florida and Alabama had to flee Dennis.

Emily The second catastrophic storm of the year, Emily, came ashore over Mexico.

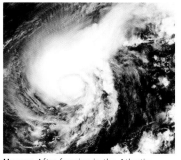

Harvey After forming in the Atlantic, Harvey made a grazing pass at Bermuda.

Irene On August 14 Irene became the fourth hurricane of the 2005 Atlantic season.

Katrina The costliest hurricane on record caused damage from Florida to Texas.

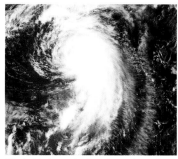

Maria Maria reached maximum intensity over the Atlantic on September 6.

Nate Growing slowly off Bermuda, Nate reached hurricane strength in September.

Ophelia Slow-moving Ophelia dumped heavy rain along the U.S. eastern seaboard.

Philippe In just one day Philippe grew into a Category 1 hurricane near the Lesser Antilles.

Rita Rita passed over the Florida Keys and the Gulf of Mexico in September.

Stan In October, Stan caused landslides, flooding, and death in Central America.

Destruction When tropical storms make landfall or approach coastal areas, damage from strong winds, storm surges, and torrential rains can be devastating. Floods can destroy sanitation infrastructure, leaving areas prone to water-borne diseases.

Washed away In June 2008 in the wake of Typhoon Fengshen, extensive flooding triggered mudslides on the island of Mindanao in the Philippines.

Submerged Situated along the path of many hurricanes, Port-de-Paix, Haiti, flooded after four storms battered the island in August 2008.

Levee failure In August 2005, Hurricane Katrina's winds pushed a 24–28 foot (7–8.5 m) storm surge into New Orleans, U.S.A., flooding 80 percent of the city.

Most intense (above) On October 19, 2005, Hurricane Wilma broke the record for the lowest pressure (882 hPa) of any Atlantic hurricane. In this false-colored composite image, orange and red indicate sea surface temperatures 82°F (28°C) and higher.

Alpha After peaking on October 23, Alpha dissipated and was absorbed by Wilma.

Beta Beta formed off Panama as a tropical depression on October 26.

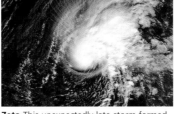

Zeta This unexpectedly late storm formed southwest of the Azores on December 30.

Timeline (left) The unusually long Atlantic hurricane season of 2005 commenced on June 8, with storm activity continuing into January 2006. This chart shows when the 15 hurricanes, 13 tropical storms, and three tropical depressions formed, their duration, and intensity.

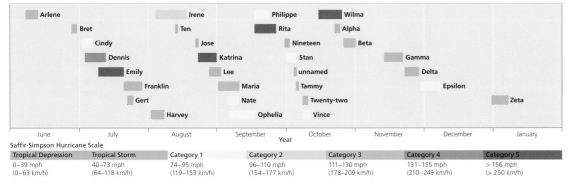

Timeline storm names: Arlene, Bret, Cindy, Dennis, Emily, Franklin, Gert, Harvey, Irene, Ten, Jose, Katrina, Lee, Maria, Nate, Ophelia, Philippe, Rita, Nineteen, Stan, unnamed, Tammy, Twenty-two, Vince, Wilma, Alpha, Beta, Gamma, Delta, Epsilon, Zeta

Months: June, July, August, September, October, November, December, January
Year

Saffir-Simpson Hurricane Scale

Tropical Depression	Tropical Storm	Category 1	Category 2	Category 3	Category 4	Category 5
0–39 mph (0–63 km/h)	40–73 mph (64–118 km/h)	74–95 mph (119–153 km/h)	96–110 mph (154–177 km/h)	111–130 mph (178–209 km/h)	131–155 mph (210–249 km/h)	> 156 mph (> 250 km/h)

 Myanmar

> ▶**REGION:** Myanmar, Southeast Asia
>
> ▶**THREATS:** Flooding, deforestation, soil erosion and sedimentation, overpopulation, poaching, earthquakes
>
> ▶**CLIMATE IMPACTS:** Rising sea levels, increased storm activity, more mosquito-borne diseases, increased CO_2 release from forest burning, drought
>
> ▶**ENDANGERED SPECIES:** Indochinese tiger, Asian elephant, clouded leopard, mangroves

The country of Myanmar, also known as Burma, experiences the monsoon climate of Southeast Asia and is particularly susceptible to damage from tropical cyclones due to its low elevation. The humanitarian disaster of Cyclone Nargis in May 2008, one of Asia's deadliest storms, sharply demonstrated the region's vulnerability. An estimated 150,000 people died because of the storm's intensity and the area's flood-prone topography.

DEVASTATING STORM

Cyclone Nargis was the first storm of the summer monsoon season in 2008, when cyclones frequently form over the seasonally warmed waters of the northern Indian Ocean. The powerful Category 3 storm ripped across the Irrawaddy Delta of southern Myanmar, leaving a broad swath of catastrophic flooding, and destroyed infrastructure in its wake.

Torrential rains and relentless monsoon winds pushed a 12-foot (3.6 m) storm surge inland up to 25 miles (40 km). Typical stilted housing structures, designed to survive regular seasonal flooding and high tides, were too flimsy to withstand the cyclone's winds. The lack of a national storm warning system meant that most coastal communities were completely unprepared, contributing to the high number of storm-related fatalities. Floodwaters covered the entire coastal lowlands, as well as

Yangon, the country's largest city, with more than four million people.

The practice of cutting down mangrove forests exacerbated the flooding. Mangroves are salt-tolerant evergreens whose dense root systems blunt the force of storm waves and protect inland areas from erosion and flooding. Vulnerable areas were inundated by the saltwater storm surge, which destroyed fields and contaminated rice crops. Hundreds of thousands of farm animals perished, adding to the food supply disaster. These losses were worsened by the government's initial inaction and blocking of international relief efforts. The United Nations estimated that 2.4 million people needed emergency food aid.

Low-lying communities, already in jeopardy due to ecologically destructive land-use practices, will be most threatened by rising seas and increased tropical storm intensity as the climate continues to warm.

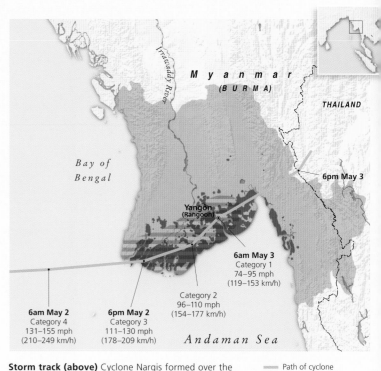

Storm track (above) Cyclone Nargis formed over the warm waters of the Bengal Sea in late April 2008. It was a Category 3 storm, with sustained winds near 130 miles per hour (210 km/h) when it made landfall on May 2. The cyclone swept northeast across the length of the low-lying Irrawaddy Delta, causing severe flooding throughout the region.

— Path of cyclone
Wind-affected areas
Flooded areas
Approximate zone of sustained wind speeds at cyclone Category 1 or greater

6am May 2
Category 4
131–155 mph
(210–249 km/h)

6pm May 2
Category 3
111–130 mph
(178–209 km/h)

6am May 3
Category 1
74–95 mph
(119–153 km/h)

Category 2
96–110 mph
(154–177 km/h)

6pm May 3

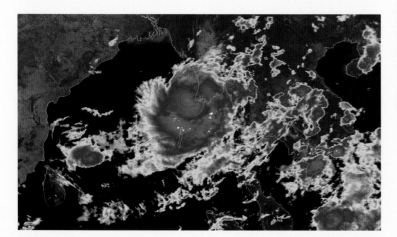

From space (left) This infrared satellite image shows the cyclone's extensive size as it hit land near the Irrawaddy River. Winds remained above 70 miles per hour (113 km/h) for most of the storm's 24-hour transit across the delta.

Damage (above) Most houses in coastal towns like Labutta (pictured on June 14, 2008) are built on stilts or platforms to protect against high tides. The fierce storm undermined these structures, leaving more than a million people homeless.

Flooded fields (above) The saltwater storm surge flooded the delta's productive agricultural region, impeding the country's ability to produce food. Mangroves, a natural defense against flooding, had been cleared to enable farming.

Run-off (far left) In the storm's aftermath, drastically diminished food and fresh water supplies threatened survivors. Here, children in a refugee camp collect rainwater to use as drinking water. Rations remained critically low for months after the cyclone.

Boats (left) The Moken people live among the islands of the Mergui Archipelago in the Andaman Sea. Families live on wooden boats called kabangs and moor along the beaches to find food. The low-lying islands are at risk of submersion by rising sea levels.

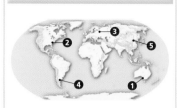

1. Australia, January 2009 During a massive heat wave in southeastern Australia, Adelaide recorded one of its hottest days ever and the Eyre Peninsula endured temperatures over 115°F (46°C).

Six days of heat of above 104°F (40°C) brought drought conditions to the region, raising the risk of wildfires.

Adelaide, Australia

2. U.S.A., June 2008 In mid-2008, the eastern U.S.A. experienced a heat wave with five successive days of temperatures over 100°F (38°C). Heat

waves are the most common cause of weather-related fatalities, with the elderly being the most vulnerable.

Eastern U.S.A.

3. Russia, May 2007 During this heat wave, temperatures in the temperate city of Moscow reached 91°F (33°C) on May 27, breaking the high-temperature record set in 1891. This was the

longest heat wave in 128 years and 1,900 square miles (5,000 km²) of spring-sown fields were wiped out.

Moscow, Russia

4. Argentina, January 2009 A summer heat wave struck subtropical South America, affecting inland and coastal areas. Temperatures in the port city of

San Antonio Oeste hit 108°F (42°C). Farther north, heat waves also struck east of the Andes mountains.

San Antonio Oeste, Argentina

5. China, 2007 North and northeast China suffered a heat wave and the worst drought in two decades, receiving 50–90 percent less rain than expected during July. This created

a water crisis for 1.2 million people. Beijing opened old air-raid shelters to help people escape the heat.

Beijing, China

Extended Heat Waves

The global average temperature has been generally increasing since 1880, with the rate of warming accelerating since about 1980. As the average temperature rises, heat waves become more frequent. Recent heat waves have starkly demonstrated the threat to human health through sunstroke and dehydration, in addition to causing crop damage and water shortages.

2008 anomaly (below) While 2008 was the coolest year since 2000, compared to the 1950–1980 baseline, much of the globe was normal (white) or warmer than normal (red). Eastern Europe, Russia, the Arctic, and the Antarctic Peninsula were exceptionally warm. No data is available for areas shaded gray.

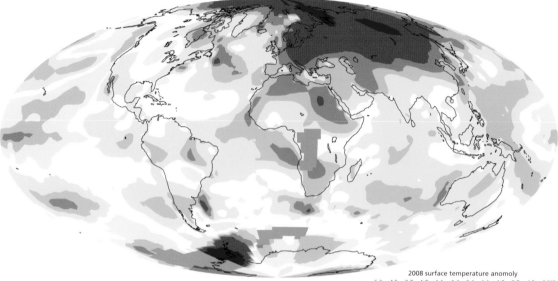

2008 surface temperature anomaly

-6.3 -4.5 -2.7 -1.8 -1.1 -0.4 0.4 1.1 1.8 2.7 4.5 6.3°F
-3.5 -2.5 -1.5 -1 -0.6 -0.2 0.2 0.6 1 1.5 2.5 3.5°C

Trends since 1880 (right) The year-to-year temperature variability can make trends harder to detect. In North America, for example, a La Niña event made 2008 cooler than past years. The five-year running average clearly shows the long-term increase in global average temperatures.

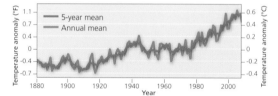

Hotter extremes (below) Heat waves become more frequent in a warming climate, threatening human health. In June 2007, Pakistan was gripped by a heat wave that killed dozens of people as temperatures reached 122°F (50°C) in parts of the country.

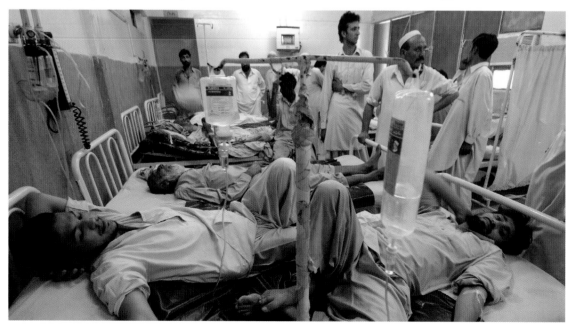

Hot spell (left) In 2003, a historic heat wave in Europe was responsible for at least 3,000 deaths in France alone. This map shows temperature differences between 2001 and 2003. A swath of red indicates where temperatures increased by up to 18°F (10°C). Melting alpine glaciers caused dangerous swelling of rivers in southern Europe.

Spanish fever (below) In Spain, the 2003 heat wave drove thousands of people to the beaches of Galicia. In southern Spain, temperatures soared to over 111°F (44°C). In addition to imperiling human health, the heat wave brought drought to southern Europe, creating a crop shortfall. More than 52,000 people died across Europe as a result of the heat.

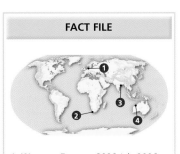 

Wildfires

Wildfires constitute an important climate feedback. Warmer temperatures facilitate more frequent and widespread wildfires, which, in turn, convert the carbon stored in trees into gaseous carbon dioxide, further warming the climate. Wildfires also emit carbon monoxide gas and aerosol particles into the atmosphere and decrease vegetative cover, leading to more erosion and drying.

Australian bushfires (below) A bushfire near Melbourne in December 2006 burned more than 4,200 square miles (11,000 km²) of bushland. High temperatures, regional drought, and the flammability of eucalyptus trees provides ideal conditions for fires.

More fires (right) The number of wildfires has increased markedly around the world in each decade since 1950. This increase has been most rapid in the Americas. However, while numbers have risen, the area burned annually has fallen in developed countries.

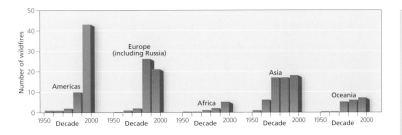

1. Greece, 2007 Forest fires broke out across Greece in August 2007. Severe drought and three consecutive heat waves with temperatures over 105°F (40°C) contributed to making

this the worst fire season on record, with 84 deaths and 1,000 square miles (2,600 km²) of forest and farmland burned.

Greece

2. South America, 2007 Vast areas of forest in the Amazon Basin regularly burn during the dry season each year. Many fires are started by farmers to

clear new pasture. In 2007, more than 10,000 fires were seen over an area of 770,000 square miles (2,000,000 km²).

Amazon Basin, South America

3. China, 2004 The Greater Khingan Range in Inner Mongolia is one of China's major forested areas. Forest fires have become more frequent in this and neighboring regions, with

lightning igniting the increasingly hot and dry taiga forests. There are two fire seasons each year, in spring and fall.

Inner Mongolia, China

ALASKAN FIRES

This satellite image captures the plume of high aerosol optical depth (in red) emitted by an Alaskan wildfire in June 2004. Optical depth is a measure of the ability of light to penetrate through to Earth's surface, a function of the concentration of particles in the atmosphere.

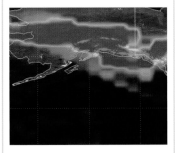

Prolonged Drought

With a warming climate, evaporation is accelerated and many parts of the world are suffering from severe drying trends. The evaporated moisture must fall somewhere, so some regions are experiencing increased rainfall. This process largely follows existing weather patterns, so that dry regions become drier while already wet regions receive the excess rainfall.

Mapping (below) The Palmer Drought Severity Index (PDSI) is a measure of drought vulnerability, derived from surface air temperatures and precipitation averaged over monthly periods. Brown denotes regions that have become drier, and green, wetter.

FACT FILE

Dust Bowl During the 1930s, the Great Plains of North America were ravaged by huge dust storms when drought, overgrazing, and harmful farming practices caused large-scale erosion and desertification. These maps show the PDSI from 1934–1936.

1934 Drought conditions were the worst in 1934, with severe drought from Canada to Texas. Browner regions indicate drier than normal conditions.

1935 Conditions in 1935 were less severe than the preceding and following years. Only small areas experienced drought; much of the east and southwest had rain.

1936 Drought conditions returned, spreading across the U.S. and Canada. Utah and southern Texas were actually wetter than usual this year.

Buried Blowing dust covered farms and equipment, killed livestock, and ruined crops, causing starvation. More than two million people abandoned farms.

PDSI CHANGES, 1950–2002

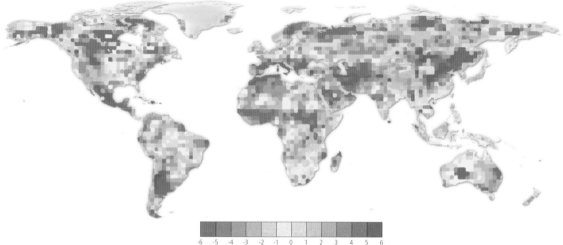

-6 -5 -4 -3 -2 -1 0 1 2 3 4 5 6
Changes in Palmer Drought Severity Index

Lower level (above) A seven-year drought and dwindling water levels in the Colorado River have lowered Lake Mead, at the Hoover Dam, in Arizona, U.S.A. A ring of mineral deposits show higher levels of water prior to this image, taken in June 2007.

Cracked earth (left) In June 2003, most of India's southern states suffered severe drought, killing at least 1,400 people. Osman Sagar Lake dried up before monsoon rains finally brought relief to the area.

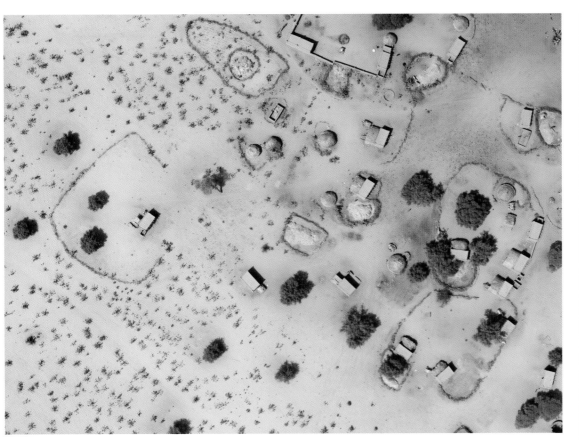

Sparse (above) Small villages dot the dry terrain of the Sahel, a region of semi-arid grasslands and savannas, which divides the Sahara desert, to the north, from the Sudanian savanna to the south.

Lake Chad (below) Recent warming, dry weather, and increased irrigation have caused the rapid evaporation of the Sahel's Lake Chad. In these satellite images, red and green indicate onshore vegetation.

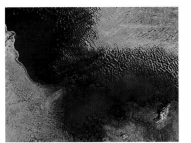

Lake Chad, 1973

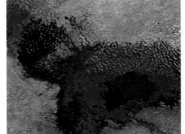

Lake Chad, 1997

Lake Chad, 2001

SAHEL PRECIPITATION ANOMALIES, 1900–2007

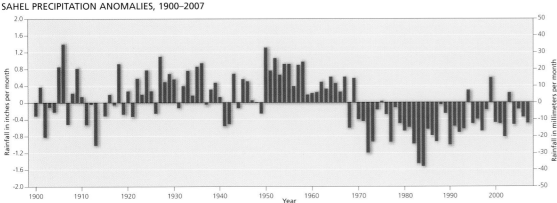

Sahel precipitation A severe decrease in monthly precipitation in the Sahel is apparent, from the late 1960s. Some recovery has occurred since the 1980s, but precipitation remains below the average for the century.

FACT FILE

Indian Ocean Changes in regional atmospheric circulation patterns can affect the distribution of droughts and flooding. Recent research suggests that Australia's drought events are driven primarily by water temperatures in the Indian Ocean.

Negative phase Cooler water in the Indian Ocean generates moisture-laden winds that carry wet air across Australia, bringing rain to the southeast.

Positive phase Warm, dry winds result from warmer Indian Ocean water. This contributes to higher temperatures and causes drought in eastern Australia.

Death Periods of severe drought have occurred repeatedly around Australia since the 1990s. Here, livestock carcasses litter a cattle station in 2005.

FACT FILE

Water inequality About 40 percent of the world's people do not have access to clean water. Disease, poor sanitation, and polluted water result from water scarcity. Lack of water is the leading cause of death in the developing world.

LACKING CLEAN WATER (IN MILLIONS)

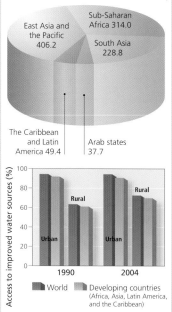

East Asia and the Pacific 406.2

Sub-Saharan Africa 314.0

South Asia 228.8

The Caribbean and Latin America 49.4

Arab states 37.7

Clean water Safe drinking water is less available in developing countries and rural areas than the world average.

LACKING SANITATION (IN MILLIONS)

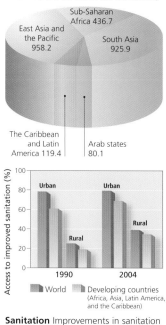

East Asia and the Pacific 958.2

Sub-Saharan Africa 436.7

South Asia 925.9

The Caribbean and Latin America 119.4

Arab states 80.1

Sanitation Improvements in sanitation since 1990 have occurred in rural areas, but are still scarce in Asia.

Water Shortage

A warming climate threatens the reliability of water supply as water evaporation rates increase and the atmosphere's hydrological cycle shifts. Rainfall, and hence freshwater availability, is redistributed. As mountain glaciers melt, less fresh water is stored on land. This results in spring flooding and fall drought, rather than a gradual meltwater supply.

World water (right) Global freshwater resources vary greatly from year to year, increasing when river runoff is high and decreasing when precipitation is low. These swings result in deviations of up to 25 percent from the mean.

Agricultural water use Globally, the agricultural sector is by far the biggest user of fresh water, accounting for 86 percent of total consumption. This is expected to decrease as farming practices become increasingly efficient.

Percentage
0–16
16–31
31–47
47–63
63–79
79–100

Industrial water use Approximately two-thirds of industrially consumed water is dammed or stored for use in hydroelectric and nuclear power generation, with the remaining third used in industrial processes.

Percentage
0–16
16–32
32–48
48–64
64–80
80–100

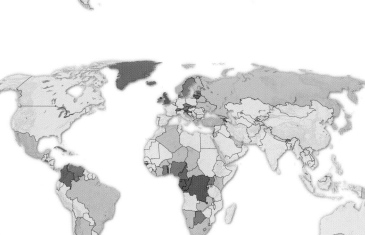

Domestic water use Domestic use accounts for the smallest fraction of water withdrawal globally. A person in the developed world consumes, on average, 10 times more water than one in a developing nation.

Percentage
0–15
15–30
30–45
45–60
60–81

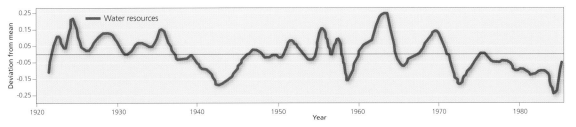

Water disputes In the war-torn Darfur region of Sudan, Africa, fighting over water and farmland has played a major role in the conflict. Each day, thousands of people line up for water in refugee camps.

FACT FILE

1. Myanmar Myanmar has a monsoon climate, receiving all of its rainfall from May to October. The rest of the year is hot and dry, and evaporation rates are high. Because Myanmar has few lakes or dams and lacks the infrastructure to use groundwater, water storage is a problem during the dry season.
Myanmar

2. Mexico Populous Mexico City, with nearly 20 million residents, presents a water provision challenge because precipitation and surface water is limited. The principal water supply for the entire metropolitan area is the Mexico City Aquifer directly beneath the city.
Mexico City, Mexico

3. Arabian Peninsula Low rainfall, high evaporation rates, and nonexistent surface water in the Arabian Peninsula means that drinking water is entirely derived from desalinated seawater and nonrenewable ancient aquifers. Access to these resources is inequitable due to poor governance of the public sector.
Arabian Peninsula

4. India Government efforts have improved water supply and sanitation in an arid country with high evaporation rates. Yet clean water provision and the treatment of wastewater remain below standard. Only one in three people in India has access to improved sanitation.
India

5. Turkey Flow in the Euphrates and Tigris rivers in Turkey determines water supply downstream in Syria and Iraq. Water is a valuable resource in this region of immense oil and gas reserves, and Turkey's proposal to dam the river for hydroelectric power generation is testing already tense regional relations.
Tigris and Euphrates rivers, Turkey

Desertification

Over one-quarter of Earth's land surface is at some risk of desertification, a process by which agriculturally productive land, generally at the edges of existing drylands, is converted to desert. The major causes are climate change, overgrazing, poor agricultural soil management, deforestation, increasing soil salinity, and local human population growth.

Desert danger (below) Desertification risk is highest at the margins of existing deserts. Warming and drying kill off the grasses, leaving soil and organic matter to be blown away by the next dust storm, diminishing agricultural productivity.

FACT FILE

Desertification risk Globally, eight percent of land is currently desert. A further four percent is at very high risk of desertification and 25 percent at moderate to high risk. The most vulnerable areas are in Australia, followed by Asia and Africa.

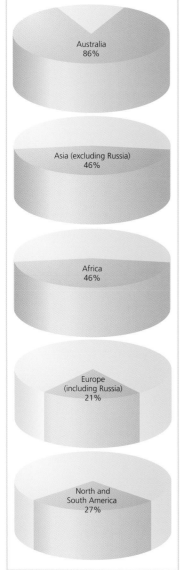

Existing desert 8%

Other land 63%

High risk 12%
Moderate risk 13%
Very high risk 4%

DESERTIFICATION RISK BY CONTINENT

Australia
86%

Asia (excluding Russia)
46%

Africa
46%

Europe
(including Russia)
21%

North and
South America
27%

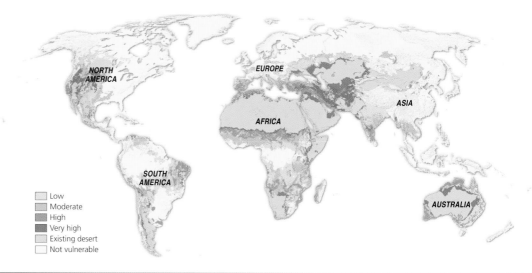

NORTH
AMERICA

EUROPE

ASIA

AFRICA

SOUTH
AMERICA

AUSTRALIA

- Low
- Moderate
- High
- Very high
- Existing desert
- Not vulnerable

Settlement in peril
The village is threatened by the shifting sand.

Forced migration Farmers abandon their now unproductive plots and search for fertile land.

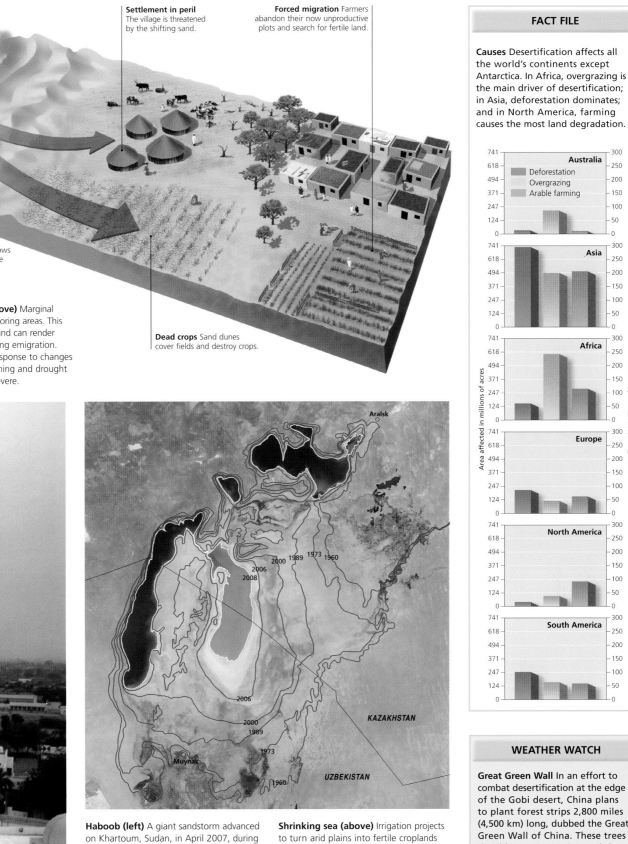

Dust storm Wind blows sand and soil from the degraded grassland.

Desert in the making (above) Marginal desert sand shifts to neighboring areas. This destroys agricultural fields and can render villages uninhabitable, forcing emigration. Deserts shift naturally in response to changes in wind patterns, but warming and drought make the problem more severe.

Dead crops Sand dunes cover fields and destroy crops.

Haboob (left) A giant sandstorm advanced on Khartoum, Sudan, in April 2007, during the seasonal monsoon. A haboob is created by an atmospheric downburst that blows up a wall of sand, promoting rapid erosion and depositing a layer of dust.

Shrinking sea (above) Irrigation projects to turn arid plains into fertile croplands have depleted water levels in the Aral Sea, turning the area into a source of huge dust storms. Contour lines indicate the receding shoreline between 1960 and 2008.

Aralsk

2000 1989 1973 1960
2006
2008

2006

2000
1989

1973

Muynak

1960

KAZAKHSTAN

UZBEKISTAN

FACT FILE

Causes Desertification affects all the world's continents except Antarctica. In Africa, overgrazing is the main driver of desertification; in Asia, deforestation dominates; and in North America, farming causes the most land degradation.

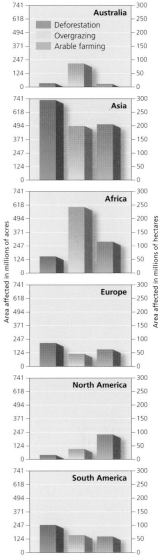

Area affected in millions of acres

Area affected in millions of hectares

Australia
- Deforestation
- Overgrazing
- Arable farming

Asia

Africa

Europe

North America

South America

WEATHER WATCH

Great Green Wall In an effort to combat desertification at the edge of the Gobi desert, China plans to plant forest strips 2,800 miles (4,500 km) long, dubbed the Great Green Wall of China. These trees should prevent erosion and slow the rate at which desert takes over marginal grassland.

SPOT ▶ Gobi Desert

▶**REGION:** Gobi desert, southern Mongolia and the Mongolia Autonomous Region of China

▶**THREATS:** Overgrazing, deforestation, soil erosion, low soil fertility, inefficient farming practices, pollution, contaminated groundwater, decreased groundwater levels, high radiation levels, overpopulation, poaching

▶**CLIMATE IMPACTS:** Desertification, dust storms and sandstorms, drought

▶**ENDANGERED SPECIES:** Snow leopard, Bactrian camel, Asian wild horse, Asiatic wild ass

Asia's largest desert—the Gobi—sits in the rain shadow of the Himalayas and extends over about 500,000 square miles (1.3 million km²). The desert is rapidly expanding along its eastern margin, largely because of deforestation, poor land management, overgrazing, and widespread erosion. Much of the region supports some grass and other vegetation, but the desert's center is arid and rocky.

PERVASIVE DUST

The Gobi desert is characterized by dryness and temperature extremes because of its relatively northerly location and high elevation. The winter temperature can plummet to as low as –27°F (–33°C) in southern Mongolia, and in inner Mongolia it rises as high as 99°F (37°C) in July.

With climate change warming the surface air, increasing water extraction for irrigation and industrial use lowering the groundwater table, and agriculture encroaching on marginal lands around the Gobi, the soil is left dry and exposed. The right combination of fierce winds and prolonged dry weather produces frequent and severe dust storms, especially in the driest spring months. Dust particles, which average about one-tenth of the size of sand grains, can rise 10,000 feet (3,000 m)

into the atmosphere and travel thousands of miles laterally. Some dust storms are so extensive that they are visible from space. The health impacts of breathing dust and pollution particulates include respiratory and cardiac diseases, and if toxic pollutants adhere to the dust particles, these can be absorbed into the body through the lungs.

Dust storms can travel as fast as 50 miles per hour (80 km/h). Near their origin, these storms may arrive with almost no warning. In 2006, nine dust storms hit the Gobi region in a two-month period. Efforts to prevent the relentless advance of the desert include the large-scale planting of trees and shrubs, but this competes with agricultural demands to farm the marginally fertile lands along the desert's perimeter.

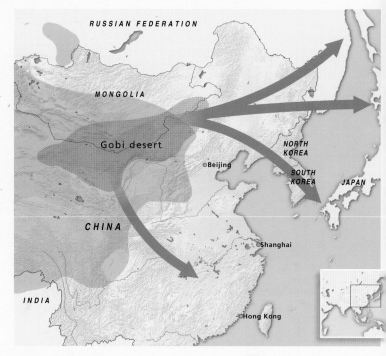

Trajectory (above) Originating mostly in spring, dust storms in the Gobi spread particles over vast distances, inundating coastal cities throughout Asia. These storms deposit over a million tons of dust on Beijing each year.

■ Barren or sparsely vegetated land
■ Scrub
■ Desert
➙ Frequent storm tracks

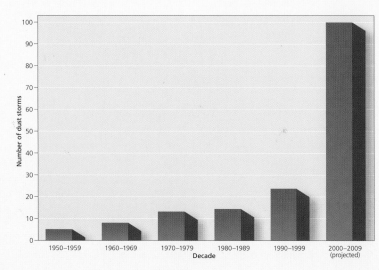

Storm numbers (left) In this decade, China is projected to have five times more dust storms than in the 1990s. In 2001, the Chinese government instigated a large-scale replanting project to restrain the encroaching desert, dubbed the Great Green Wall.

Slim pickings (above) In addition to coating cropland, desertification diminishes the land available for livestock. Here, a herd of sheep graze near Long Baoshan village, on the outskirts of Beijing, indicating that the desert is creeping ever closer to urban centers.

Aloft (above) This composite satellite image, from April 2001, shows a dust plume from Asia stretching across the Pacific Ocean. When dust enters rising air, the particles can travel great distances.

Stopping erosion (below) To prevent the desert's advance, a grid of straw was woven into the dunes along a train line in inner Mongolia. Farmers also use straw as a barrier against windblown sand.

Red sky (right) Airborne dust, mixed with pollution, created an eerie haze at a train station in Tianjin, China, in 2007. The air can appear yellow, red, brown, or black, depending on the type and thickness of the dust.

THREAT TO LIFE

The Earth's biosphere, the global aggregate of all ecosystems, including humans, has evolved to thrive under the current moderate, interglacial Holocene climactic conditions. The abrupt disruption to our climate caused by rapid CO_2 release from human industrialization challenges the adaptive ability of plants and animals. As global average temperatures warm, climate zones shift geographically, and some may disappear altogether. The rich biodiversity that has developed over many millennia of evolution will diminish, because some species are unable to adapt as rapidly as the climate is changing.

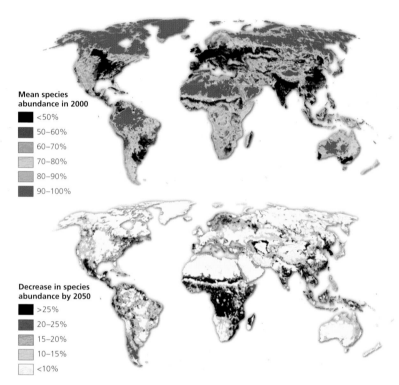

Mean species abundance in 2000
- <50%
- 50–60%
- 60–70%
- 70–80%
- 80–90%
- 90–100%

Decrease in species abundance by 2050
- >25%
- 20–25%
- 15–20%
- 10–15%
- <10%

Biodiversity loss (left) Global biodiversity had already significantly diminished by 2000, relative to the natural state, particularly in populous regions (upper map). By 2050, the United Nations anticipates further dramatic decreases in biodiversity (lower map). The greatest losses are predicted for Africa and Latin America, driven by land use changes.

Cyclical flooding (right) Botswana's Okavango Delta is a rare inland delta with no outlet to the sea. Previously part of an ancient lake, this delta now experiences seasonal flooding only in mid-summer, irrigating portions of the Kalahari Desert. A giraffe gets most its water from food, but will drink water when it is available.

Lush biome (right) With copious rainfall and nearly constant 100 percent humidity, rain forests are home to much of Earth's biodiversity. Costa Rica's Monteverde Cloud Forest Reserve is protected by the World Wildlife Fund, preserving the habitats of over 100 mammal, 400 bird, and 2,500 plant species from the threat of deforestation.

Endangered (below) The snow leopard lives in the remote mountains of Central Asia, a threatened ecosystem. Its dense coat and small, round ears minimize heat loss, and its large, furry paws act as snowshoes. The IUCN estimates that the total wild population numbers less than 6,600 individuals.

ECOSYSTEM IMPACT

Conservation scientists warn that massive global loss of species is underway. Current extinction rates are up to 1,000 times greater than during the past 65 million years. Human activity speeds biodiversity losses by promoting habitat reduction, especially when a warming climate displaces existing arable land or usable forest. A warming climate promotes poleward and upslope mountain range shifts of flora and fauna, altering breeding and hatching seasons, and changing migratory patterns.

Natural distributions of plants and animals provide valuable ecosystem benefits, such as producing clean drinking water, regulating nutrient cycles, decomposing wastes, and pollinating crops. While a warming climate could make some regions with increased rainfall or lengthened growing seasons more hospitable to agriculture, many arid and semi-arid regions will dry out and no longer be able to be cultivated. Although increased atmospheric carbon dioxide acts as a fertilizer to crops, expanding production if other nutrients are available in sufficient quantities, the overall risks to the environment of extra greenhouse gas emissions will outweigh the positive effects.

Climate change is also predicted to intensify the spread of certain diseases, as warmer temperatures increase the range of insect carriers, such as mosquitoes. More frequent flooding in wetter climates stresses urban sanitation systems, aiding the spread of bacterial and viral diseases that thrive in aquatic environments. To confront the many ecosystem and human health threats, societies must adopt a combination of climate mitigation—reducing warming by decreasing energy consumption to minimize CO_2 emissions—and adaptation to adjust agricultural practices and lifestyles to cope with the consequences of climate change.

Plants and Trees

Plants and trees that are adapted to a certain climate zone are capable of some migration, but climate change is predicted to outpace the adaptation capabilities of many flora. Additionally, the range of many destructive insects is projected to expand. Many plant ecosystems, such as coastal mangrove forests, provide valuable climate protection and need to be safeguarded.

Beetle damage Pine beetles are a threat to evergreen forests across western North America **(below)**. With a warming climate, the altitude at which cold kills off the beetles also rises. A satellite image **(bottom)** shows severe beetle damage (red) in Canada.

FACT FILE

Adaptation Plant species evolve to adapt to their environment. This results in some marvelous traits that enable them to survive fires, extreme dryness, or long, cold winters. In some regions, climate change is too rapid for successful adaptation to occur.

Fire Banksia, an Australian genus, is well adapted to its bush habitat. Fire ruptures the seed case and releases the seeds, which fall to the ground and germinate.

Drought The quiver tree of the Kalahari Desert is an African species of aloe that has adapted to extremely dry conditions by storing water inside its spongy trunk.

Cold Spruce trees are adapted to cooler climates. With climate warming, large tracts of dead spruce have been found at high elevations in North America.

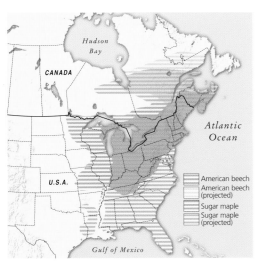

Current climate zones

- Permafrost desert
- Alpine wet tundra
- Subalpine moist forest
- Montane steppe
- Lower montane thorn steppe
- Premontane thorn woodland

Projected climate zones after 6.3°F (3.5°C) increase and 10% precipitation change

- Permafrost desert
- Alpine wet tundra
- Subalpine moist forest
- Montane steppe
- Lower montane thorn steppe
- Premontane thorn woodland

Higher latitudes (left) As climate warms, optimal conditions for plant species move toward the Poles. Some models project that American beech and sugar maple will disappear from the U.S.A. by 2050.

Higher altitudes Climate zones and their plant species are moving to higher elevations in mountainous regions, forcing current high-altitude flora distributions **(above left)** into smaller areas **(above)**.

FACT FILE

Root systems Mangroves have evolved complex aerial buttress root systems to support the trees in coastal mud. This root system allows mangrove forests to thrive despite saltwater inundation and drainage of the coastal wetlands.

Knee roots So-called knee roots are pneumatophores, or aerial roots, that take in oxygen from the air above the oxygen-poor, saltwater mud.

Stilt roots Some mangroves require stilt roots to prop up the trunks above the water level. This helps them remain stable in the shifting mud.

Peg roots Another type of breathing root, or pneumatophore, is the peg root. These poke up through the mud like snorkels, harvesting oxygen from the air.

Mountains (top) The shifting ranges and seasons of wildflowers, such as these lupines, are being monitored in the European Alps. This helps to gauge how ecosystems respond to climate change.

Mangroves (above) These natural coastal barriers are threatened by increasingly stormy waters caused by warming oceans. Mangroves are disappearing, along with the rich biodiversity they support.

Insects, Amphibians, and Reptiles

As climate zones shift geographically, insects, amphibians, and reptiles can partially adapt by altering their ranges. However, as certain ecosystems—especially rain forests—reduce in size, the species that inhabit them are threatened with extinction. Climate change can also give false cues for breeding, dispersal, and even sex determination, affecting the distribution of genders in certain reptile species.

FACT FILE

Insects Climate change affects the migration of many species. Insects and spiders are particularly impacted because they often have reproductive processes that are very temperature sensitive. This makes them more vulnerable to local extinctions.

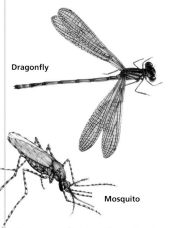

Dragonfly

Mosquito

Signs Dragonflies are affected by air temperature, making them good climate indicators. Mosquitoes have already adjusted their hibernation patterns.

Butterflies The shifting range of butterfly species is influenced by climate. In the Italian Alps, the Apollo butterfly has moved to higher altitudes.

BALLOONING SPIDERS

Many spiders undergo seasonal aerial dispersions, called ballooning. One study observed that two annual balloonings had collapsed into a single midsummer event.

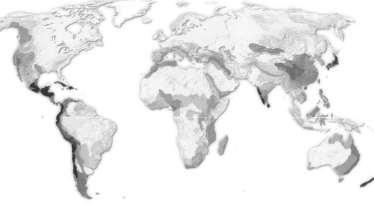

Amphibians under threat (below) A larger percentage of amphibian species—such as frogs, toads, and salamanders—is threatened than bird or mammal species. This map shows the percentage of threatened freshwater amphibians by region. In 2006, it was found that 1,356 of the 4,025 species were threatened.

Percentage of threatened freshwater amphibian species

- 0
- <2
- 2–5
- 5–9
- 9–15
- 15–25
- 25–33
- 33–50
- 50–80
- 80–100
- No amphibians

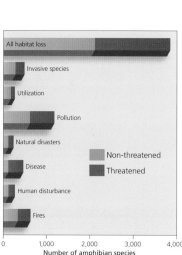

All habitat loss
Invasive species
Utilization
Pollution
Natural disasters
Disease
Human disturbance
Fires

Non-threatened
Threatened

0   1,000   2,000   3,000   4,000
Number of amphibian species

Amphibian perils (above) Habitat loss is by far the greatest cause of global decline in amphibians, threatening a large percentage of species. This is largely due to the rapid deforestation of rain forests, the primary amphibian habitat.

Assisting nature (left) Corroboree frog tadpoles are being raised at Taronga Zoo in Sydney, Australia, to help replenish the diminishing wild population. The frogs are endangered by shorter winters, which affect their breeding patterns, and a deadly fungus.

Race to the sea (right) Sea turtles are particularly threatened by climate change. Because egg incubation temperatures determine the sex of turtle offspring, a warmer climate results in a disproportionate number of females.

Reptiles under threat (left)
Reptiles—including lizard, snake, crocodile, and turtle species—are imperiled in many of the same geographic regions as amphibians, but more severely in Southeast Asia and the southeastern U.S.A. A major threat to these semi-aquatic species is habitat loss and damage, which is exacerbated by climate change.

Percentage of threatened reptile species

0–11	33–44
11–12	44–77
12–33	77–100

FACT FILE

Red List The International Union for Conservation of Nature (IUCN) maintains a Red List of Threatened Species to monitor the conservation status of plants and animals. The charts below do not include species that have already become extinct.

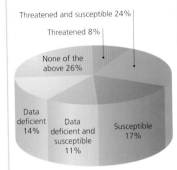

Threatened and susceptible 24%

Threatened 8%

None of the above 26%

Data deficient 14%

Data deficient and susceptible 11%

Susceptible 17%

Amphibians Almost one-third of amphibian species are classified as threatened, or threatened and susceptible, by climate change.

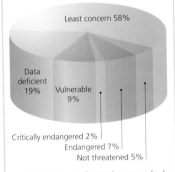

Least concern 58%

Data deficient 19%

Vulnerable 9%

Critically endangered 2%

Endangered 7%

Not threatened 5%

Reptiles Most reptile species are ranked as being of least concern for extinction. Reptiles are less dependent on vulnerable rain forest ecosystems.

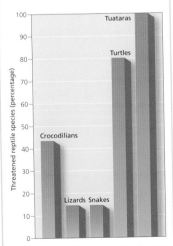

Rate of threat Tuataras and turtles are the most threatened species of reptile. The two kinds of tuataras are endemic to New Zealand, with very limited range.

SP^H_TOT ▶ The Galápagos Islands

▶**REGION:** The Galápagos Islands, Ecuador

▶**THREATS:** Introduced plant and animal species, overfishing, pollution, development, tourism

▶**CLIMATE IMPACTS:** Rising sea levels, warming seas, coral bleaching, increased intensity of El Niño and La Niña events, changes in water salinity, ocean acidification

▶**ENDANGERED SPECIES:** Galápagos sea lion, Galápagos fur seal, Galápagos petrel, Galápagos penguin, Galápagos stringweed, flightless cormorant; marine iguana and Galápagos giant tortoise vulnerable

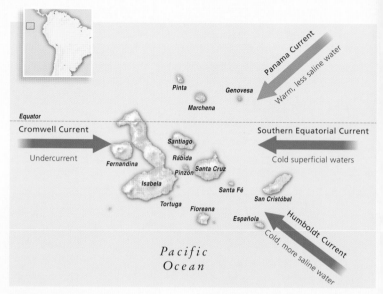

The rich volcanic archipelago and marine reserve of the Galápagos Islands is located about 600 miles (965 km) off the coast of Ecuador. Here, Charles Darwin alighted from the HMS *Beagle* in 1835 and catalogued the exceptional diversity of animal life—observations that inspired his theory of evolution by natural selection. In 1959, the centenary year of Darwin's book *On the Origin of Species,* all land in the archipelago not already colonized was declared a national park.

Intersection (above) The Galápagos Islands are situated at the confluence of three major currents: the warm Panama Current, cold Southern Equatorial Current, and cold Humboldt Current.

Visitors (below) Annually, more than 100,000 tourists explore the Galápagos. Tragically, tourism introduces invasive plant species, threatening the very wildlife these travelers come to see.

SHOWCASE OF EVOLUTION

A cluster of 19 islands, the Galápagos is relatively new on a geological timescale, with some parts of the archipelago forming by volcanic activity as recently as a million years ago. Volcanic and seismic activity continues on the islands, and scientists have observed species distribution that attests to changes in their topography. The world that Darwin observed in the 19th century had evolved over eight million years, since the volcanic mantle plume emerged from the ocean, fostering such unique endemic species as the marine iguana, 13 species of finches, fur seals, giant tortoises, and the only living tropical penguin.

Galápagos' biodiversity faces threats from invasive species and disease, reduced marine food sources, and damage to animal habitats. The region is imperiled by climate change in the form of rising and warming seas, and increasingly frequent and intense El Niño events. In 1998, a strong El Niño year, the surrounding ocean experienced unprecedented coral reef bleaching and death, spurring repercussions up the marine food chain, with dramatic biodiversity loss. With ocean temperatures steadily rising, such events will become more common and more severe.

Fortunately, the high international profile of the Galápagos Islands has raised the alarm about environmental threats, and conservation efforts are underway. UNESCO designated the Galápagos a World Heritage in Danger site in 2007, leading to greater regulation of development, tourism, and fishing in the park.

Effects Warm El Niño waters suppress plankton growth in the Galápagos **(below)** and cause coral reef bleaching. After El Niño subsides, upwelling waters replenish vital nutrients, producing a phytoplankton bloom **(below right).**

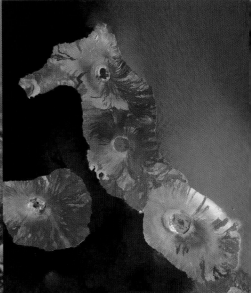

Volcanic (above) Wolf Volcano (top) is home to the unique pink iguana, discovered in January 2009. This species diverged from other iguanas and evolved independently, when the sinking of the island isolated Wolf Volcano. Volcanic uplift later recurred.

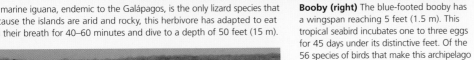

Feeding (above) The marine iguana, endemic to the Galápagos, is the only lizard species that feeds underwater. Because the islands are arid and rocky, this herbivore has adapted to eat algae. Adults can hold their breath for 40–60 minutes and dive to a depth of 50 feet (15 m).

Booby (right) The blue-footed booby has a wingspan reaching 5 feet (1.5 m). This tropical seabird incubates one to three eggs for 45 days under its distinctive feet. Of the 56 species of birds that make this archipelago home, 27 are found only in the Galápagos.

Basking (left) Endangered Galápagos sea lions rest ashore at Gardner Bay, on Española. The sea lion population fluctuates from 20,000–50,000 due to predation by killer whales and sharks, as well as El Niño events that limit their marine food source.

Giant tortoise (below) The largest living tortoise, the Galápagos giant tortoise, can weigh up to 660 pounds (300 kg). Its life expectancy in the wild is 100–150 years. This vulnerable fauna is threatened primarily by poachers and is now strictly protected.

Marine Life

A warming ocean is detrimental to marine life. Phytoplankton, at the base of the marine food chain, is highly temperature dependent. In the tropics, warming isolates plankton from nutrients; in colder regions, warmer seas disperse plankton. Fish, crabs, sharks, mollusks, and turtles are further threatened by habitat loss and changing currents and weather patterns that disrupt migration.

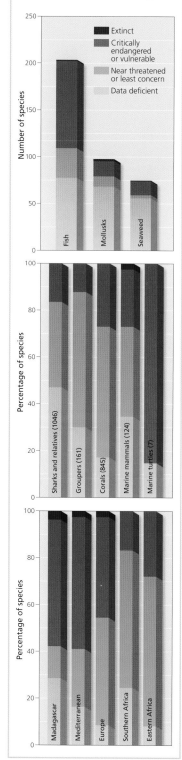

Washed up (right) In May 2002, a warm El Niño event beached thousands of pelagic red crabs in San Diego, California. Warming seas will cause more frequent El Niño events, shifting ocean currents and altering the seasonal migration of many marine species.

Ocean changes (left) Satellite analysis shows that increased water temperatures (red, upper map) in the already warm tropics cause a reduction (pink, lower map) in the primary biological productivity of the oceans—the growth of phytoplankton. Warmer surface waters increase ocean stratification, isolating phytoplankton on the surface from the nutrient-rich, cold, deeper water.

CHANGE IN OCEAN TEMPERATURE

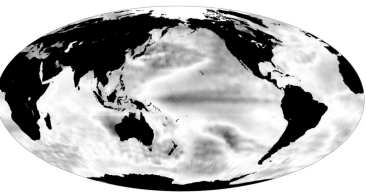

CHANGE IN OCEAN PRODUCTIVITY

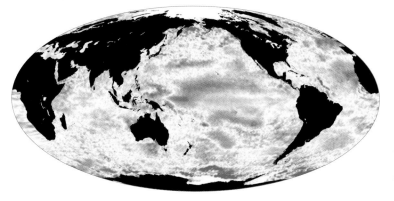

Change in ocean temperature

°F	°C
+5.4	+3
+3.6	+2
+1.8	+1
0	0
-1.8	-1
-3.6	-2
-5.4	-3

Change in ocean productivity

+60%
+30%
0%
-30%
-60%

Safe haven (below) Mangrove roots make an ideal habitat for baby sharks. The many species of crabs, shrimp, and smaller fish that shelter there provide an abundant food supply for the pups, and the tangled root systems create a hiding place that larger predators cannot negotiate.

FACT FILE

Plankton A clear indicator of climate change is the shifting distribution of marine plankton species, markers of primary biological productivity. In the latter half of the 20th century, dramatic changes in plankton range were observed in the North Sea.

1958–81 Monitored by the Continuous Plankton Recorder (CPR) survey, high plankton concentrations (red) were only observed south of the U.K.

Northward In the 1980s and 1990s, melting Arctic ice began to warm the North Atlantic, allowing the northward expansion of plankton.

Over the top By 2000, the disappearing Arctic ice cap allowed a Pacific plankton species to migrate through polar waters into the Atlantic Ocean for the first time.

Birds and Mammals

Birds and mammals have varying abilities to adjust to climate change. Modifications have been observed in hatch timing, migration patterns, and the ranges of various species, demonstrating their adaptations to climate change. However, where species cannot sufficiently adapt, their populations are observed to decline. Conservation organizations document these changes to monitor the level of threat posed to various species.

FACT FILE

Birds Most species of bird are very mobile, with a capacity to alter their range. However, 38 percent of species are listed as threatened or susceptible to threat. Their food sources may not be as mobile, or new breeding sites may contain new predators.

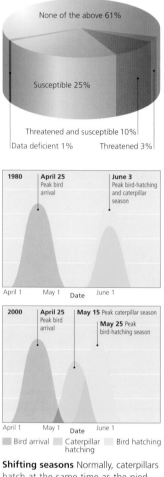

None of the above 61%

Susceptible 25%

Threatened and susceptible 10%

Data deficient 1%   Threatened 3%

1980	April 25 Peak bird arrival	June 3 Peak bird-hatching and caterpillar season

April 1   May 1   **Date**   June 1

2000	April 25 Peak bird arrival	May 15 Peak caterpillar season
		May 25 Peak bird-hatching season

April 1   May 1   **Date**   June 1

■ Bird arrival   ■ Caterpillar hatching   ■ Bird hatching

Shifting seasons Normally, caterpillars hatch at the same time as the pied flycatchers. Warmer springs have led to earlier caterpillar hatches. The birds now lag behind the peak of their food source.

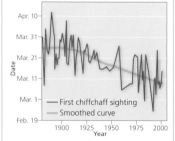

Apr. 10
Mar. 31
Mar. 21
Mar. 11
Mar. 1
Feb. 19

— First chiffchaff sighting
— Smoothed curve

1900   1925   1950   1975   2000
Year

Migration The timing of bird migration is linked to temperature. Over the past century, chiffchaffs have been observed returning to southern England from their breeding grounds progressively earlier.

Waterfowl (left) Geese, swans, and other waterfowl are vulnerable to smoke and air pollution, which eventually pollutes rivers and streams. Climate change caused by industrial emissions also threatens birds with reproductive failure and the extinction of species unable to shift to new habitats.

Flamingo (below) The lesser flamingo, pictured at Lake Nakuru National Park in Kenya, is the most numerous flamingo species. Nevertheless, it is classified as near threatened due to pollution, disease, and habitat loss. These threats are magnified by climate change and drought.

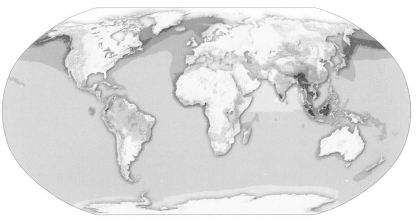

Mammal and marine species (left)
The largest number of threatened land mammal species occurs in Asia, especially Indonesia. The main threats to marine species are habitat loss and human utilization, both of which are intensified by a warming climate.

Threatened marine species	Threatened terrestrial species
1	1–3
2	4–6
3	7–10
4	11–20
5	21–29
6	30–45
7–8	

Large cats (above) The cougar ranges from northern Canada to the southern Andes, in South America, and is known by various names: puma, mountain lion, and panther. Cougars can adapt to different habitats, and a warming climate allows them to occupy an expanding range.

Reindeer (left) A warming climate is already threatening the food supply of reindeer in Norway. Rain is replacing snow, a phenomenon that is expected to increase by 40 percent over the next century. When frozen, this moisture coats the lichens and mosses that reindeer eat with an impenetrable layer of ice.

FACT FILE

Mammals In an assessment of the 5,487 known mammal species, the IUCN found that nearly 21 percent are globally threatened: classified as vulnerable, endangered, or critically endangered. Since 1500, 76 species are known to have become extinct.

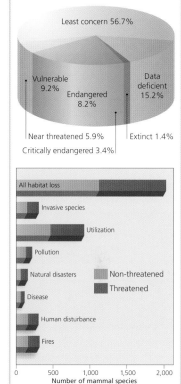

Least concern 56.7%

Vulnerable 9.2%

Endangered 8.2%

Data deficient 15.2%

Near threatened 5.9%

Extinct 1.4%

Critically endangered 3.4%

All habitat loss
Invasive species
Utilization
Pollution
Natural disasters
Disease
Human disturbance
Fires

Non-threatened
Threatened

0   500   1,000   1,500   2,000
Number of mammal species

Threats The primary threats to mammals include habitat loss, human utilization, invasive species, human disturbance, and fires. Many threats are exacerbated by a warming climate.

PIKA DECLINE

The pika is a mountain-dwelling relative of the rabbit, found in Asia and North America. As the climate warms and ecosystems shift to higher altitudes, the pika follows, but its population has declined by around 30 percent as its range is squeezed.

Failing Human Health

Extreme heat waves and shifting precipitation patterns reduce food production and diminish water supply. Beyond these straightforward consequences, climate also influences the spread of illness. Waterborne disease is promoted by flooding, and the population of disease vectors—insects and rodents that transmit viruses—is affected by local climate.

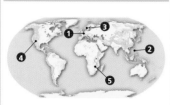

Sanitation (below) Improved sanitation dramatically reduces the spread of disease. However, in sub-Saharan Africa and parts of Southeast Asia, less than half the population has access to public sewers, septic systems, or ventilated pit latrines.

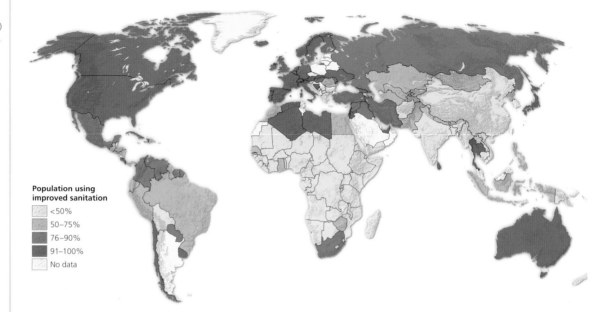

Population using improved sanitation
- <50%
- 50–75%
- 76–90%
- 91–100%
- No data

Cold wave (left) Cold weather extremes threaten human health primarily through hypothermia, when the body temperature drops below the level necessary for normal metabolism. In January 2007, several people died when temperatures in Moscow dropped below –2°F (–19°C).

Vaccination (below) A warming climate increases the range of vector-borne diseases, spread by insects and rodents that thrive in new areas. This has led to increasing efforts to develop vaccines for diseases like malaria. In the Kintampo Health Center, in Ghana, a child is weighed before receiving a malaria shot.

Water (above) Many of the public health threats exacerbated by climate change involve water. Flooding causes wastewater treatment problems that spread disease. Scarcity of water can lead to the use of polluted water for drinking and food preparation. Here, a child collects water from a muddy pool during a water shortage in Mexico.

Heat wave (below) A scorcher hit Lahore, Pakistan, in June 2007, driving many people to cool themselves in canals as temperatures soared to 124°F (51°C). Heat waves can be lethal as they increase the incidence of heat stroke, and respiratory and cardiac problems, particularly in the elderly.

FACT FILE

1. Floods Flooding can cause sewage contamination, enabling the spread of bacteria like the Giardia parasite. An outbreak in Bergen, Norway, in August 2004, was attributed to heavy rains.

 Giardia thrives in warmer water, making outbreaks more common in warm climates and during summer.

Bergen, Norway

2. Heat Prolonged heat waves can be extremely hazardous, causing wildfires, crop failures, and power outages, as well as mortality from hyperthermia.

 In Chicago in 1995, 600 people—mainly the poor, infirm, and elderly—died in five days of extreme temperatures.

Chicago, Illinois, U.S.A.

3. El Niño A study of cholera epidemics in Matlab, Bangladesh, found that disease transmission correlated with population density and climate patterns. Monsoonal rains and Brahmaputra

 River water levels were determining factors, while El Niño conditions directly influenced the size of outbreaks.

Matlab, Bangladesh

LOST YEARS

The health impact of climate change can be assessed in terms of Disability Adjusted Life Years. DALYs are the years of productive life lost due to disability or mortality resulting from climate-related diarrhea, flooding, malnutrition, and malaria.

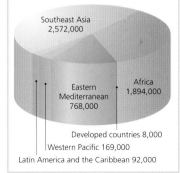

Southeast Asia
2,572,000

Africa
1,894,000

Eastern
Mediterranean
768,000

Developed countries 8,000

Western Pacific 169,000

Latin America and the Caribbean 92,000

Disease

The World Health Organization predicts that the first signs of climate change affecting human health will be alterations to the range and seasonality of infectious diseases. Vector-borne diseases, transmitted by insects or rodents, peak in warmer periods, which are lengthening. Heavy rainfall and flooding provide breeding grounds for mosquitoes and bacteria, such as cholera.

Malaria risk (right) The December 2004 Indian Ocean tsunami displaced millions of people. In the hot, damp weather, malaria outbreaks were feared. Health workers in Aceh, Indonesia, fumigate tents to kill disease-spreading mosquitoes.

1. Rift Valley fever The earliest identification of the Rift Valley fever (RVF) virus was in Kenyan livestock in 1915. Although primarily an animal disease, a series of unexpected human

fatalities during 2006–07 were diagnosed as an RVF epidemic, eventually causing 118 deaths in Kenya and Somalia.

Kenya and Somalia

2. West Nile virus This mosquito-borne disease, first reported in the U.S.A. in 1999, has expanded its range as climate warms. In California, the mosquito

species that carries the virus is now found at higher elevations and coastal regions, making outbreaks in humans more likely.

California, U.S.A.

3. Ross River fever An outbreak of Ross River fever in Perth, Australia, in 2007 was attributed to heavier spring rains. This flooded the breeding sites of the mosquitoes that transmit the

virus from native wildlife to humans. Australian cases of the disease have increased since initial identification in 1963.

Perth, Australia

4. Dengue fever Typically a tropical and subtropical disease, dengue fever is regaining a foothold in southern Texas, U.S.A., where a 2005 outbreak

led to 25 hospital cases. Dengue has been kept in check by mosquito controls; there is no vaccine or drug to treat it.

Texas, U.S.A.

Cholera spread Increased flooding in normally arid regions provides ideal conditions for the spread of cholera, especially in areas without wastewater management. In 1950, cholera had almost disappeared **(below)**, but by 1990–2004 it made a global resurgence **(right)** due to a warming climate and worsening socio-economic conditions.

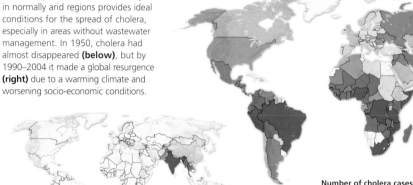

Number of cholera cases declared per country

1–30	1,000–10,000
30–100	10,000–100,000
100–1,000	>100,000

Cholera risk (right) Massive flooding struck Dhaka, Bangladesh, in August 2007, dislocating millions and leading to outbreaks of cholera and other water-borne diseases. Cholera is spread by bacteria that thrive in water and attack the lining of an infected person's small intestine, causing the disease's characteristic diarrhea.

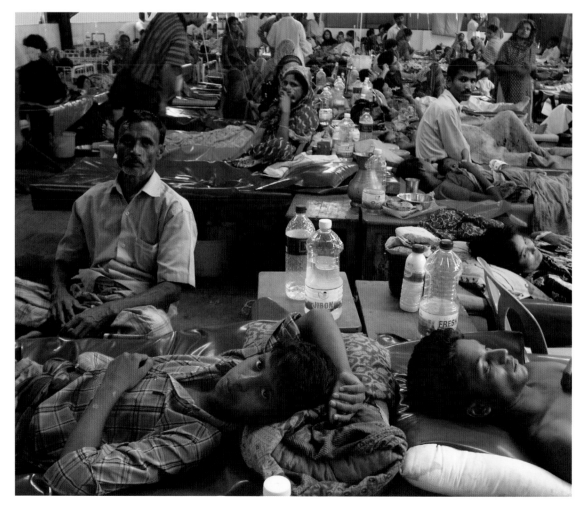

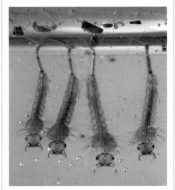
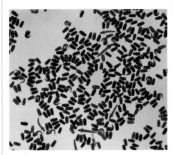

Change in malarial risk from a 5.4°F (3°C) temperature rise

- >2
- 1.7–2.0
- 1.4–1.7
- 1.1–1.4
- Areas newly at risk
- No change
- Decrease
- No significant risk

Malaria risk (left) The mosquito-borne parasites that spread malaria are dependent on temperature. The growing malarial risk predicted to accompany a 5.4°F (3°C) increase in average global temperature could result in millions more infections.

SP̂OT ▶ Indonesia

▶**REGION:** Indonesia, Southeast Asia

▶**THREATS:** Flooding, deforestation, land use changes, soil erosion and sedimentation, water pollution from industrial waste and sewage, air pollution, overpopulation, tsunamis, earthquakes, volcanoes, overfishing

▶**CLIMATE IMPACTS:** More mosquito-borne diseases, increased CO_2 release from forest burning, increased atmospheric temperature, drought, increased storm activity, increased intensity of El Niño and La Niña events, rising sea levels, warming seas, coral bleaching, ocean acidification

▶**ENDANGERED SPECIES:** Orangutan, proboscis monkey, Pagai macaque, silvery gibbon, Asian elephant, Sumatran tiger, Javan rhinoceros, black-spotted cuscus

Incidence rate
- <5 (per 100, 000)
- 5–9 (per 100, 000)
- >10 (per 100, 000)

Infected (above) Dengue fever cases plague Indonesia, especially during the rainy season from January–March. This map shows the incidence rate per 100,000 inhabitants by province in 2004, with particularly bad outbreaks in Java, Kalimantan, and Sulawesi.

Orangutans (below) These apes are almost exclusively tree-dwelling; their name means "man of the forest" in Malay. With their arboreal habitat decreasing rapidly due to deforestation, orangutan numbers have dropped 30–50 percent in the past decade.

Indonesia is a hot spot for observing the effects of climate change, as well as one of its major contributors. As the world's fourth-most-populous country, which comprises an archipelago of 17,000 islands, Indonesia is vulnerable to drought, tropical diseases, storm activity, and sea-level rise. This relatively poor country with vast forest resources is the world's third largest CO_2 emitter when including deforestation and fire emissions.

FEVERISH CRISIS

Timber export is a major source of income for Indonesia. Deforestation and the associated burning, as well as land conversion, are proceeding sufficiently rapidly that the country's forestry-related emissions equal about half of the energy-related greenhouse gas emissions of the U.S., the world's largest CO_2 emitter. With climate change causing the rain forest to become drier, uncontrolled fire emissions will continue to increase.

While the average temperature increase in Indonesia is predicted to be moderate, projected changes in precipitation patterns will have more severe effects. Rainfall is expected to become more substantial and occur over a shorter rainy season, significantly heightening the risk of flooding. This flooding creates prime conditions for the spread of mosquito-borne diseases such as

dengue fever because standing water is the ideal breeding ground for the mosquitoes that transmit the virus. Dengue is prevalent in tropical rural and urban areas alike, infecting as many as 50 million people worldwide every year. Malaria, diarrhea and dysentery also spread more readily in regions that experience flooding. As many as 2,000 small islands are threatened with complete submersion by mid century with the sea-level rise that is projected to accompany a warming climate.

The shift to shorter, more intense rainy seasons will be accompanied by increasingly frequent periods of drought, as global climate patterns cause Indonesia to be especially warm and dry between monsoonal rainy seasons. Drought will cause water shortages and diminish crop output, decreasing food security in this densely populated nation.

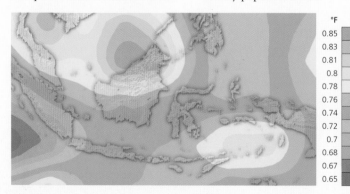

°F	°C
0.85	0.47
0.83	0.46
0.81	0.45
0.8	0.44
0.78	0.43
0.76	0.42
0.74	0.41
0.72	0.4
0.7	0.39
0.68	0.38
0.67	0.37
0.65	0.36

Warming (above) By 2020, average surface air temperatures in Indonesia are projected to warm by 0.65–0.85°F (0.36–0.47°C) relative to 2000 climate. The greatest increases are predicted for Kalimantan and the Molucca Islands.

Parched (above) During El Niño events, which may become more frequent or more severe in a warming climate, much of Indonesia is devastated by drought. Here, boys carry buckets over dry rice paddies after collecting water near Cibarusah, east of Jakarta, in 2002.

Fouled water (left) Rivers near Jakarta Bay are polluted with garbage and plastic waste. Rising sea levels will inundate Indonesia's many coastal farming areas with salt water, threatening food security, and flooding rivers will bring polluted waters inland.

Epidemic (below) An April 2007 outbreak of dengue fever infected 11,000 in Jakarta alone. Close monitoring is critical during the early stages of the disease, which can cause hemorrhaging. A four-year-old receives fluids in a Jakarta hospital to prevent dehydration.

Food Shortage

FACT FILE

Production Food production shifts in response to changing incomes, but also to climate threats. Globally, cereal grains are the main source of calories, while richer nations consume a larger share of more economically and environmentally expensive meat.

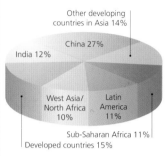

Other developing countries in Asia 14%
China 27%
India 12%
West Asia/ North Africa 10%
Latin America 11%
Sub-Saharan Africa 11%
Developed countries 15%

Cereal demand The increase in demand for cereal grain in 1997–2020 is projected to total 720 million tons (653 million tonnes). The largest consumer is China.

Cereal production Europe is the world's leading cereal producer, with the fastest rate of increase; China's production has increased to match the world average.

Meat vs cereal As per-capita incomes rise, consumers switch from mainly cereal grains to more expensive meat products as significant sources of calories.

ANCHOVY RISE AND FALL

Peruvian anchovy catches crashed in the 1970s, caused by overfishing and warming ocean waters off the South American coast in the wake of a strong El Niño event in 1972.

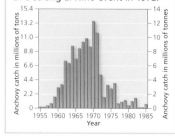

Weather and climate are vital to food production. Hotter, drier climates are less hospitable for many crops, causing food crises when combined with water scarcity. While some colder regions will benefit from longer growing seasons, and increased atmospheric carbon dioxide enhances plant productivity, these benefits will be more than offset by crop losses at lower latitudes.

Agriculture (below) A model of change in food production by 2080 incorporates the climate effects of heat, water scarcity, shifting precipitation, and carbon fertilization, which is an increase in crop productivity due to higher carbon dioxide concentrations.

Projected changes in agricultural production by 2080

- -50 to -15%
- -15 to 0%
- 0 to +15%
- +15 to +35%
- No data

Rice (right) Japan's 20th century climate showed considerable temperature variability. During cool years, when the temperature anomaly was lower than −2.7°F (−1.5°C), rice harvests failed. In the later decades, year-to-year fluctuations were large.

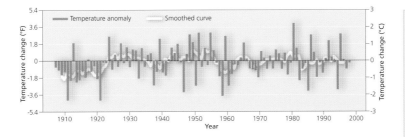

Designer food (above) Genetically modified (GM) crops can be engineered to require less fertilizer and pesticide, or to tolerate a drier climate, addressing climate-induced food shortages. However, they remain controversial. Protesters rip out GM crops in Toulouse, France.

Fishing (left) Longer dry seasons and water scarcity are major concerns for the fishing industry, particularly in areas like central Java, Indonesia, which experiences severe drought linked to El Niño events. A fisherman fixes his net near a drought-stricken dam.

Cloud seeding (below) Some countries attempt to take climate into their own hands, deploying airplanes with wing-tip cannons to inject water-absorbing chemicals into clouds to coax precipitation. The effectiveness of these cloud-seeding experiments is uneven.

FACT FILE

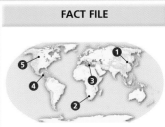

1. China The productivity of China's staple food crops of wheat and rice, and animal-feed maize crops is projected to diminish 10 percent by 2050 if current agricultural practices are maintained.

This would result from increased drought and heat waves in the North China Plain and more flooding in China's south.

China

2. Kalahari Desert The continent most threatened by food shortages is Africa, and climate change will worsen this situation. The dunes of the Kalahari

Desert are projected to spread, shifting sands across precious arable land in Angola, Botswana, Zambia, and Zimbabwe.

Kalahari Desert, southern Africa

3. Mediterranean Desertification is occurring around the shores of the Mediterranean Sea. In Southern Europe, the key drivers are the expansion of tree crops over marginal lands and

sheep overstocking. In North Africa, desertification is accelerated by the encroachment of agriculture.

Mediterranean Sea

4. Honduras Over the past 40 years, food production increases in Central America have not kept pace with population growth. Future changes in

rainfall may reduce maize crop yields by 22 percent in Honduras if current practices are maintained.

Honduras

5. Canada A warmer climate may increase agricultural productivity in the northern Canadian prairie. A longer growing season and CO_2 fertilization combine to generate more productive

agriculture, partially offsetting productivity losses closer to the Equator. However, poor soil may not sustain crop growth.

Canadian prairie

SP⊕T ▶ Africa

> ▶**REGION:** Africa
>
> ▶**THREATS:** Famine, shortage of drinking water, localized flooding, extreme poverty, civil unrest, poaching
>
> ▶**CLIMATE IMPACTS:** Drought, desertification, increased atmospheric temperature, more mosquito-borne diseases, rising sea levels
>
> ▶**ENDANGERED SPECIES:** Black rhinoceros, pygmy hippopotamus, leopard, 30 shrew species, 15 lemur species, 7 monkey species, 7 frog species, 3 gazelle species, riverine rabbit, cherry, mahogany, ebony

Malnutrition and famine, perennial problems in Africa, are exacerbated by climate change, as increasingly erratic weather patterns make subsistence farming difficult. Arid and semi-arid regions throughout Africa are becoming drier, and many areas are currently undergoing desertification or degradation due to overfarming. Annually, over 25 million people in sub-Saharan Africa face food crises.

GOING HUNGRY

With widespread poverty limiting the ability to adapt, Africa is deemed to be the inhabited continent most vulnerable to the impacts of climate change. Increasing temperatures, drought, desertification, and lack of water for irrigation combine to threaten crops across Africa. While the average temperature increase in Africa thus far is 0.9°F (0.5°C), temperatures have risen considerably higher in some regions, causing unprecedented stresses on agriculture.

Beyond the humanitarian crisis of acute food shortages, climate-induced threats to farming attack the mainstay of Africa's economy: agriculture contributes more than half of the total value of African exports and 20–30 percent of the gross domestic product in sub-Saharan Africa. The average number of food emergencies per year in Africa has tripled since the

mid 1980s. For example, Zimbabwe was known as southern Africa's "bread basket" until the late 1980s because of its bountiful agricultural production. But by 2007 half that country's population was thought to be malnourished. Corruption and poor government policies contributed to this crisis, but food emergencies around Africa are driven, in large part, by water scarcity. Of the 19 countries classified as water-stressed around the world, more are in Africa than any other region. African countries receive the most international food aid, through programs such as the United Nations' World Food Program.

Within Africa, land management can help mitigate climate change. Control of deforestation, improved rangelands management, expansion of protected areas, and sustainable forestry would all ensure better agricultural productivity.

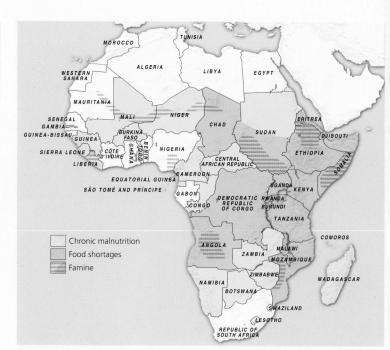

Lack of food (above) Food shortages are frequent in several sub-Saharan countries and large swaths of Africa have populations that suffer chronic malnutrition. Famine crisis zones are depicted since the 1970s.

Withered crops (below) Drought struck Kenya in 2009, following two years without a rainy season and severe drought in 2006. Here, children harvest corn from their family's parched field in Kwale, Kenya.

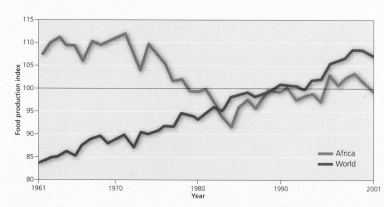

Food index Over the past several decades, food production in Africa has not kept pace with population growth. Drought in the 1980s and 1990s caused domestic consumption to exceed production, necessitating food aid.

Dead cattle (above) Prolonged drought wipes out pasture plants, starving livestock. In the aftermath of the 2006 East African drought, one Kenyan farmer's herd plummeted from 85 cows to five.

Primitive technique (right) A woman removes husk from millet near Maradi, Niger. This painstaking method of food processing limits yield. Food crises threatens two million people in Niger every year.

Starving (below) Malnutrition is a serious problem for thousands of Ethiopian children. In 2008, a malnourished two-month-old baby received treatment at a medical center run by Médecins Sans Frontières.

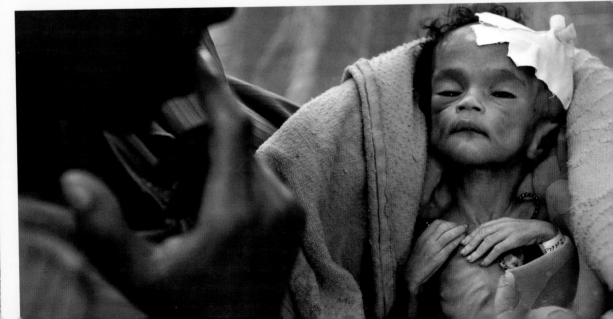

REVERSING THE TREND

The climate change problem is coming into ever sharper focus. Scientists are filling in the details of climate models as field observations provide evidence of the scope and impacts of a warming planet. An international clearinghouse has been established—the IPCC. It strives to synthesize the science into a format useful to policymakers, whose constituents globally are more and more interested in solving the climate problem. The approaches to slowing and reversing climate change are as myriad as its impacts: public education, collective political actions, technological innovation, and individual lifestyle choices.

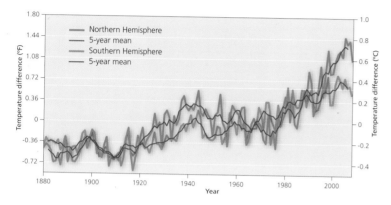

Reverse this! (left) This is the scope of the climate problem. In the Northern Hemisphere, where most emissions are produced, a temperature increase of around 1.8°F (1°C) has been observed since preindustrial times. The increase is lower in the Southern Hemisphere.

Earth Hour (right) To draw attention to energy use and climate hazards, Earth Hour was conceived in 2007 by the World Wildlife Fund (WWF) in Sydney, Australia. More than two million people and businesses turned off their lights **(bottom)** for one hour. The event has since become global.

POWER TO THE PEOPLE

Efforts by activists to draw attention to ecological degradation became a political force relatively recently. With environmentalism now solidly ensconced in the political mainstream, activists still have a crucial role to play in raising public awareness and educating citizens about environmental problems and solutions. This is especially important in the case of climate change, the impacts of which are gradual, variable, and spatially dispersed.

Laws to reduce environmental degradation began appearing around the world in the 1950s, and most countries now have legislation regulating pollution of water, soil,

and air. International cooperation on the global climate problem began with the United Nations' formation of the Intergovernmental Panel on Climate Change (IPCC) in 1988.

The high point of international climate policy thus far has been the Kyoto Protocol, a 1997 treaty committing the signatories of the developed world to reduce greenhouse gas emissions to five percent below 1990 levels by 2012. Ratified by 174 nations, with the notable exception of the historically largest CO_2 emitter, the U.S.A., this treaty provided the framework for national emissions reduction programs. The Protocol is widely recognized as being too weak,

and negotiations are underway to develop a more effective post-2012 international climate agreement.

The business world has responded creatively and diversely to demands to reduce emissions. By far the cheapest and easiest way is to improve energy efficiency. Many companies have found that efficiency improvements have helped their environmental image and their profits.

On the supply side, a variety of breakthrough low-carbon energy generation technologies are in development. Hydroelectric power from dams has long been the cleanest form of energy production and remains the largest renewable energy resource in use today. Solar, wind, tidal, and geothermal electricity generation all provide ways to harness Earth's weather. As materials and designs improve, a portfolio of renewable energy sources, tailored to each region's resource strengths, may compete sufficiently well economically to supplant CO_2-emitting fossil fuels. Biofuel crops are being developed to provide a renewable source of liquid fuel. Burning any liquid hydrocarbon fuel releases CO_2, but as plants also absorb CO_2 from the atmosphere as they grow, the life cycle of these fuels can have lower net carbon emissions than petroleum.

Finally, simple lifestyle changes made by individuals can collectively have a great impact on reducing global CO_2 emissions and slowing down climate change.

Algae in action Algae can be genetically engineered to consume CO_2 from the atmosphere and could be deployed in the world's oceans, or to turn power plant emissions into biofuel feedstock.

Slowing Down Climate Change

The first phase of the international climate policy agreement, the Kyoto Protocol, will expire in 2012. The Protocol has spurred many actions to reduce carbon dioxide emissions, though many signatories have not met its targets. Further, it was not ratified by the greatest historical emitter, the U.S.A., and had set no emissions reduction targets for developing countries such as China and India.

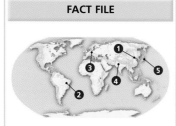

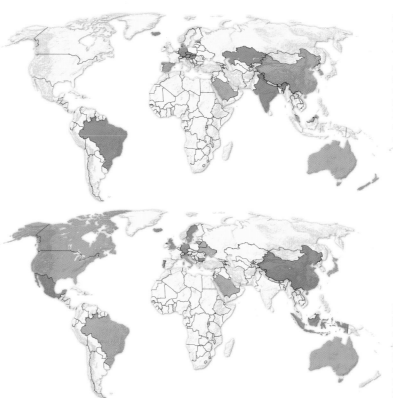

Scorecard (left) A 2007 study assessed the climate change performance of the top 56 carbon-dioxide-emitting nations. The CO$_2$ emissions trends (upper map) and climate policy actions (lower map) are shown, with the best- and worst-performing countries in each category. China's rapidly increasing emissions, and North America's lack of engagement in international climate policy development, stand out as areas for improvement.

■ Top 10 countries
▨ Bottom 10 countries

Reforestation (below) Governments are increasingly recognizing the importance of forests and mangroves as carbon sinks and protection against rising seas and intense coastal storms. The government of the Philippines has initiated a mangrove reforestation program to replace native forests illegally cut for firewood.

Recycling (above) For the production of plastics and paper products, using recycled materials requires much less energy than does the extraction, processing, and transport of virgin raw materials.

Packaging (right) The recycling of packaging materials removes waste from incinerators and landfills—both major greenhouse gas emitters. European countries are world leaders in waste reduction and recycling.

Desalination (below) Converting salt water to fresh water is a method used in arid areas and where demand outstrips freshwater resources. Desalination plants, such as this one in Spain, use the process of reverse osmosis.

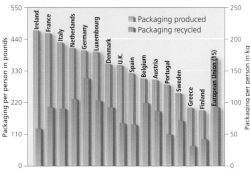

FACT FILE

Trading To reduce emissions, many countries set targets for specific industries, with companies allocated emissions quotas. They can often trade these between themselves, and also purchase carbon offset credits from projects that reduce emissions.

KINDS OF TRADING PROJECTS

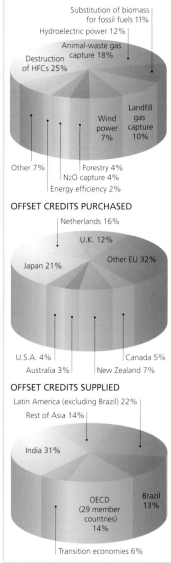

Substitution of biomass for fossil fuels 11%
Hydroelectric power 12%
Animal-waste gas capture 18%
Destruction of HFCs 25%
Landfill gas capture 10%
Wind power 7%
Other 7%
Forestry 4%
N2O capture 4%
Energy efficiency 2%

OFFSET CREDITS PURCHASED

Netherlands 16%
U.K. 12%
Other EU 32%
Japan 21%
U.S.A. 4%
Canada 5%
Australia 3%
New Zealand 7%

OFFSET CREDITS SUPPLIED

Latin America (excluding Brazil) 22%
Rest of Asia 14%
India 31%
Brazil 13%
OECD (29 member countries) 14%
Transition economies 6%

WEATHER WATCH

Carbon sequestration CO_2 is emitted in massive amounts at point sources like coal-fired electricity generation plants. Engineers are exploring technologies to sequester the CO_2 before it is emitted. As these sources are large and localized, a small number of facilities could have a significant impact on emissions.

WORKING TOGETHER

Climate change is a global environmental problem requiring global solutions. The Kyoto Protocol, the first international treaty with binding greenhouse gas emissions reductions for industrialized countries, came into force on February 16, 2005. It has been ratified by 184 countries. While an important first step in international climate cooperation, it is widely recognized that Kyoto's emissions reduction target is too weak.

Protest (right) Environmental activists find creative ways to draw attention to the perils of climate change. In a Greenpeace campaign to raise awareness of man-made CO₂ emissions causing Alpine glacial melt, naked volunteers lined up on the Swiss glacier Aletsch, the largest in the Alps.

GLOBAL SOLUTIONS

The rise of environmentalism in the 1970s brought international public attention to the effects of human activity on the planet, including the climate system. With global trends appearing, scientists engaged in international cooperative research. As climate models developed and predicted the impacts of droughts, storms, and rising seas, politicians began to recognize the policy ramifications of climate change.

The adoption of the United Nations Framework Convention on Climate Change in 1992 launched a new age of international cooperation on climate research and policy. Since then, climate scientists have published consensus documents on the status of climate change. An agreement to follow the first phase of the Kyoto Protocol, which ends in 2012, was discussed in Copenhagen in December 2009.

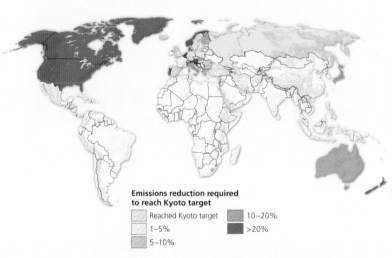

Emissions reduction required to reach Kyoto target

- Reached Kyoto target
- 1–5%
- 5–10%
- 10–20%
- >20%

Kyoto targets (above) The Protocol set emissions reduction targets for industrialized countries. The U.S.A. never ratified the treaty and is not bound by it. Future agreements will set targets for developing countries.

Kyoto in force (below) The primary Kyoto mechanisms are the carbon market, in which emissions can be traded, and the clean development mechanism, which sponsors projects that offset CO₂ emissions.

Montreal Protocol (below) In effect since 1989, this 1987 treaty is an international success story. After recognizing that CFCs deplete stratospheric ozone, 191 countries agreed to phase out these chemicals.

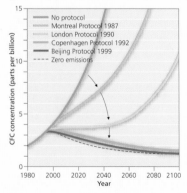

- No protocol
- Montreal Protocol 1987
- London Protocol 1990
- Copenhagen Protocol 1992
- Beijing Protocol 1999
- Zero emissions

CFC concentration (parts per billion)

Year

April 22, 1970 The first Earth Day was celebrated, marking the beginning of the modern environmental movement. Earth Day promotes good environmental citizenship.

February 12–23, 1979 The First World Climate Conference was held in Geneva, Switzerland. This was one of the first international scientific conferences on climate change.

1988 The World Meteorological Organization (WMO) and the United Nations Environment Program (UNEP) established the Intergovernmental Panel on Climate Change (IPCC).

1990 The First Assessment Report of the IPCC published, concluding that Earth is warming. The report concluded that this was likely due, in part, to human-induced emissions of carbon dioxide.

1992 An international treaty, the United Nations Framework Convention on Climate Change (UNFCCC), was produced in Brazil. It aimed to stabilize greenhouse gases in the atmosphere.

1995 The IPCC's Second Assessment Report was released, making a stronger statement that the evidence points to a discernable human influence on the global climate system.

1997 The Kyoto Protocol, an international agreement linked to the UNFCCC, was written and adopted. It set emissions reduction targets for industrialized countries, to be met by 2012.

2007 The IPCC and Al Gore, environmental activist and former U.S. vice president, shared the Nobel Peace Prize for their work to disseminate knowledge about man-made climate change.

December 7–18, 2009 Parties to the UNFCCC convene for the UN Climate Change Conference in Copenhagen, Denmark, to negotiate international policy actions to follow Kyoto.

2012 The Kyoto Protocol's first commitment period ends. Emissions reductions for industrialized nations were set at five percent below 1990 levels. New targets are likely be more stringent.

Alternative Energy

Replacing fossil fuels, which emit large amounts of CO_2, with alternative energy sources will require strong international and national policies and great technological advances. Harnessing weather—with developing wind, solar, and wave-power technologies—accounts for a small but rapidly growing portion of global energy production.

Usage (below) A 2008 study surveyed the percentage of renewable energy use based on total consumption. With heavy fossil fuel use, most developed nations have the lowest ranking. Countries that obtain much of their energy from hydroelectricity rank the highest.

Solar arrays (bottom) The Sun's energy is collected and converted to electricity using photovoltaic devices, usually made of silicon. Scientists are developing cheaper materials for solar power, which is presently more expensive than coal-generated electricity.

Wind farm (above) Wind turbines capture the kinetic energy in wind using blades, or rotors. Windy areas, such as this wind farm near Palm Springs, California, U.S.A., can generate enough energy to power an entire town. As winds are intermittent, wind power requires backup electricity sources.

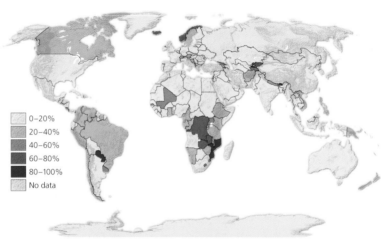

- 0–20%
- 20–40%
- 40–60%
- 60–80%
- 80–100%
- No data

Sizes (right) To capture the maximum amount of wind energy, modern wind turbines are typically constructed with three rotors, each longer than the wingspan of a 747 aircraft. Rotors are built of extremely lightweight materials, such as carbon fiber.

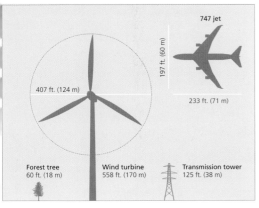

747 jet		
407 ft. (124 m)	197 ft. (60 m)	
	233 ft. (71 m)	
Forest tree 60 ft. (18 m)	Wind turbine 558 ft. (170 m)	Transmission tower 125 ft. (38 m)

Waves (above) Tidal turbines harness the energy in currents. Built to move with the direction of the tide, these rotors power a generator. Because tides are regular, they produce a predictable amount of electricity.

Alternative Energy continued

In addition to harnessing sunlight, wind, and waves, renewable energy sources include geothermal, biomass and waste gas, and hydroelectric power. Hydroelectricity makes 16 percent of the world's electricity. Nuclear power, six percent of world energy supply, is not renewable, though its CO_2 emissions are low. Safety concerns have all but stalled nuclear energy development. Globally, non-hydroelectric renewable sources account for about 10 percent of total energy production, but all are on the rise.

Garbage power (below) Bioreactors process methane harvested from decaying buried waste to burn in a generator for electricity production. This operation has the dual benefits of generating electricity and preventing the release of methane, a much more potent greenhouse gas than carbon dioxide.

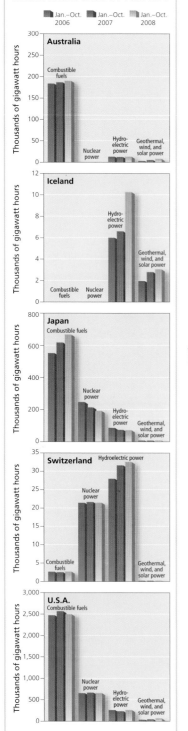

ENERGY PRODUCTION BY FUEL TYPE

Jan.–Oct. 2006 · Jan.–Oct. 2007 · Jan.–Oct. 2008

Australia — Combustible fuels, Nuclear power, Hydro-electric power, Geothermal, wind, and solar power (Thousands of gigawatt hours)

Iceland — Combustible fuels, Nuclear power, Hydro-electric power, Geothermal, wind, and solar power (Thousands of gigawatt hours)

Japan — Combustible fuels, Nuclear power, Hydro-electric power, Geothermal, wind, and solar power (Thousands of gigawatt hours)

Switzerland — Combustible fuels, Nuclear power, Hydroelectric power, Geothermal, wind, and solar power (Thousands of gigawatt hours)

U.S.A. — Combustible fuels, Nuclear power, Hydro-electric power, Geothermal, wind, and solar power (Thousands of gigawatt hours)

Collection Leachate—liquid released by rotting garbage—is collected with perforated pipes, refined in a treatment tank, then recirculated through the landfill to increase methane production.

Methane collection Methane gas is collected from deep within the landfill using perforated pipes and pumped up to the generator.

Generator Methane is burned in a generator, converting it to electrical power, which is added to the power grid.

Buried waste As waste buried in a landfill slowly decays, it releases methane gas (yellow arrows).

Leachate Liquid from the decomposing waste (blue arrows) contains chemicals that speed its decay.

Monitoring Environmental scientists monitor the gaseous emissions from the landfill and downstream water quality to ensure ecological safety.

Geothermal power (above) Extensive underground geothermal reserves at Iceland's Blue Lagoon are exploited by pumping water through hot rocks at high pressure. The resulting hot water drives a steam-powered generator to produce electricity and fills a pool of soothing water for a nearby spa.

Hydroelectric power (below) The world's largest hydroelectric power station, China's Three Gorges Dam, generates electricity via water-driven turbines. However, the controversial dam has caused significant erosion and water quality problems.

FACT FILE

Biofuel Rising oil prices and a focus on reducing CO_2 emissions have fed the demand for biofuels. Biofuels, such as ethanol from soybeans or corn and biodiesel from oilseeds, currently make up about one percent of global transportation fuels.

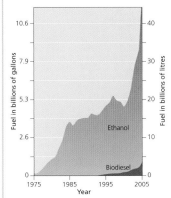

Quantities Biofuel production has increased rapidly since 2000. The U.S. is the largest producer and consumer of ethanol; Brazil is the second largest.

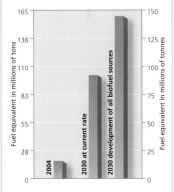

Consumption At current rates, world biofuel consumption is predicted to rise 500 percent by 2030. Pro-biofuel policies may result in a seven-fold increase.

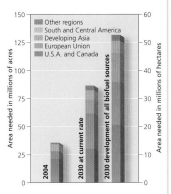

Land required Oilseed crops such as soybean, sunflower, and rapeseed will need an increasing share of the world's arable land to support biofuel production.

Slowing Down Change: Urban

Urban planning has a critical role to play in reducing the greenhouse gas emissions that cause climate change. More than half of the world's population lives in urban areas, which are large emitters of carbon dioxide through industrial activity, domestic energy use, and transportation. Urban "smart growth" policies can improve energy efficiency of the built environment, and public transit reduces vehicle miles driven.

1. Reykjavik Sitting atop an extensive geothermal resource that provides heat for all its buildings and renewable hydropower to make up the balance of the city's electricity needs, Reykjavik, in Iceland, is the "greenest" city in Europe. Its fleet of buses are currently being converted to run on hydrogen.

Reykjavik, Iceland

2. Portland In addition to its excellent public transportation system, Portland, Oregon, is the most bike-friendly city in the U.S.A. It has over 70 miles (113 km) of dedicated bike trails within the city and bike lanes are planned as attentively as those for vehicular traffic.

Portland, Oregon, U.S.A.

3. Curitiba Three-quarters of the two million residents of Curitiba, in southern Brazil, rely on public transport, using a well planned bus rapid transit system that is considered one of the world's best. The city's green space is "mowed" by a flock of sheep, and residents recycle 70 percent of household waste.

Curitiba, Brazil

4. Vancouver City planners in Vancouver have encouraged dense development, leading to a compact urban center interspersed with 200 parks. It uses 90 percent renewable power sources, and its sustainability plan includes developing coastal wind, tidal, and wave energy.

Vancouver, British Columbia, Canada

5. Sydney Australia was a world leader in making the transition from inefficient incandescent lightbulbs to compact fluorescents, and in 2007 Sydney drew global attention to the contribution of electric lighting to global warming by staging Earth Hour. In 2008, the city's government became carbon-neutral.

Sydney, Australia

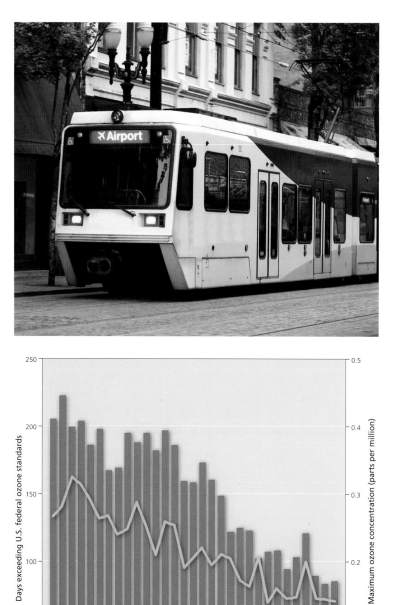

Air quality (above) California has some of the world's best air-quality monitoring infrastructure. Its vehicle emissions standards are stringent, helping to reduce ozone levels in southern California over the past 30 years.

Public transit (left) One-third of carbon dioxide emissions in the U.S.A. come from transportation. Cities like Portland, Oregon, which invest in mass transit, eliminate a large amount of vehicle emissions. Portland limits the number of downtown parking spaces, while offering free use of buses and trains.

LA smog (below) Los Angeles developed with a car culture, producing a sprawling city connected by a web of congested freeways. Traffic emissions, combined with the bright Californian sun, created a massive smog problem. This has been improving over the past 30 years due to better fuel efficiency.

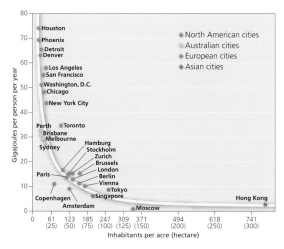

Energy consumption (above) Densely populated urban areas encourage heavier use of mass transit, resulting in lower per-capita transportation-related energy consumption. More vehicle miles are driven in North American and Australian cities than in Europe or Asia.

Pedal power (right) Many city governments are launching schemes to encourage cycling, such as Barcelona's "Bicing" program in Spain, which places bicycles for hire around the city. Some bike-sharing programs are free, while others charge a nominal fee.

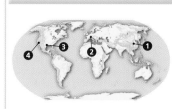

Slowing Down Change: Industrial

Scientists and industry researchers around the world are investigating ways to reduce damaging atmospheric emissions from power stations and motor vehicles, which produce more than 15 percent of fossil-fuel-based emissions. Companies that consider greenhouse gas emissions in business planning have the potential to improve the global CO_2 balance sheet while securing economic gains.

Saplings (below) Because trees provide the predominant terrestrial carbon dioxide sink, some paper manufacturers seek to reduce their carbon footprint by replanting seedlings after timber is harvested, thus reversing some of the environmental damage.

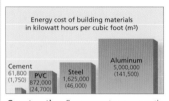
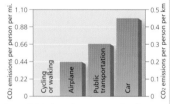
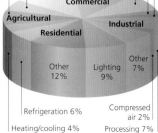

Phone recycling (right) Personal consumer electronics have high energy costs associated with their manufacture and are often treated as disposables. Cell phones are typically used for only 18 months before replacement, with the majority discarded in landfills. Recycling cell phones reduces the need for new raw materials and, hence, cuts industry greenhouse gas emissions.

Cogeneration plant (below) The electricity generation industry can dramatically reduce its carbon footprint by developing creative ways to squeeze more useful energy out of its facilities, or by collaborating with other industries. Here, a waste incineration plant in Herten, Germany, doubles as a cogeneration plant, producing both electricity and heat.

FACT FILE

1. Voluntary actions In many countries, industrial CO_2 emissions reductions are accomplished in part through "voluntary actions," joint government-industry performance standards. The Danish

agreement on energy efficiency is binding once a company enters into it, with tough fiscal sanctions for non-compliance.

Denmark

2. Emissions trading In Switzerland, companies adopt voluntary CO_2 limits and trade emissions among themselves to be compliant with the European

Union's Kyoto Protocol commitments. If these voluntary caps prove insufficient, a tax on fossil fuel usage is invoked.

Switzerland

3. Labels Not having ratified the Kyoto Protocol in the first commitment period, the U.S. has thus far relied on voluntary actions to reduce industrial emissions. The highly successful, voluntary Energy

Star labeling program promotes the purchase of energy-efficient products, particularly household appliances.

U.S.A.

4. Cement Cement production makes up about five percent of global CO_2 emissions, and about 45 percent of the world's cement is produced in China.

Representatives from Chinese cement companies recently formed the Cement Sustainability Initiative to monitor emissions.

China

5. Semiconductors The semiconductor industry is a major CO_2 emitter. At a recent symposium in Japan, industry leaders evaluated the many ways semiconductors can be used to mitigate

climate change, from more efficient use of electric motors to "smart metering" networks that enable energy savings.

Japan

Slowing Down Change: Domestic

Home energy use and building practices have a significant impact on global greenhouse gas emissions. There are many simple ways that individuals can reduce their domestic carbon dioxide output. The choice of building materials, insulation, and heating and cooling systems, and lowered water use can all affect a home's carbon footprint, as well as installing renewable power sources.

Standby mode (right) The standby mode on household electronics, such as computers, DVD players, televisions, and stereo systems, contributes up to two percent of some countries' CO_2 emissions. Plug these devices into a power strip that can be turned off.

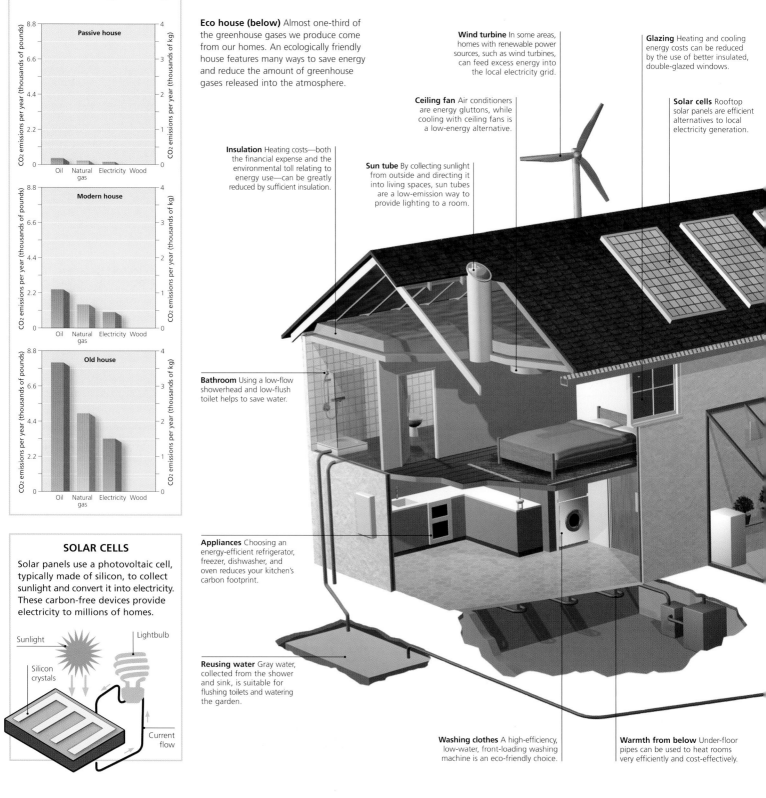

Eco house (below) Almost one-third of the greenhouse gases we produce come from our homes. An ecologically friendly house features many ways to save energy and reduce the amount of greenhouse gases released into the atmosphere.

Wind turbine In some areas, homes with renewable power sources, such as wind turbines, can feed excess energy into the local electricity grid.

Glazing Heating and cooling energy costs can be reduced by the use of better insulated, double-glazed windows.

Ceiling fan Air conditioners are energy gluttons, while cooling with ceiling fans is a low-energy alternative.

Solar cells Rooftop solar panels are efficient alternatives to local electricity generation.

Insulation Heating costs—both the financial expense and the environmental toll relating to energy use—can be greatly reduced by sufficient insulation.

Sun tube By collecting sunlight from outside and directing it into living spaces, sun tubes are a low-emission way to provide lighting to a room.

Bathroom Using a low-flow showerhead and low-flush toilet helps to save water.

Appliances Choosing an energy-efficient refrigerator, freezer, dishwasher, and oven reduces your kitchen's carbon footprint.

Reusing water Gray water, collected from the shower and sink, is suitable for flushing toilets and watering the garden.

Washing clothes A high-efficiency, low-water, front-loading washing machine is an eco-friendly choice.

Warmth from below Under-floor pipes can be used to heat rooms very efficiently and cost-effectively.

SOLAR CELLS

Solar panels use a photovoltaic cell, typically made of silicon, to collect sunlight and convert it into electricity. These carbon-free devices provide electricity to millions of homes.

Sunlight

Lightbulb

Silicon crystals

Current flow

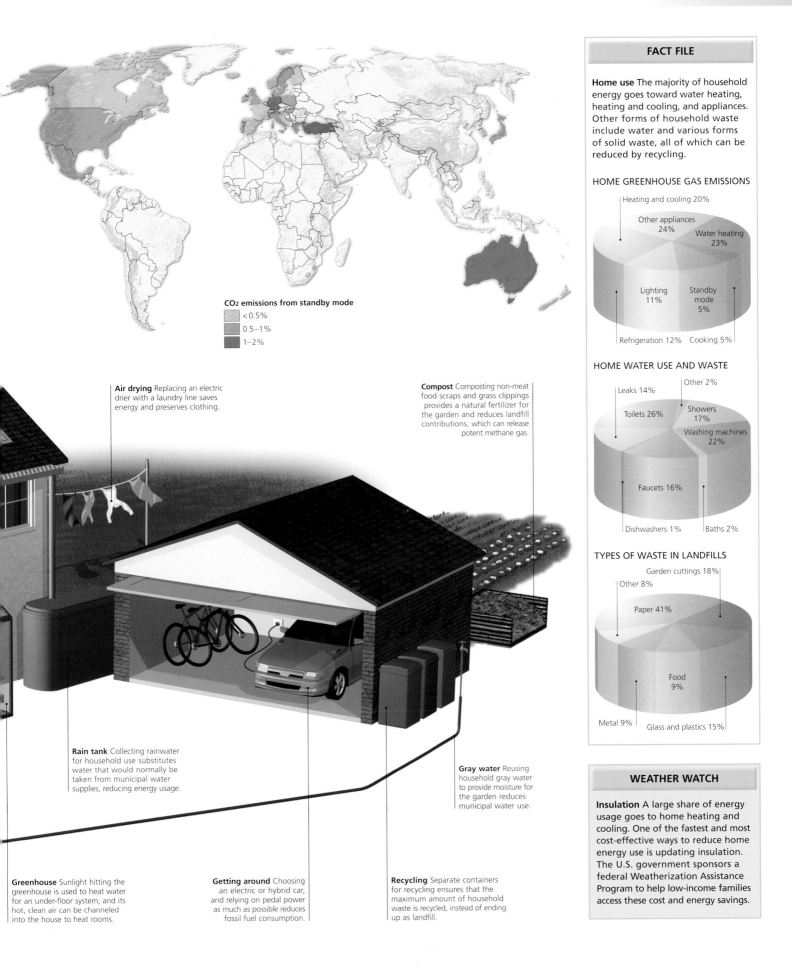

CO₂ emissions from standby mode
- <0.5%
- 0.5–1%
- 1–2%

Air drying Replacing an electric drier with a laundry line saves energy and preserves clothing.

Compost Composting non-meat food scraps and grass clippings provides a natural fertilizer for the garden and reduces landfill contributions, which can release potent methane gas.

Rain tank Collecting rainwater for household use substitutes water that would normally be taken from municipal water supplies, reducing energy usage.

Gray water Reusing household gray water to provide moisture for the garden reduces municipal water use.

Greenhouse Sunlight hitting the greenhouse is used to heat water for an under-floor system, and its hot, clean air can be channeled into the house to heat rooms.

Getting around Choosing an electric or hybrid car, and relying on pedal power as much as possible reduces fossil fuel consumption.

Recycling Separate containers for recycling ensures that the maximum amount of household waste is recycled, instead of ending up as landfill.

FACT FILE

Home use The majority of household energy goes toward water heating, heating and cooling, and appliances. Other forms of household waste include water and various forms of solid waste, all of which can be reduced by recycling.

HOME GREENHOUSE GAS EMISSIONS

- Heating and cooling 20%
- Other appliances 24%
- Water heating 23%
- Lighting 11%
- Standby mode 5%
- Refrigeration 12%
- Cooking 5%

HOME WATER USE AND WASTE

- Leaks 14%
- Other 2%
- Toilets 26%
- Showers 17%
- Washing machines 22%
- Faucets 16%
- Dishwashers 1%
- Baths 2%

TYPES OF WASTE IN LANDFILLS

- Garden cuttings 18%
- Other 8%
- Paper 41%
- Food 9%
- Metal 9%
- Glass and plastics 15%

WEATHER WATCH

Insulation A large share of energy usage goes to home heating and cooling. One of the fastest and most cost-effective ways to reduce home energy use is updating insulation. The U.S. government sponsors a federal Weatherization Assistance Program to help low-income families access these cost and energy savings.

TOMORROW'S FORECAST

Public awareness of the climate and energy crisis is at an all-time high, with an increasing number of consumers choosing hybrid vehicles, compact fluorescent bulbs, and energy-efficient appliances. The Kyoto Protocol and other treaties have shown us that global cooperation can help mitigate the long-lasting environmental challenge of climate change.

Carbon trading allows countries and companies to seek the most economically efficient ways to reduce greenhouse gas emissions, but progress is slow. Policies to regulate CO_2 emissions have spurred remarkable technological developments in energy efficiency and sources of renewable energy, but coal and oil remain too cheap and available to be made obsolete quickly.

Although some of the damage may be irreversible, much can still be done to slow and reverse climate change. Every person has a role to play in tackling the problem, to leave the planet more hospitable for future generations. By spreading the word about the positive impact of energy conservation and reducing emissions, choosing to conserve energy in our daily lives, and electing political representatives who champion climate-protection policies, we can collectively prevent the worst-case projections from becoming reality.

Rising to the challenge Climate change has taught us how vulnerable Earth's atmosphere is to human impact. The delicate balance of gases that provide optimal conditions for life on Earth has been disrupted by the recent rise of industrial activity. Fortunately, this also suggests that human efforts can reverse the trend, if we act quickly enough to avoid crossing an irreversible climate tipping point. Tackling climate change will require a major overhaul of our global energy infrastructure, and it is essential that the international community addresses this challenge together.

GLOSSARY

ablation The removal of snow and ice by melting and direct vaporization.

absolute zero The lowest temperature possible, 0 Kelvin (-460°F or -273°C), at which all motion ceases.

acid rain An acidic form of rain that occurs when chemicals produced by the burning of *fossil fuels* mix with water vapor in the air.

advection The transfer of energy or heat by the movement of a substance, such as the atmosphere or ocean. For example, advection fog is made up of water droplets moved by winds.

aerosol A suspension of liquid or solid droplets in a gas.

albedo The reflectivity of an object when illuminated by sunlight.

algae (singular alga) Simple, plant-like organisms lacking a stem, leaves, or roots, but containing chlorophyll and performing *photosynthesis*.

alluvial fan *or* plain A fan-shaped, level area covered by river sediment deposited where the gradient of the river bed decreases abruptly.

amplitude Of an ocean wave, the vertical distance between the midpoint and the peak or the trough (bottom).

anemometer An instrument that measures wind speed.

anthracite coal The type of coal with highest carbon content and fewest impurities.

anthropogenic Caused by human activity.

anticyclone A system of rotating winds that spiral out from a high-pressure area. Generally associated with stable weather.

aphelion The point in orbit when a planet is farthest from the Sun.

apparent temperature The temperature a person senses when the effects of windchill and humidity are taken into account.

archipelago A group of islands or an area that contains many small islands.

atmosphere A layer of gas and suspended material held around a planet or moon by gravity.

atmospheric pressure Also called air pressure or barometric pressure, the weight of the atmosphere over a unit area of Earth's surface. Changes of weather are usually accompanied by air pressure fluctuations.

atoll A coral reef surrounding a central lagoon, often around the rim of an extinct underwater volcano.

atom The smallest unit that can be called a chemical element.

aurora Colored streaks of light in the sky seen in high latitudes, caused by *solar wind*.

barometer An instrument that uses air, water, or mercury to measure air pressure, making it possible to anticipate weather changes.

barrier reef A coral reef around islands or along coasts, with a deep lagoon between the reef and the coast.

Beaufort scale A scale devised by William Beaufort in 1805, used to estimate wind speeds.

bedrock The solid mass of rock lying beneath the ground or water surface.

Big Bang According to our best cosmological theory, the event that marked the birth of the *universe* about 13.7 billion years ago.

biodegradable Describes material that can be broken down by bacteria, insects, or other natural substances.

biodiversity The variety of plant and animal species found in a habitat.

biofuel A fuel that is produced from biological sources, usually plants, rather than from *fossil fuels*.

biomass Biological material derived from living organisms, such as wood and plant waste. Biomass is the term applied to a class of renewable energy derived from such materials.

biome A large area defined by climate and vegetation and having a distinctive community of plants and animals.

bituminous coal A type of coal of lower quality and purity than *anthracite*; bituminous coal has 60-80 percent carbon content and significant sulfur and water impurities.

blizzard A severe snow storm with strong winds and reduced visibility.

blocking A sustained high-pressure system that disrupts the prevailing atmospheric flow, such as a *jet stream*.

canyon A deep, steep-sided valley formed by river erosion.

carbon dioxide A molecule made up of one carbon atom and two oxygen atoms. Because of its molecular structure, carbon dioxide (CO_2) is a strong absorber of infrared radiation, making it a potent greenhouse gas.

carbon fertilization An increase in crop productivity due to higher carbon dioxide concentrations.

carbon footprint An estimate of the impact of one person's activities on the environment, measured by the amount of CO_2 emissions.

carbon monoxide A molecule composed of one carbon atom and one oxygen atom. The product of incompletely combusted fossil fuels or biomass, carbon monoxide (CO) is a toxic gas.

carbon sink Any process that removes carbon dioxide from the atmosphere.

carbonic acid The acid formed when carbon dioxide dissolves in water.

Celsius Scale of temperature in which the melting point of ice is 0° and the boiling point of water is 100°.

chaparral A type of vegetation found in parts of California and other regions with a similar climate, dominated by plants adapted to hot, dry summers.

chinook The warm, downslope winds descending from various mountain ranges in the western United States.

chlorofluorocarbons (CFCs) Molecules composed of carbon and chlorine and/or fluorine atoms. CFCs are both potent greenhouse gases and destroyers of stratospheric ozone.

chlorophyll The green pigment present in all green plants and some bacteria. It absorbs sunlight and releases the energy that drives the process of *photosynthesis*.

chromosphere The thin layer of the Sun's atmosphere just above the *photosphere* and below the *corona*.

climate The pattern of weather that occurs in a region over an extended period of time.

climate variability Statistical changes in climate on short timescales.

cloud forest Tropical forest on mountainsides at an elevation where atmospheric water vapor condenses, shrouding the vegetation in mist.

cogeneration The use of a power station to generate both electricity and usable heat.

cold front A boundary between two air masses of different temperatures. As the front passes, cold air replaces warmer air.

compact fluorescent lamp (CFL) A fluorescent lightbulb lined with phosphor that uses much less energy than traditional incandescent lightbulbs.

condensation The formation of liquid water from water vapor; occurs when moist air reaches its dew point and comes into contact with a solid surface or with condensation nuclei.

continental drift The slow movement of continental plates, caused by convection currents in the *mantle*.

continentality The degree to which a location's climate is influenced by a landmass, which typically experiences a greater annual range in temperature than sea-influenced regions.

convection A heat-driven process that causes hotter material or air to move upward, while lighter matter sinks.

convergence A flow of air moving from different directions toward a central point.

coral bleaching The loss of color affecting coral reefs when the algae that live in them are killed or forced out.

coral polyps Cylindrical, elongated organisms that join together to form a coral reef. Each polyp is fixed on one end to the firm carbonate structure and bears a mouth on the other end.

cordillera A system of mountain ranges, with their plateaus and basins.

core The centermost part of Earth, consisting of a solid inner core, surrounded by a liquid outer core.

Coriolis effect The tendency of a freely moving object to follow a curved path in relation to the rotating surface of Earth, as shown in the direction that weather systems rotate (clockwise in the Northern Hemisphere and counterclockwise in the Southern Hemisphere).

corona The high-temperature, outermost atmosphere of the Sun.

coronal mass ejection A massive eruption of material from the Sun's *corona* over a period of several hours.

crest The highest point of a wave.

crevasse A deep fissure in a *glacier*.

crust The outer layer of Earth, consisting of solid rock averaging 3 miles (5 km) thick beneath the oceans and up to 40 miles (64 km) thick beneath mountain ranges.

crustacean A mostly aquatic animal, such as a lobster, crab, or shrimp, that has a hard external skeleton.

crystal A regular, repeating geometric structure found in frozen liquids, such as hexagonal or cubic ice crystals. Also, a solid mineral with a definite geometric shape.

cumuliform The shape of a cloud formed by strong *convection*, resulting in a "heaped" appearance, which is an indicator of atmospheric instability.

cyclone The name given to a *hurricane* in Australasia and countries around the Indian Ocean.

deciduous A plant or tree that drops its leaves in fall, winter, or the dry season, and grows new leaves in spring.

deforestation The logging and/or burning of native trees, typically for fuel or to enable farming.

delta A layer of sediment deposited at the mouth of a slow-moving river and protruding beyond the coastline.

dendrochronology A technique for dating, based on analysis of tree ring growth patterns.

deposition The direct formation of ice from water vapor; a type of *condensation*.

desertification The process by which fertile land turns into desert as a result of decreasing rainfall.

dew point The temperature to which air must be cooled at constant pressure for water *saturation* to occur, followed by *condensation*.

dike An embankment, or artificial earthen wall, built to prevent nearby lowland from flooding.

divergence A flow of air that moves outward from a central point.

dominant Of a species, one that has the greatest influence on the composition and character of a certain plant or animal community.

downdraft Draft caused by the descent of a cooling column of air. It can result in powerful wind gusts and heavy rain.

erg An extensive area in a hot desert that is covered by gently sloping sand dunes, also known as a sand sea.

erosion The breaking down and removal of rock and soil by wind, water, and ice, as well as by mass movements, such as landslides.

estuary A partly enclosed area of coastal water that is open to the sea and into which a river flows.

eutrophication An increase in the nutrients in an *ecosystem*, most often a body of water. This can result in enhanced algal growth that depletes the water's oxygen content, harming fish and other animal populations.

evergreen A type of plant that has leaves year round.

exosphere The outermost region of the *atmosphere*, 300–450 miles (480–725 km) above Earth's surface.

extratropical cyclone A cyclonic storm occurring in the midlatitudes. It is powered by the interaction of warm and cold air masses.

eye The clear center of intense low pressure within a hurricane or cyclone.

Fahrenheit Scale of temperature in which the melting point of ice is 32° and the boiling point of water is 212°.

Ferrel cell A large-scale atmospheric circulation feature at the midlatitudes.

front In meteorology, a boundary between two air masses of different temperatures. See also *cold front, occluded front,* and *warm front.*

frontal system The interface between air masses of different temperatures and humidities, where the most significant weather tends to occur.

fugitive emissions Unintended and unmonitored emissions of harmful gases due to leaks, primarily from industrial activities.

fynbos Drought-resistant vegetation that is adapted to hot, dry summers, found in Africa's Mediterranean climate zone, notably Western Cape Province, South Africa. It is similar to *chaparral,* but its plant species are different.

gale A wind between forces 7 and 10 on the Beaufort scale, blowing 38–63 miles per hour (60–100 km/h).

genus (plural genera) A low-level taxonomic rank used to classify living and fossil organisms. The full, binomial name of every species is formed from a genus name (capital initial), followed by the species name.

geographic pole One of the two points where Earth's rotational axis intersects its surface.

geothermal warming Warming of water by contact with the hot rocks beneath Earth's surface.

geyser An opening at Earth's surface, from which a fountain of hot water periodically spouts upward.

glaciation The effect of a moving ice mass on a landscape, resulting in erosion, the gouging of U-shaped valleys and fjords, and the deposition of ridges and sheets of rock debris.

glacier A mass of ice that moves over the underlying surface.

global warming An observed increase in Earth's average surface temperature.

Gondwana The supercontinent fragment comprising New Zealand, Antarctica, Australia, South America, Africa, and India. It began to break up 130 million years ago.

graupel Frozen precipitation that forms when water condenses onto a snowflake, making 0.08–0.2 inch (2-5 mm) diameter balls of ice.

gray water Used water drained from baths, showers, dishwashers, or washing machines that is recycled in order to reduce water use in the home.

great ocean conveyor belt See *thermohaline circulation.*

greenhouse effect The retention of heat in the *atmosphere*. Sunlight reaching Earth's surface is re-radiated as heat, which is then absorbed, trapped, and re-radiated back to Earth by certain gases, such as *carbon dioxide.*

greenhouse emissions Substances released into the air by machines or natural processes, contributing to the greenhouse effect.

groundwater Water that moves through the spaces between rocks below the ground.

Gulf Stream An ocean current that carries warm water from the Caribbean Sea to the North Atlantic Ocean.

gyre A circular motion in a body of water.

haboob An intense sandstorm, commonly observed in northern Africa and the Middle East.

Hadley cell A large-scale atmospheric circulation feature dominating the tropics, with rising motion near the Equator, poleward flow aloft, and then descending motion in the subtropics.

Haines index A weather index used to assess the potential for rapid forest fire growth.

hamada A type of desert landscape characterized by rocky, barren plateaus with little sand.

harmattan A dry and dusty West African trade wind, blowing south from the Sahara.

heat wave A period of at least one day, but more usually several days or weeks, during which the air temperature is unusually high for a certain location at that time of year.

helium The second most common element after hydrogen. It was created in the *Big Bang* and continues to be made inside stars by nuclear reactions.

hemisphere One half of Earth. Asia, Europe, and North America are in the Northern Hemisphere; while Africa, Antarctica, Australia, and South America are located in the Southern Hemisphere.

high latitudes The regions near Earth's Poles, north of the Arctic Circle and south of the Antarctic Circle, which receive the least energy from the Sun and so have cold climates.

high-pressure system An area of high-pressure that rotates clockwise in the Northern Hemisphere and counterclockwise in the Southern Hemisphere.

humidity The amount of water vapor in air.

hurricane The term used in North America and the Caribbean for an intense low-pressure cell of tropical origin in which mean wind speeds are greater than 74 miles per hour (119 km/h).

hydroelectric power A form of renewable energy, based on converting the movement of water into electricity. Dams capture river water and direct it past turbine generators at high speed.

hydrogen The most common and lightest substance in the *universe*. Stars and gas-giant planets are made mostly of hydrogen gas and *helium*.

hydrological cycle The continuous cycling of water between land, ocean, and *atmosphere*.

ice age A cold phase in the climatic history of Earth, during which large areas of land were covered in ice.

ice cap A layer of ice stretching over land less than 19,300 square miles (50,000 km²) in area, but thick enough to bury underlying landscape features.

ice core A column of ice retrieved by drilling into glaciers, typically to analyze for prehistoric climate and atmospheric composition.

ice field A layer of ice that develops where the land surface is high enough, or level enough, for ice to accumulate.

ice floe A large, flat expanse of floating sea ice.

ice sheet A layer of ice extending over land greater than 19,300 square miles (50,000 km²) in area. There are currently two ice sheets: one in Antarctica and one in Greenland.

ice shelf An area of floating ice, once part of a *glacier*, that is still attached to land.

ice storm A winter storm characterized by *freezing rain*.

iceberg A mass of ice that has become detached from a *glacier* or *ice shelf*, and floats in the sea.

Icelandic low A center of low pressure found between Iceland and southern Greenland, the center of Northern Hemisphere atmospheric circulation.

International Union for Conservation of Nature (IUCN) A multinational environmental organization that maintains the Red List of Threatened Species to monitor the conservation status of plant and animal species.

Intertropical Convergence Zone (ITCZ) The *convergence* of the trade winds from the Northern and Southern Hemispheres, forming a band of convective clouds, usually with thunderstorms, at the Equator.

inversion Of temperature, a region in the atmosphere in which temperature increases with height, not decreases.

ionization The loss or gain by an *atom* of one or more *electrons*, resulting in the atom having a positive or negative electrical charge.

isobar A line drawn on a weather map that connects points of equal air pressure. When close together, isobars indicate areas of strong wind.

isotopes *Atoms* of a single chemical element with varying numbers of neutrons, resulting in different atomic masses.

jet stream Currents of fast-moving air at upper levels of the atmosphere. In midlatitudes, jet streams are more pronounced in winter.

K-T extinction event The geological dividing line, approximately 65 million years ago, when a mass extinction of animal and plant species, including the extinction of the dinosaurs, occurred over a geologically short period of time.

katabatic wind A type of wind carrying high-density air from higher elevations down a slope, due to gravity. Also known as a "fall wind."

Kelvin Scale of temperature on which all molecular motion ceases at 0 ("absolute zero"). The melting point of ice is 273°K and the boiling point of water is 373°K. There are no temperatures below 0°K, which is equal to -273°C.

Köppen climate classification Developed by Russian climatologist Wladimir Köppen, this classification system designates climate zones based on the native vegetation, driven by temperature, precipitation, and seasonality of precipitation.

Kyoto Protocol A 2005 agreement between world governments that aims to reduce *greenhouse emissions* to slow climate change.

La Niña Periods of unusually cold ocean temperatures in the equatorial Pacific that occur between *El Niño* events.

lake-effect snow High snowfall that occurs on the *lee* side of a large, unfrozen lake that is entirely enclosed by land.

land bridge A connection between two large landmasses.

landslide The rapid movement down a slope of material that has become detached from the underlying surface.

latent heat Heat either released or absorbed when water changes form. Latent heat is absorbed, cooling the environment, when water changes from ice to liquid, liquid to vapor, or ice to vapor. Latent heat is released, warming the environment, when water changes from vapor to liquid, liquid to ice, or vapor to ice.

latitude A measurement of distance from the Equator.

Laurasia The ancient supercontinent fragment comprising most of the landmasses that make up today's Northern Hemisphere continents.

Laurentian Ice Sheet A thick layer of ice that covered eastern Canada and the northeastern U.S. during the Pleistocene ice ages.

lee The side sheltered from the wind.

lightning A flash of light in the sky produced by atmospheric electricity generated in a thundercloud.

lithosphere The outermost solid layer of Earth, comprising the *crust* and the brittle top part of the *mantle*.

Little Ice Age A period of relatively low average temperatures that affected the whole world and lasted from the 16th century to the mid-19th century.

low latitudes The regions nearest the Equator, south of the Tropic of Cancer and north of the Tropic of Capricorn, where the Sun is almost directly overhead and its heat is most intense. These regions have hot or warm climates.

low-pressure system A weather system in which air pressure decreases toward the center. This is usually caused by a mass of warm air being forced upward by cold air. Such systems are usually associated with unsettled weather.

magnetic field A region surrounding a magnetic object, within which any magnetic body will experience a force.

magnetic pole The North or South Pole of Earth's *magnetic field* with which magnetic compass needles align. The magnetic pole is some distance from the *geographic pole*.

magnetosphere A region of space around a planet or star, dominated by that body's *magnetic field*.

mangrove Flowering shrubs and trees tolerant of salt water, found on low-lying tropical and subtropical coasts and estuaries.

mantle The part of Earth that lies between the underside of the *crust* and the outer edge of the outer *core*.

megacity An urban area with a population greater than 10 million.

meltemi Strong, dry north winds of the Aegean Sea.

mesopause In Earth's *atmosphere*, the boundary, located about 53 miles (85 km) above Earth's surface, between the *thermosphere* above and the *mesosphere* below.

mesosphere The layer of the *atmosphere* above the *stratosphere*, from about 31–53 miles (50-85 km) above sea level.

meteorite The name given to any piece of interplanetary debris that reaches Earth's surface intact.

meteorology The scientific study of weather.

methane A molecule composed of one carbon atom and four hydrogen atoms. Methane (CH₄) is a more potent greenhouse gas than carbon dioxide on a per-molecule basis, but there is less of it in the atmosphere.

microclimate A local variation in the normal climate of the region, with differences in temperature and moisture caused by topography, vegetation, or proximity to bodies of water or urban areas.

microorganism An organism too small to be visible to the naked eye.

midlatitudes The regions located between the Arctic Circle and the Tropic of Cancer (including much of North America and Europe) and between the Antarctic Circle and the Tropic of Capricorn. These regions tend to have moderate climates.

midnight sun The natural, annual phenomenon occurring north of the Arctic Circle and south of the Antarctic Circle during local summers, when the Sun is visible for a continuous 24 hours.

mineral A naturally occurring substance with a characteristic chemical composition and crystal structure. Rocks are made of minerals.

mirage An optical phenomenon in which light rays bend through warmer air to produce a displaced image of distant objects.

molecule The smallest particle into which a chemical substance can be divided without it becoming something else.

monsoon A sustained seasonal wind that produces heavy rain in tropical and subtropical zones. Any region receiving most of its precipitation during one season is classified as a monsoon climate. Monsoon regions include North America, sub-Saharan West Africa, and, most dramatically, South and East Asia.

Montreal Protocol An international treaty written in 1987, which went into effect in 1989, to protect the *ozone layer* by phasing out the production of the *CFCs* that deplete *ozone*.

moraine Rock and gravel that has been removed and ground by glacial scouring and then deposited at the side or terminus of a *glacier*.

NASA The National Aeronautics and Space Administration, the U.S. government's agency for space exploration. NASA operates several satellites that provide valuable information about the atmosphere.

neap tide A tide with a much smaller range than a *spring tide*. This occurs when the gravitational pulls of the Sun and Moon on the oceans work against each other.

nitrogen oxides (NOx) Molecules composed of one nitrogen atom and one or two oxygen atoms, emitted by combustion engines and forming a primary ingredient of urban air pollution, in the form of photochemical smog.

NOAA The National Oceanic and Atmospheric Administration is the U.S. scientific agency charged with assessing the condition of Earth's atmosphere and oceans.

North Atlantic drift The less well defined portion of the *Gulf Stream,* that moves eastward from Cape Hatteras across the North Atlantic.

North Atlantic Oscillation (NAO) A large-scale climatic phenomenon in which fluctuations in the difference in sea-level atmospheric pressure between the Icelandic Low and the Azores High control the strength and direction of westerly winds and storm tracks across the North Atlantic.

North Atlantic Oscillation Index An index proportional to this sea-level pressure difference is used to describe the atmospheric state.

oasis An area in a desert where the water table is close enough to the surface to allow plants to flourish.

occluded front An amalgam of two fronts, produced when a *cold front* catches up with a *warm front;* usually associated with a *low-pressure system.*

optical depth The measure of the ability of light to penetrate through to Earth's surface, a function of the concentration of particles in the atmosphere.

ore A mineral or rock that contains a particular metal in a concentration that is high enough to make its extraction commercially viable. Iron ore is an example.

orographic precipitation Rainfall produced when a mass of air pushed up the side of a mountain or elevated land formation cools, causing condensation.

outlet glacier A *glacier* that drains ice from an *ice sheet* or *ice cap.*

ozone A type of oxygen in which the molecule consists of three atoms (O_3) rather than two (O_2).

ozone hole A region of the *ozone layer* over Antarctica, and to a lesser extent over the Arctic, where the amount of *ozone* is depleted.

ozone layer A region of Earth's *stratosphere,* 12–25 miles (20–40 km) above sea level, where the *ozone* concentration is higher than elsewhere.

pack ice Drifting sea ice that has become packed together to form a large mass.

paleoclimatology The study of ancient climates.

Palmer Drought Severity Index A measurement of dryness and drought susceptibility, based on recent precipitation and temperature.

pampas A vast area of level grassland in Argentina and Uruguay, extending from the Atlantic coast to the Andean foothills and bounded by the Gran Chaco and Patagonia.

Pangea The ancient supercontinent that once contained all of Earth's continents. It began to break up, about 200 million years ago, into *Gondwana* and *Laurasia.*

pannus An "accessory" cloud beneath or attached to another cloud.

peat A soil that contains at least 65 percent organic material by dry weight and that has a surface horizon at least 16 inches (40 cm) thick.

peninsula A strip of land that is bordered on three sides by the sea.

periglacial lake A lake that has formed where the natural drainage of the topography has been blocked by *moraine* or a *glacier.*

perihelion The point in orbit when a planet is nearest the Sun.

period In geology, a specific division of geological time. Periods are further broken down into epochs and ages.

permafrost Ground that has remained frozen for at least two successive winters and the intervening summer.

photosphere The visible surface of the Sun or any other star.

photosynthesis The process by which plants produce their own nutrients, in the form of sugars, using daylight, water, and carbon dioxide. Plants give off oxygen during *photosynthesis.*

photovoltaic cell The basic unit of most solar energy systems.

plankton The plant (phytoplankton) or animal (zooplankton) organisms that float in the open sea.

plate tectonics The theory that Earth's *crust* consists of a number of plates that float on top of the *mantle* and move in relation to one another.

plateau An area of high, level ground.

pneumatophore A type of aerial root structure in some aquatic plants, such as mangroves.

Polar cell A large-scale atmospheric circulation feature at the Poles, in which warm air rises at about 60°N and 60°S, moves poleward, then descends near the Pole.

prairie An area of level or rolling grassland, especially found in central North America, that has few trees, and generally a moderately moist climate.

precipitation Water that falls from the sky as dew, fog, mist, drizzle, rain, hail, frost, or snow.

predator An organism that obtains energy by consuming and usually killing another organism (the prey).

pressure gradient The difference in pressure over some spatial dimension.

proton An elementary particle that carries a positive charge.

pyronado A tornado-like column of wind that sometimes forms in intense fires.

radiation Energy radiated from a source as wavelengths or particles.

radiative forcing The change in net solar energy received at the top of the *troposphere* relative to preindustrial times. The "forcing" from atmospheric constituents, such as greenhouse gases and aerosols, is a useful metric for comparing the climate-changing ability of these components.

rain forest A type of forest that develops in regions with high rainfall or frequent fog throughout the year.

rain shadow A certain reduction in precipitation that eventually produces a dry climate on the *lee* side of a mountain barrier.

recycle To keep, process, and reuse materials in order to save energy and reduce waste.

refraction Bending of a light ray or sound wave when it passes into a medium with a different density.

reg A type of desert, consisting of rock pavements.

relative humidity The amount of water vapor present in air expressed as a percentage of the amount needed to saturate air at a prevailing temperature.

renewable energy Energy from sources such as sunlight or wind that does not run out.

Sahel A semi-arid tropical savanna region south of the Sahara desert in Africa.

salinity A measure of the amount of dissolved salts in water.

sandstorm A wind storm that raises sand grains from the surface, often to a considerable height, and transports them, often for a long distance.

Santa Ana winds Strong, dry offshore winds that sweep southern California in late fall to winter.

satellite Any small object in orbit around a larger one. The term is most often used for rocky or artificial objects orbiting a planet.

saturation The condition in which air (or another medium) contains as much water vapor as it is capable of holding.

savanna A type of tropical vegetation, dominated by grasses with varying numbers of bushes and trees, that is adapted to an annual dry season. Sometimes there are two dry seasons (and two wet seasons), because the *Intertropical Convergence Zone* passes over the area twice.

season Distinct periods of a year, characterized by recurring weather changes. Generally, four seasons that differ mainly by temperature are observed in mid and polar latitudes. In the tropics and subtropics, changing precipitation levels produce a dry and a wet season.

seasonal precipitation *Precipitation* associated with seasonal changes.

sediment Fine particles of mud, sand, and organic debris that are carried by rivers and currents, and settle at the bottom of a pond, river, or ocean.

sensible heat Potential energy of an atmospheric air mass in the form of heat. As opposed to *latent heat*, this form is not produced by a change in state.

Siberian high A high-pressure area of very cold, dry air accumulated over the Eurasian terrain.

sirocco A Mediterranean wind from the Sahara that reaches hurricane speeds in North Africa.

sleet A type of precipitation either characterized by a mix of snow and rain, or by ice pellets.

smog A form of urban air pollution formed by the combination of combustion emissions and sunlight initiating photochemistry.

snow field An extensive, level area covered by snow or ice.

Snowball Earth A hypothesis in *paleoclimatology* that the Earth was once nearly or entirely covered in glaciers.

snowline The lower edge of the snow that remains on a mountain throughout summer.

snowmelt Freshwater surface runoff caused by melting snow.

snowpack Naturally accumulated snow that melts during warmer months.

solar cycles Variations in sunspot activity that affect solar energy output.

solar storm A violent explosion in the solar atmosphere, usually occurring close to a *sunspot*, that emits *radiation* and a stream of charged particles.

solar wind A ceaseless, but variable stream of high-energy and charged particles emitted by the Sun, traveling hundreds of miles per second.

solstice An astronomical event that occurs twice a year, when the orbit puts the Earth's axis maximally inclined toward or away from the Sun, causing its apparent position to reach its northernmost or southernmost limit.

South Pacific Convergence Zone A band of clouds and precipitation located at part of the *Intertropical Convergence Zone*, extending west from the Pacific warm pool toward French Polynesia.

spring tide A tide that occurs when the gravitational forces of the Sun and Moon act together.

squall line A line of simultaneously occurring storms along a cold front.

stack A pillar of rock that stands offshore, formed when wave action erodes a cave through a headland, producing an arch that later collapses.

stalactite A long mineral deposit that descends from the roof of a cave.

stalagmite A pinnacle of a mineral deposit that rises from the floor of a cave.

star A globe of gas that shines of its own accord because of energy released by nuclear reactions in its core.

steppe A climate region characterized by semi-arid grasslands without trees.

stoma (plural stomata) A pore in a leaf surface through which the plant absorbs carbon dioxide and releases oxygen and water vapor.

storm surge A mound of ocean water drawn up by low pressure below a hurricane; it can cause enormous waves and widespread damage if the hurricane reaches the coast.

stratopause The boundary, about 31 miles (50 km) above Earth's surface, between the *stratosphere* below and *mesosphere* above.

stratosphere An atmospheric layer that lies above the *troposphere* and below the *mesosphere*.

stromatolites Layered structures formed in shallow water by the accretion of microorganisms into sedimentary rock, providing excellent paleoclimate records in the form of ancient fossils.

sublimation The changing of any gaseous vapor into its solid form, such as steam to ice, without it passing through the liquid phase.

subtropical high Semi-permanent regions of high atmospheric pressure centered around 30°N and 30°S, associated with the subsidence of the Hadley cell circulation.

subtropics The region that lies approximately between latitudes 23° and 35° or 40° in both hemispheres.

sulfur oxides (SOₓ) Molecules composed of one sulfur atom and one or two oxygen atoms, emitted primarily by coal combustion.

sumatras An eastward-moving line of thunderstorms that typically form in tropical maritime Southeast Asia.

sunspot A dark, highly magnetic region on the Sun's surface that is cooler than the surrounding area.

supercell thunderstorm A type of thunderstorm characterized by a deep, rotating updraft.

supercooled droplets Water droplets cooled to below freezing but still in liquid form.

supersaturation The condition of air that contains more water vapor than would typically be supported at a given temperature. Supersaturated air will readily condense to form water droplets.

symbiosis The close beneficial feeding relationship between two species.

synoptic forecasting Weather forecasting based on the preparation and analysis of a chart that records surface weather observations taken simultaneously over as wide an area as possible.

synoptic map A chart that shows the weather at a particular time.

taiga The Russian name for the belt of coniferous forest that lies across northern Eurasia.

teleconnections Connected climate anomalies that appear over extremely large distances. One example is the *El Niño-Southern Oscillation*.

temperate Neither very hot nor very cold. Temperate areas have four distinct seasons.

temperature inversion The atmospheric condition in which cold air at the surface is overlaid by a deck of warmer air. This inhibits vertical mixing of the air masses and often gives rise to *smog* in urban areas.

thermocline A zone in the *water column* of a lake or ocean, where the present temperature decreases rapidly with depth.

thermohaline circulation Water movement caused by differences in density produced by changes in temperature and/or salinity.

thermopause The boundary, at an altitude varying from about 310–620 miles (500–1,000 km), between the *thermosphere* below and *exosphere* above.

thermosphere The layer of the *atmosphere* that extends from the *mesopause* about 53 miles (85 km) above the surface, to the *thermopause* at 310–620 miles (500–1,000 km).

thunder A rumbling shock wave created as lightning heats the air.

tide The regular rise and fall of the sea due to the gravitational attraction of the Moon and, to a lesser extent, the Sun.

tornado A spinning column of air that can measure more than 1 mile

(1.6 km) in diameter, move at up to 65 miles per hour (105 km/h) and generate winds of up to 300 miles per hour (482 km/h).

trade winds The winds that blow toward the Equator from the northeast in the Northern Hemisphere and from the southeast in the Southern Hemisphere.

trade-wind inversion An *inversion* associated with subsiding air on the side of the Hadley cells farthest from the Equator.

transpiration The evaporation of water from plants, transporting water from soil into the *atmosphere*.

treeline A line marking the latitudinal or altitudinal limit of tree growth.

tropical cyclone A storm system characterized by a low-pressure center, around which strong winds circulate, accompanied by cloud formation, heavy rain, and thunderstorms. These storms typically form over warm tropical waters, where moist air rises, then condenses. Depending on its region of origin, cyclones are also alternately known as hurricanes (western Atlantic Ocean) or typhoons (western Pacific or Indian oceans).

tropical storm A class of storm less severe than a tropical cyclone.

tropics The equatorial regions of Earth, between the Tropic of Cancer and the Tropic of Capricorn, at 23°N and 23°S, respectively.

tropopause The boundary, about 10.6 miles (17 km) above Earth's surface, between the *troposphere* below and *stratosphere* above.

troposphere The lowest layer of the *atmosphere*, up to approximately 4.4–10.6 miles (7-17 km) above sea level. This is the layer in which most of Earth's weather occurs.

trough An elongated region of low atmospheric pressure, or the lowest point of a wave.

tsunami A huge wave caused by an earthquake, landslide, or volcanic eruption, not by weather.

tundra A treeless plain found in the Arctic and some Antarctic regions, where the predominant vegetation consists of grasses, herbs, shrubs, and trees.

turbulent mixing The mixing of fluids (atmospheric or oceanic) in regions characterized by chaotic flow.

typhoon The name given to a hurricane in the western North Pacific and China Sea.

ultraviolet (UV) radiation *Electromagnetic radiation*, with a wavelength of 10 to 400 nanometers, located in the spectrum between visible light and X-rays.

UNESCO An agency of the United Nations established to contribute to peace and security through international collaboration. The World Heritage Committee fosters international cooperation to secure world cultural and natural heritage sites from destruction.

universe Everything that physically exists around us, including space, time, energy, and matter.

updraft Any movement of air away from the ground. The strongest form is found within thunderstorms.

upslope wind A wind that moves from lower to higher elevation.

upwelling The rising of deep, cold nutrient-rich waters into the surface layers, close to continental coasts.

virga Rain that evaporates before it reaches the ground. Virga is often visible as streaks in the sky.

volcano A landform created by the buildup of *lava* flows and ash. Volcanoes are typically cone-shaped.

vortex A spinning mass of air or fluid, especially the funnel of a tornado.

warm front A boundary between two air masses at different temperatures that advances with the warmer air behind it.

warm pool The deep layer of warm ocean water that lies in the region of Indonesia, except during *El Niño* events, when the pool is depleted.

wave height The difference between the elevations of a wave crest and a neighboring trough.

wavelength The distance between the crest or trough of one wave and another wave's crest or trough.

westerlies Prevailing winds, from the west, in the midlatitudes.

wetland Land that is covered for part of the year with fresh or salt water. It has vegetation adapted to life in saturated soils.

wildfire An uncontrolled fire occurring in the wild. In Australia, also known as a bushfire.

wind shear Motion caused by one layer of air sliding over another layer that is moving at a different speed and/or in a different direction.

windward The direction from which a wind is blowing.

world ocean The interlinked network of oceans and seas covering 71 percent of Earth's surface.

zooanthellae Single-celled, plant-like organisms that live in a symbiotic relationship inside coral polyps.

INDEX

Page numbers in *italics* indicate illustrations of the topic.

t=top; l=left; r=right; tl=top left; tcl=top center left; tc=top center; tcr=top center right; tr=top right; cl=center left; c=center; cr=center right; b=bottom; bl=bottom left; bcl=bottom center left; bc=bottom center; bcr=bottom center right; br=bottom right

BBoc = Bolot Bochkarev/YakuitaToday.com, BOM = Bureau of Meteorology, CCD = Corel Corp, DSCD = Digital Stock, ESA = European Space Agency, GI = Getty Images, GRIDA = GRIDA Ardenal, iS = istockphoto.com, N = NASA, N_EO = NASA Earth Observatory, N_ES = NASA Earth From Space, N_G = Great Images In NASA, N_GS = NASA Goddard Space Flight Center, N_H = NASA Hubble Space Telescope, N_ISS = NASA International Space Station, N_JPL = NASA Jet Propulsion Laboratory, N_MI = NASA Missions, N_NEO = NASA Earth Observations, N_S = NASA Solar and Heliospheric Observatory, N_SF = NASA Space Flight/Human Space Flight, N_V = NASA Visible Earth, NGS = National Geographic Society, NIRD = National Institute of Rural Development, NIWA = National Institute of Water and Atmospheric Research, NOAA = National Oceanic and Atmospheric Administration, NLN = National Library of Norway, PDCD = Photodisc, PUB = Public Domain, SH = Shutterstock, USDA = United States Department of Agriculture, USGS = United States Geographical Survey, USN = U.S. Navy

PHOTOGRAPHS

Front cover GI

1c iS; **2**c GI; **4**c GI; **6**tc, tr GI; tl N; **7**tr GI; tl N_SF; tc SH; **8**c NOAA; **12**c GI; **14**cl N_EO; cr N_GS; **15**r GI; **16**cl DSCD; **18**tl GI; bl iS; cl N_EO; **19**tc GI; cr, tr iS; tl N_GS; **20**b GI; tr N_G; cl N_GS; bl, tl N_H; **22**c DSCD; bc GI; bl SH; **23**br, cr, cr, tr GI; **24**cl N; tl N_GS; bl N_S; **25**t N_GS; **26**b, br iS; **27**b, bc, bl iS; **29**c, cr, t, tr GI; bc SH; **30**bc GI; bl, c, tl iS; cl N; tc N_G; **31**cl GI; **32**br, cl GI; **33**c SH; **34**cl, tl GI; bl N; **35**br iS; bc N_EO; **36**b GI; **37**t GI; b iS; **38**bl, tl GI; cl SH; **39**tr GI; br, cr iS; **40**bc N_EO; **41**c DSCD; **42**t GI; **43**b GI; tc SH; **44**b, t GI; **45**cc iS; **46**t GI; **47**b iS; br, cr, tr NOAA; **48**tl GI; br N; cr N_EO; bl NGS; **49**bc, t, tr GI; **50**br Ann Woolf/BOM; tr N_GS; **51**bc GI; **53**c GI; **54**b GI; t N_EO; **55**t GI; b iS; cr, tr N_EO; cr N_MI; **56**bc N_EO; br, cc, tl N_G; bl N_V; **57**bc, bl, cc, tc, tl N_G; **58**cl, tl GI; **59**br, cl GI; **60**c GI; **62**t, tr GI; **63**tl iS; **64**bl GI; cl N_EO; cr N_G; **65**br GI; t SH; **66**br, tc GI; **67**bc iS; **68**b, bl GI; **69**bc, cr, tc GI; **70**cc, tc, tl, tr N_EO; bl PUB; **71**tl N_EO; **72**bc, t iS; **73**cr ESA; b GI; tr N_EO; **74**b, cl GI; cr PUB; **75**bl iS; t NLN; cc NOAA; br PUB; **76**b GI; t N_G; **77**bc GI; **78**tr PDCD; tl USDA; **79**bc GI; tl iS; cr, tr SH; **80**bl, bl, tc GI; bc N_V; **81**tr GI; b, br iS; **82**br, r GI; bc, bl iS; **83**cl GI; **84**cc N_EO; **85**bc iS; t SH; **87**bc iS; **88**b N; bl N_EO; **89**br, cc GI; tc N_G; **90**cl, bcl CCD; bl, br, cl, tl iS; cr N; **92**b SH; **93**br, br, tc SH; **94**b SH; **95**tr GI; bc SH; **96**b GI; **97**br, cr, tr GI; **98**b GI; tr SH; **99**br GI; tr N_EO; tl N_JPL; **100**bc iS; tr SH; **101**br, tc GI; **103**b, br, tc GI; **104**b iS; **105**br, tr GI; **106**b, tl GI; **107**tc GI; bc iS; **108**br NOAA; **109**cl, cr, tr GI; br N; **110**br SH; **111**br, tc GI; **112**b GI; **113**br GI; **114**br, cr GI; **115**c GI; **116**br GI; tr N_EO; **117**cc GI; **118**b, cl GI; tl SH; **119**br, tr GI; t iS; **120**b GI; tr N_ISS; **121**tc SH; **122**bc, bc GI; cl NOAA; **123**b, cr, t, tr GI; **124**tr GI; bc SH; **125**tl GI; tr iS; b N_V; cr USN; **126**cc iS; **127**bc, t GI; **128**bc, br GI; **129**br, br, tl GI; tr iS; cr, r SH; **130**t, tr GI; tr SH; t USDA; **131**tl SH; **132**cc, tr GI; **133**bc iS; br NOAA; **134**br, cc GI; **135**t GI; br SH; **136**b, cc iS; **137**t GI; **138**br GI; bl PUB; **139**bc, tl GI; cr iS; tr PUB; **140**c GI; **142**tl, tr GI; **143**tr GI; **145**b, tc GI; **147**br N; tr N_ISS; cr N_V; **148**b iS; **149**bc N; **150**bl NOAA; SH; **151**tr GI; br NOAA; cr SH; **152**b, cl, tl GI; bl NOAA; **153**br, t NOAA; **154**b, cl GI; **155**tr GI; bc iS; t SH; **156**tr GI; bl, br NOAA; **157**c GI; **158**bl GI; **159**tc GI; bc NOAA; **160**bl, bl GI; tr, tr NGS; **161**bc, bc, cr, cr GI; tl, tl NGS; cr, cr, tr, tr NOAA; **162**bc, bl GI; **163**bl, tr GI; **164**bl PUB; t SH; **165**bc, tc GI; **166**bl, br GI; **167**c N_V; **168**tc, tc, tl, tr N_EO; bl N_G; tl N_S; cl N_V; cl NOAA; **170**br GI; t PUB; **171**bc N_V; **172**br GI; cc PUB; **173**bc, t, tr GI; br, tc PUB; **174**bc GI; cl N; tl N_EO; t N_G; bl, br NOAA; **175**b, bc, br, cr, bcr, tl, tr GI; bc SH; **176**bl, t GI; **177**bc GI; tr USGS; **178** GI; **179**b, br GI; cc, tl N_V; **180**bc, cl GI; **181**tc GI; **182**t GI; bl PUB; **183**cr, t, tr GI; br, cr iS; bc N_EO; **184**br, tr GI; **185**bc, br, tr GI; **187**br, t GI; **188**bc GI; c N_GS; **189**b, br, t, tr GI; bc, bc N_V; **190**b GI; **191**t GI; cr, tr N; br, cr N_V; **192**cr GI; **193**t iS; **194**tr GI; bc, br N_NEO; **195**tl BBoc; bc GI; **196**b GI; tc N_NEO; **197**bc, t GI; br iS; **198**br GI; **199**bc, t GI; **200**b, bl GI; **201**bc, br, cr, tl, tr GI; **202**c CBT; **204**tl, tr GI; **205**t GI; **206**br, cl GI; **207**t GI; bc, br iS; **208**bl, tl GI; t iS; **209**b, br, cr, tl, tr GI; **211**br GI; tr N; c SH; **212**b, tr GI; **213**cl, tr GI; **214**cr, tl, tr GI; b NOAA; **215**br, c, cr, tr GI; **216**bl, cl, tr GI; tl N_MI; tl PUB; **217**cl GI; tr N; **218**br GI; l N_SF; **219**t N_V; **220**b BOM; bl, t GI; **221**b, tr iS; cr SH; **222**tc, tl, tr N_JPL; br SH; **223**cr, tr iS; cl N_EO; br N_JPL; **224**bc BOM; c, r GI; **226**tc GI; bl, cl, tl iS; bc NOAA; **227**br, cr, tr iS; b NOAA; **228**b, t GI; **231**cr, r BOM; c, tr GI; br N_S; **232**c GI; **234**t, t GI; **235**tl iS; **236**bl GI; br N_G; **237**bc N_G; **238**b iS; **239**b, tc GI; br SH; **240**tc GI; bl SH; **241**b GI; t SH; **242**b GI; **243**bc, tc GI; cc SH; **244**bc GI; tc iS; **245**c SH; **246**bc, tc GI; **247**b GI; tl iS; **248**b, t GI; **249**tc iS; **250**bl GI; tr iS; br SH; **251**bl GI; cr SH; **252**b, bl GI; **253**bc, cc GI; **254**bc GI; c SH; **255**br, c iS; **256**b SH; **257**cc, tc GI; bc SH; **258**bc iS; br SH; **259**cl iS; **260**bl, br GI; cr SH; **261**cc iS; **262**cr GI; br iS; b SH; **263**bl GI; **264**br, cr, tr GI; **265**tl iS; br SH; **266**tl GI; l SH; **267**tc GI; c SH; **268**bc SH; **269**b GI; br, cl, cr, tc SH; **270**br, tl GI; cc iS; tc SH; **271**bc GI; **272**bc GI; r N_L; **273**b, t SH; **274**bl, c SH; **275**bc, t GI; b SH; **276**bc, cc iS; **277**b GI; cr NOAA; **278**b GI; **279**tl SH; **280**bc GI; tc SH; **281**bc, cr iS; tc N_V; cr, tr SH; **283**t GI; cc NOAA; b, tc SH; **284**tl, tr N_EO; **285**bc, bl GI; tl iS; **287**bc, cl, tr GI; tl iS; **288**tl, tl iS; cl, t SH; **289**bc GI; tc iS; **290**tr GI; br SH; **291**bc GI; cr, tr iS; tl N_EO; cr SH; **292**br GI; tc iS; cc, tr SH; **293**bc SH; **294**b GI; **295**tc, tl iS; bc SH; **296**br GI; cl, tc, tr iS; **297**tl GI; **298**t SH; **299**bl, tr GI; br SH; **300**tr GI; bc SH; **301**tl GI; b, bc SH; **302**tl SH; **303**bc, bc, cc GI; tc, tl SH; **304**bc SH; **305**bc, cr, tl GI; **306**cr, t GI; cl N; bc, cl, tl SH; **307**tc GI; **308**b, bc, t SH; **309**bc GI; **310**bc GI; tc, tl N; tr SH; **311**bc, tl GI; cc iS; **313**cr, tl GI; bc SH; **314**bc GI; cr iS; **315**bl GI; cr, tr iS; tl N_V; bc SH; **316**bc GI; t SH; **317**cc GI; b, tc SH; **318**bc, tc GI; **319**b SH; **320**bl iS; br, cl, cl N_EO; **321**bc, bl GI; tl iS; **322**b GI; tr iS; **323**t iS; bc N; br SH; **324**b, cl, l, tc GI; t SH; **325**bl GI; **326**bc GI; tr SH; **327**tl iS; tc N; bc, cr, cr SH; **329**cr, t GI; cl, tr SH; **330**tc GI; **331**bc, t GI; b iS; **332**bc, br GI; **333**bc GI; t N; **334**br, cl iS; cl N_V; **335**bc GI; tc iS; l NIRD; **336**bc, br GI; **337**bc, tl GI; **338**b, tr GI; **339**b GI; t iS; **340**b iS; **341**t GI; bc N_EO; br SH; **342**br iS; **343**cc GI; cl, tl iS; bl SH; **345**br, t GI; bl, cl, tr SH; **346**cl GI; b SH; **347**t GI; cc iS; bc, tc SH; **348**tr GI; b SH; **349**bc, tl GI; cr, cr SH; **350**br GI; **351**bc GI; tc iS; bc, tl SH; **352**b GI; c SH; **353**bc, t GI; **354**bc, c GI; **355**b, tc GI; cl iS; c SH; **356**tr NIWA; b, c SH; **357**cc, tl iS; bc SH; **358**b, c, tc GI; **359**bc N_L; **360**c GI; **362**t, tl GI; **363**t iS; **364**t iS; **365**b, tc GI; **366**b, bl, cl, tl GI; tc iS; **368**br GI; bl, tl iS; cl SH; **369**cr GI; br iS; tr SH; bc SP; **370**cc GI; **371**t GI; **372**b GI; **375**r, tl GI; bc SH; **376**t GI; **377**tr GI; br iS; **378**c GI; br N_V; **379**c GI; **380**b GI; **381**bc iS; tl N_G; **382**bl, cl, cl, tl N_GS; **383**c GI; br SPL; **384**br GI; bl, bl N_JPL; **385**bc, cr, tl GI; br, tr iS; **386**c GI; bc N_G; **387**c GI; **388**cr GI; **389**c iS; **390**bl, bl, cl N_G; **391**br, tr GI; cr N_G; **392**br iS; **393**cl iS; **394**cl, tl iS; bl, bl N_L; b N_V; **395**br GI; cr, tr iS; cr SH; **396**b, bl GI; tl iS; cl PDCD; **397**b GI; **398**br GI; cr N; **399**bc, br GI; tl, tr iS; **400**tc GI; **401**br iS; **402**b GI; **403**cc GI; br, cr, tr iS; **404**bc N; **405**cr GI; br iS; tr SH; **406**b GI; bl iS; **407**tl N_EO; bc, bl N_V; **408**br GI; bl iS; cl N_EO; **409**bc, tc iS; **410**bl, br N_V; **411**tc N_EO; bc N_T; tl, tl N_V; **412**b GI; **413**b, t GI; **414**br GI; **415**br, cc, cr GI; **416**b N_EO; bl N_G; **417**cr, tl N_EO; cr N_G; br N_JPL; **418**br iS; **419**bl, br, tl, tr GI; **420**br GI; **421**bc, tl, tr GI; **422**br GI; **423**bc, cc GI; **424**b, br GI; **425**bc, tl GI; cr, tr iS; br SH; **426**br iS; **427**br, tc GI; bc N_EO; **429**c, tr GI; cr iS; br SP; **430**bc GI; **431**b, cc GI; tr N; cr N_MI; **432**br SP; **433**bc, cl, tl GI; r iS; **434**b, t N_EO; **435**bc, l, tl GI; **436**bl N_EO; bc, bcl, br, c, cc, cl, cr, l, t, tcl, tr, r N_V; **437**br, cr, tr GI; t N_G; b, bc, bl N_V; **438**br GI; bl NOAA; **439**bc, bl, tc GI; **440**br GI; tr N_EO; **441**bc GI; tl N_EO; **442**b GI; **443**br N_EO; **444**bc, br GI; bl NOAA; **445**br, tl GI; cc, cc, cl N_V; **447**bl GI; **448**b GI; **450**br GI; **451**bc, tr GI; tl N_GS; **453**b GI; t iS; **454**tr GI; bl, cl, tl iS; b N_EO; **455**bc, cc iS; **456**bc GI; cl iS; **457**bc SP; **458**cr GI; bl, br N_G; **459**tl GI; bl, cr iS; tr N_EO; br SH; **461**cl GI; br, cr, tr GRIDA; **462**b, tc GI; **463**bl, c iS; **464**b, br GI; **465**b, t GI; **466**br GI; **467**br, cr, tl, tr GI; **468**br GI; **469**bl, br, tc GI; **470**b GI; **471**bc, tl GI; **472**b GI; **473**br, r, tl GI; **474**bl GI; **475**b, t GI; **476**b GI; **477**bc, tl GI; **478**bc GI; **479**t GI; **480**b GI; **481**br, cr, tc, tr GI; t iS; **483**t iS; bc SH; **484**c SH; **485**b, tc GI; **486**br GI; **487**tc GI; bc iS; **488**bl GI; **491**c GI

ILLUSTRATIONS

Mike Atkinson 243cr, **Richard Bonson/The Art Agency** 21bl, 69tr, 86c, 91cl, **Leonello Calvetti** 73br, **Karen Carr** 370cl, 370tl, 370cl, 370cl, 370bl, 373cr, **Robin Carter/The Art Agency** 79br, 79br, **Barry Croucher/The Art Agency** 26c, 67t, 71tr, 81t, 83tr, 83cr, 83br, 131c, 132br, 132tl, 153tr, 212bl, 212tl, 212bl, 212cl, 269tr, 269tr, 277tr, 277tr, 382cc, 382tc, **Damien Demaj** 309tl, **Jane Durston/The Art Agency** 28br, 28cr, 29cc, 29bc, 295r, **Simone End** 373tr, **Christer Eriksson** 375tr, **Chris Forsey** 367tl, **Lloyd Foye** 316tl, **John Francis** 274cl, **Mark A. Garlick** 17b, 18b, 25cc, 25tr, 25cr, 25r, 47tl, 47cl, 47tl, 92r, 94bl, 95tl, 96bl, 98bl, 100bl, 102cr, 104bl, 106bl, 108, 110bl, 112bl, 114bl, 116bl, 120bl, 124bl, 168c, 181bc, 230, 277tl, 373tl, 373cc, 373bc, **Mike Gorman** 133tl, 158c, **Jon Gittoes** 239cr, 239tr, **Godd.com** 179cr, 179cr, 179tr, 197tr, 197cr, **Malcolm Godwin/Moonrunner Design** 217cr, **Ray Grinaway** 32bl, 265r, **Tim Hayward** 245cr, 252tl, **Robert Hynes** 430cl, 430bl, 430tl, 430l, **Ian Jackson/The Art Agency** 265cr, **David Kirshner** 71br, 71cr, 71cr, 240tl, 240tl, 243tr, 245tr, 246cl, 255tr, 255cr, 257tr, 257r, 258cl, 265tr, 274cl, 292cl, 292cl, 295cr, 300cl, **KJA-artists** 164br, 169cr, 182b, 371cc, 371cc, 371cl, 374r, 375l, 395bc, 397t, 397c, 404c, 404cr, 405c, 405bc, 406c, 426cr, 482b, 489b, **Frank Knight** 246tl, 292tl, 316cl, 338l, 372tl, 372cl, 372l, 372bl, 452br, **John Mac** 243br, **Rob Mancini** 252l, **MBA Studios** 87br, 100br, 101bc, 106tr, 107br, 111bc, 118cl, 118bl, 121br, 139br, 139bc, 148tl, 148bl, 148cl, 148cl, 149tr, 149cr, 149cr, 149br, 149tc, 150b, 151cc, 155br, **Iain McKellar** 220tl, **Ed Merritt** 279bc, 281c, **Yvan Meunier/Contact Jupiter** 28br, 29bl, **Nicola Oram** 371cr, 371tr, 371br, **Terry Pastor/The Art Agency** 455br, 455cr, 455tr, 456tl, **Peter Bull Art Studio** 28bc, 34r, 39bc, 39cc, 39tc, 40tc, 43br, 43tr, 43cr, 45tr, 50tl, 50cl, 58br, 58tr, 59cr, 59cr, 59tr, 66cl, 66cl, 66tl, 68tl, 68cl, 69br, 72bl, 72bl, 72cl, 72cl, 72tl, 72tl, 73tc, 77br, 77cr, 77tr, 89tr, 89bc, 129cl, 129cc, 129cc, 130tl, 130cl, 130cl, 134tc, 134tl, 134cl, 134cl, 135bl, 138cr, 138tc, 138cc, 165br, 165cr, 165tr, 165cr, 170cl, 195tr, 195br, 195cr, 195cr, 201cr, 216br, 222bl, 222l, 230bl, 230cl, 230tl, 240cl, 251tl, 251tr, 253tc, 257cr, 265bc, 266c, 274bc, 275tr, 282cc, 284b, 286br, 288br, 293tr, 296tl, 296cl, 300tc, 304t, 307cc, 307tr, 309tr, 309cr, 309cr, 312tr, 313cl, 314tc, 314tr, 316tc, 318tl, 320r, 320tl, 322tc, 325bc, 325tc, 327tr, 329bc, 329br, 330bc, 332tc, 334tl, 336tr, 336tc, 340tc, 340cr, 341tr, 341cr, 342tl, 342cl, 342cl, 342tr, 342tl, 345cc, 345cr, 346tc, 346cl, 348tc, 349tr, 350tl, 350cl, 350tl, 350tr, 352tc, 353cr, 353tr, 354tr, 358cl, 358l, 358tl, 417bc, **Lionel Portier** 66bl, 488bl, **Mick Posen/The Art Agency** 38b, 51tl, 51tl, 51tc, 57b, 57cr, 57cr, 57tr, 70b, 78br, 146br, 146tl, 146bl, 146cl, 159tr, 159cr, 159cr, 159br, 258r, 266cl, 267tr, 272t, 400bc, 400bc, 400bc, 400br, 460br, **Tony Pyrzakowski** 246cl, 292cl, **Oliver Rennert** 300cl, **Edwina Riddell** 338cl, **Barbara Rodanska** 375cr, **Claudio Saraceni** 456cl, **Michael Saunders** 258cl, 258bl, 401bc, 401bc, 401bc, 401bl, **Peter Schouten** 239cr, **Guy Troughton** 245br, 249cr, 249tr, 258tl, 300tl, 338tl, **Glen Vause** 208b, 210b, 210c, 210t, 210c, 220cl, **Genevieve Wallace** 274tl, 316l, **Trevor Weeks** 255cr, **Rod Westblad**e 41cr, 41tr, 41cr, 46bl, 46cl, 46cl, 46tl, **Ann Winterbotham** 252cl, 295tr

MAPS/GRAPHICS

All charts and graphics by **Andrew Davies/Creative Communication**.
All additional maps by **Andrew Davies/Creative Communication**, adapted from **Map Illustrations**.
229c redrawn from map by Center for Ocean-Land-Atmosphere Studies, **270**tr redrawn from map by Global Forest Watch Canada, supplied by International Boreal Conservation Campaign, **304**t redrawn from map supplied by the Met Office, adapted from British Crown copyright data.